A GUIDE BOOK OF THE
UNITED STATES
MINT

COLONIAL, STATE, PRIVATE, TERRITORIAL, AND FEDERAL COINING FACILITIES

A GUIDE BOOK OF THE
UNITED STATES
MINT

COLONIAL, STATE, PRIVATE, TERRITORIAL, AND FEDERAL COINING FACILITIES

The first coin federally
minted for wide
circulation—the silver
half disme of 1792.

Q. David Bowers

Foreword by
Kenneth Bressett

Whitman
Publishing, LLC
PUBLISHING SINCE 1934
Whitman.com

A GUIDE BOOK OF THE
UNITED STATES MINT

© 2016 Whitman Publishing, LLC
1960 Chandalar Drive, Suite E, Pelham, AL 35124

THE OFFICIAL RED BOOK is a trademark of Whitman Publishing, LLC.

Correspondence concerning this book may be directed to
Whitman Publishing, Attn: U.S. Mint book, at the address above.

ISBN: 0794843972

Printed in China

About the cover: The Denver Mint's impressive Second Renaissance Revival granite structure is pictured at the bottom. Mintage of coins for circulation began there in 1906. Today it is one of Denver's most popular tourist attractions. The coin and medals pictured at top are all collectibles related to the history of the United States Mint. Clockwise from top left: A 1993 silver medal celebrating 200 years of coinage at Philadelphia; a 1920 silver medal marking the opening of the Manila Mint in the Philippines (then a territory of the United States); a 1902 U.S. Assay Commission medal showing the third Philadelphia Mint, built in 1901; a 1955 Assay Commission medal depicting the first coining press at Philadelphia; and a 2006 silver dollar commemorating "The Granite Lady," San Francisco's Old Mint, which survived the earthquake and fires of 1906. See appendix C for more information on each of these collectibles and many more.

Other books in the Bowers Series include: *A Guide Book of Morgan Silver Dollars; A Guide Book of Double Eagle Gold Coins; A Guide Book of United States Type Coins; A Guide Book of Modern United States Proof Coin Sets; A Guide Book of Shield and Liberty Head Nickels; A Guide Book of Flying Eagle and Indian Head Cents; A Guide Book of Washington and State Quarters; A Guide Book of Buffalo and Jefferson Nickels; A Guide Book of Lincoln Cents; A Guide Book of United States Commemorative Coins; A Guide Book of United States Tokens and Medals; A Guide Book of Gold Dollars; A Guide Book of Peace Dollars; A Guide Book of the Official Red Book of United States Coins; A Guide Book of Franklin and Kennedy Half Dollars; A Guide Book of Civil War Tokens; A Guide Book of Hard Times Tokens; A Guide Book of Mercury Dimes, Standing Liberty Quarters, and Liberty Walking Half Dollars; A Guide Book of Half Cents and Large Cents; A Guide Book of Barber Silver Coins; A Guide Book of Liberty Seated Silver Coins,* and *A Guide Book of Modern United States Dollar Coins.*

Whitman Publishing is a leader in the antiques and collectibles field.
For a complete catalog of numismatic reference books, supplies, and
storage products, visit Whitman Publishing online at www.Whitman.com.
If you enjoy the U.S. Mint, join the American Numismatic Association.
Visit www.Whitman.com/ANA for membership information.

WCG™   Whitman®   OCG™

CONTENTS

Foreword

Few people ever give much thought to the coins that jingle in their pockets or how or why they came to be there. They serve a necessary purpose and that is all that seems to matter. In fact, if it were not for the convenience of that handy medium of exchange, the commercial world would still be operating in the dark ages. It seems likely that the invention of money is one of the five greatest inventions of mankind, exceeded only by writing, the wheel, mathematics, and computers. Yet the origin of money as we know it today is largely ignored or forgotten.

This book holds an account of the various United States mints that have produced our coins since the beginning of the nation, and the colorful stories behind the trials and tribulations they faced to supply the country with the necessary "hard cash" that is so vital to day-to-day transactions.

Beyond the intrigues and factual descriptions of how and why our mints function as they do are many real-life anecdotes and events of the day—sometimes tragic, sometimes humorous—that add to the allure of coins from the vantage point of numismatists who have devoted years to chronicling these events. This book however is a story that does not simply begin or end with the mechanical business of making our federal coins. It will take you back in time to the days when the first settlers arrived in colonial America and their struggle to find ways to carry on trade and survive in a harsh land devoid of the kinds of coins they knew in their former homeland.

From the very beginning of commerce in America "hard money" was needed for trade with overseas nations. European merchants were eager to purchase dried fish and pelts from the Americans, who could obtain quantities of them by bartering with Native Americans who eagerly sought their knives, hatchets, beads, and blankets. The settlers established trade with Europeans following the same procedure to obtain guns, powder, shot, and other essentials. Out of necessity, barter became the prime means of transactions among settlers where there was no mint or local coinage.

The struggle to find a convenient means of making trades and settling debts was not a new dilemma. In fact it likely began in prehistoric times when individuals and clans found that it was to their advantage to exchange surplus items with others for needed things that they were lacking. That was the beginning of problems in agreeing on the value of diverse items—for instance, how to make change for a goat given in trade for two containers of wheat, or what to do with the goat once you owned it. Solutions only came with the establishment of a common medium of exchange in the form of something desired, useful, and recognized by all, and a governing authority to control and warranty the agreed-upon value. As it turned out, gold, silver, and copper were and still are the lucky winners. Yet, as you will see in this account, the rules were later changed with the advent of tokens and paper currency.

The ancient kings and rulers who produced the first metallic coins made it known that they authorized their use by including in the design something symbolic of the nation, their portrait, or one of a local god to insure the value and respectability of their money. In ancient Athens some of the first mints were set up in the Agora, their most public place of business, to insure that their coins were being produced with the greatest of care, and were of full value.

The ancient Romans went a step further to instill confidence in their coins by establishing their first mint in the temple of Juno Moneta in the very heart of the city. The goddess Juno was a popular favorite among Romans as the Advisor of Good Counsel, and inspiration to those about to be married. Her temple stood at the citadel on the

Capitoline Hill overlooking the old Roman Forum. It was the place where the books of the magistrates were deposited, and dedicated to Juno as the Protector of Funds. As such it was so well suited to the production of money that her very epithet "Moneta" became the origin of our words "mint" and "money."

Today, more than 2,500 years after those first coins were minted, we continue to rely on a ready supply of those shiny, round objects for daily transactions and rarely consider where they come from or why we place our faith in the government mints that produced them. The coins we use today are not so different from those of the ancients either in form or substance. They are backed by faith in the governing authority, three very modern and efficient mints, and that familiar motto "IN GOD WE TRUST."

Here now is the story of how it all came to be, starting with the first mint in the colonies—in Boston where Pine Tree shillings and other coins were struck—continuing down to the newest federal facility, the West Point Mint. Dave Bowers takes you on a "you are there" experience with over three dozen mints, federal, state, and private—a unique experience. I know you will enjoy this delightful narrative!

Kenneth Bressett
Colorado Springs, Colorado

Kenneth Bressett, long-time editor of the popular *Guide Book of United States Coins*, has actively promoted the study and hobby of numismatics for over 50 years. His published works on various subjects cover a wide range of topics and extend from short articles to standard reference books on such diverse areas as ancient coins, paper money, and English and American coins.

U.S. Assay Commission medals (shown enlarged) from 1869 and 1881.

AUTHOR'S PREFACE

Writing this book was an exceptional pleasure. Over a long period of time I have written many books, including quite a few that have become standard references. Many of my important titles can be found at www.whitman.com. When a book has been about certain series—say colonial coins, half cents and large cents, Morgan silver dollars, or whatever, I have had a lot of numismatic material to consult. Large copper cents, for example, have been studied by numismatist starting with Jeremiah Colburn in the late 1850s down to books and auction catalogs of the present century. I dare say that a comprehensive library of books and auction offerings of cents dated from 1793 to 1857 would more than fill a bookshelf.

Being of the old school, so to speak—my first printed sales catalog was in 1955, first auction catalog in 1958, first hardbound book *(Coins and Collectors)* in 1964—I wrote my earliest books the hard way. After consulting publications in and out of print, including issues of the *American Journal of Numismatics* from 1866 onward, *The Numismatist* from the first issue turned out on a hand press by Dr. George F. Heath in Monroe, Michigan, in 1888, *Annual Reports of the Director of the Mint* since the 1790s, *American State Papers*, the *Congressional Record*, and my complete file of *Niles' Weekly Register* from 1811 onward, my search turned to trying to track down information that would be new to readers. I filled out my share of call slips in the Library of Congress in Washington, in the New York Public Library, and elsewhere. On occasion I camped among books at the American Numismatic Society in their temple-like building on Audubon Terrace at 155th Street and Broadway in New York City.

The *Dictionary of American Biography* and old and new sets of the *Encyclopædia Britannica* saw constant use. I also remember turning out letters on my Royal typewriter, then an IBM Selectric, then an IBM Executive typewriter. Today I have none of the tools mentioned in this paragraph! I could not have imagined this back in 1988 when I purchased my first computer, a Macintosh Plus.

Since then the world has changed. While I still consider myself a bit old-fashioned and certainly conservative, I have embraced the computer and the Internet. The point of this is the book you are holding, with much information that is not available in any publication currently in print, involved electronic research. At my fingertips have been newspapers dating back to colonial times, thousands of pages of congressional reports, a cornucopia of books printed in the United States that are out of copyright (generally from the early 1920s and before), and more. The end result is that new (adding to my existing archives) research for this book took less than a half year!

As to the subjects of this book—the various mints in the United States past and present—I have visited and done in-depth research, courtesy of the U.S. Mint and the Treasury Department at Philadelphia, Denver, and San Francisco, on multiple occasions and at the West Point Mint once (accompanied by Tom Jurkowsky, of the public-relations department of the Mint). I have never been to Dahlonega, as the mint that was in operation from 1838 to 1861 burned to the ground on one winter night in December 1878. I've been to the Charlotte Mint, which was also in service from 1838 to 1861, and since has moved to another location where it is used as an art museum. I've also visited the buildings that used to house the operations of the New Orleans Mint and the Carson City Mint. I've also been to the mint that never was, in The Dalles, Oregon, discussed in chapter 13.

I was able to serve the Mint when, in 1992, Senator Alan Cranston of California was trying to convince Congress that the sleepy Mint could not market the forthcoming

Los Angeles Olympics Coins, but that Occidental Petroleum, run by his buddy Armand Hammer, could do a great job. At the time Occidental was famous, a darling of the stock market. At the request of the Treasury Department I went to Washington and, with a few others, testified that the Mint should market its own coins, this was tradition, and until the government decided to rent out the White House it should not turn coin marketing over to the private sector. We won, and I got a warm handwritten letter of thanks from United States Treasurer Bay Buchanan.

Also in that year, on April 2, 1992, the Treasury asked me to be the keynote speaker when in Philadelphia the Mint observed its 200th anniversary. Many times the Treasury Department or the Mint has asked me for historical information, advice, and more. All of this has been very enjoyable.

The present book is mostly about coin factories—the federal mints—and not about the coins they produced, but many cross references have been added, including a synopsis of the different denominations and designs produced by each mint over the years.

It is my hope that as you read these pages you will have an "I am there" feeling—at the Dahlonega Mint in the midst of the Georgia Gold Rush, or at the Philadelphia Mint in the uncertain times of the 1790s.

This study could have filled an entire *set* of volumes! As such, although there are thousands of words for you to read, it is distilled from what it could have been. The Internet beckons, and any one of the topics can be explored in greater depth with a few keystrokes.

The above said, I dearly love the printed page. I give a nod to Whitman Publishing for not only preserving the printed page but, remarkable, turning out more numismatic reference books by more authors than any other publisher in history—and this is in the electronic age!

Q. David Bowers
Wolfeboro, New Hampshire

 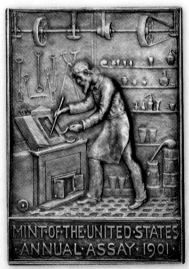

U.S. Assay Commission medals (shown enlarged) from 1887 and 1901.

PUBLISHER'S PREFACE

The United States Mint is at the center of American coin collecting. It is the well-spring, the mother lode, of the little copper, silver, and gold objects of material culture that numismatists collect, study, and catalog. The Mint is the sole manufacturer of the nation's legal-tender coinage, and its products are used every day by millions of Americans nationwide. As a repository its facilities safeguard $311 billion in national assets. Its programs are self-sustaining and operate at no cost to taxpayers. It employs nearly 2,000 people, including its own police force.

The Mint is, quite simply, a big and important part of American life.

Naturally anything this big attracts attention. Most collectors today love the Mint. (At times some love to critique it; opinions in the hobby community, unlike 1804 silver dollars and 1792 half dismes, are not scarce.) Hobbyists discuss the latest coins, programs, and developments in blogs, editorials, online forums, and print articles. We eagerly await the latest announcements from the Mint's Office of Corporate Communications. Entire industries surround the Mint and its products—book-publishing firms, weekly newspapers, monthly magazines, coin conventions, makers of folders and albums, bullion wholesalers and retailers, and rare-coin dealers. Museums and scholarly societies archive and display U.S. coins. Hobbyists join affinity groups that focus on specific topics (like Liberty Seated silver coins, or early American coppers) as well as larger groups that cover the entire expanse of coinage (like the American Numismatic Association, online at www.money.org).

Still, even longtime collectors can misunderstand how the Mint operates, how coinage designs come to be, how coins are distributed, and the million moving parts that make our national mint facilities run.

Meanwhile, outside the hobby community the Mint seems perpetually misunderstood. Non-numismatists sometimes think the Mint makes our nation's paper currency. Hollywood's 1967 madcap romp *Who's Minding the Mint?* includes the spectacle of George Washington dressed as Milton Berle (or vice versa) surrounded by printing presses cranking out $100 bills. Some people think that it's the Federal Reserve, not the Mint, that makes coins. Few non-numismatists could name all the American cities that are home to the modern Mint's coinage factories. And most don't realize that the famous gold depository at Fort Knox is part of the Mint, or that a huge amount of silver is stored in scenic West Point.

Prior to this volume many books had been written about American coinage, and about individual mint facilities. However, at Whitman Publishing we've seen the need for a single authoritative history of not just the

Whitman publisher Dennis Tucker, author Q. David Bowers, and Mint Director of Corporate Communications Thomas Jurkowsky in front of the Denver Mint.

modern (post-1872)[1] U.S. Mint, but also of the early Mint, plus each of its branches, and the many private and territorial mints and assay offices that provided local and regional "coins" (from colonial days up to the 1860s) in places where official money was scarce.

Here are just a few of the predecessors to the book you're now reading:

- Colonial coins and tokens and early American patterns (dating from before 1792, when the federal mint was established under the management of the Department of State) were studied in classic texts such as Sylvester S. Crosby's *The Early Coins of America*, published in 1875. Q. David Bowers's *Whitman Encyclopedia of Colonial and Early American Coins*, 2009, discussed the colonial minters in a single-source reference containing the latest research.

- Philadelphia coin collector Andrew Madsen Smith published an *Illustrated History of the U.S. Mint*, which was released in later editions (over the course of nearly 20 years starting in the 1880s) by bookstore and gift-shop owner George G. Evans, who sold it to tourists and visitors. Frank H. Stewart published his *History of the First United States Mint, Its People and Its Operations*—an important reference about the Philadelphia Mint—in 1924. Stewart, his book, and the first U.S. mint were explored in great depth by Leonard Augsburger and Joel Orosz in *The Secret History of the First U.S. Mint*, published in 2011 and subtitled *How Frank H. Stewart Destroyed—And Then Saved—A National Treasure*.

- Neil Carothers (*Fractional Money*, 1930) and Don Taxay (*U.S. Mint and Coinage*, 1966) also wrote about the early Philadelphia Mint and its coins.

- Clair M. Birdsall wrote two studies of the history and coinage of the U.S. branch mints at Dahlonega (Georgia) and Charlotte (North Carolina), published in 1984 and 1988 respectively.

- Coin dealer and collector Rusty Goe has researched and written extensively on the history of the Carson City Mint.

- Numismatic historians Richard G. Kelly and Nancy Y. Oliver have written on many Mint-related topics, including the branch mint at San Francisco.

- Michael F. Moran's books *Striking Change* (2007) and *1849: The Philadelphia Mint Strikes Gold* (2016) explore the inner workings of the Philadelphia Mint from its early days to the "Renaissance" era of President Theodore Roosevelt and sculptor Augustus Saint-Gaudens.

- Richard Doty's *America's Money, America's Story* provides colorful background to the nation's coins and its mints.

- Roger W. Burdette's work, shared in his "Renaissance of American Coinage" trilogy and the *Journal of Numismatic Research* among other places, shines light on many previously unpublished aspects of the Mint.

- David W. Lange wrote the *History of the United States Mint and Its Coinage*, a richly illustrated and engaging study published in 2006—perhaps the closest in scope to the present work.

- My own *American Gold and Silver: U.S. Mint Collector and Investor Coins and Medals, Bicentennial to Date* (published 2016) offers behind-the-scenes coverage of modern Mint production, in particular at Philadelphia and West Point.

Each of these books, and others, focused on one or several parts of the giant tapestry of the United States and its mints. Before the *Guide Book of the United States Mint*, though, there was no single book that brought together all of these subjects and more, in vibrant full color and in fascinating depth, connecting the water-, horse-, and man-powered machines of the 1600s and 1700s with today's massive Gräbener presses that strike three-inch silver coins with 500 metric tons of pressure.

Our choice of author for this complex undertaking was easy. There is no better craftsman for such a volume than Q. David Bowers himself—Whitman Publishing's numismatic director, after whom the popular Bowers Series of references is named. For more than 60 years Bowers has made the study of American coinage his life's work. He's known many of the Mint's directors, superintendents, sculptors, and chief engravers as personal friends. In recent years I've had the honor of traveling with Dave on research missions to each of the Mint's operating facilities, in Philadelphia, Denver, San Francisco, and West Point, as well as its headquarters in Washington, D.C., with the blessing of its directors and public-affairs officers. The insight we've gained from these recent studies is icing on the cake for Dave's readers. His already remarkable knowledge is constantly expanding, and he shares that knowledge generously. His first-hand interviews with Mint personnel going back to the 1960s, plus decades of archived research at his fingertips (accessible with a photographic memory), are updated by frequent communication with Mint staff and ongoing study of new issues. The result is a historical record that lives and breathes. It is a landscape painting of the U.S. Mint stretching back to the colonial era but also as close as today's dimes and quarters jangling in our pockets.

The *Guide Book of the United States Mint*, like those coins, is for every American—artistic but also useful, and a tangible expression of our nation's history.

Dennis Tucker
Publisher, Whitman Publishing

A 1963 U.S. Assay Commission
medal showing commissioners
at the Philadelphia Mint.

"Ye Olde Mint" (the first federal
mint, in Philadelphia), on a
1952 Assay Commission medal.

EARLY AMERICAN MONEY AND COMMERCE

COMMERCE IN THE COLONIES

COMMODITIES AND BARTER

In the area that later comprised the United States of America, the first permanent English settlement was Jamestown (also called James City), in what became known as the colony of Virginia, on the James River off of Chesapeake Bay. It was established in 1607. New Amsterdam (renamed New York in 1664) was settled in the years after the 1609 arrival of Henry Hudson in the mercantile ship *Halve Maen* (*Half Moon*) by subsequent Dutch-employed explorers such as Afro-Dominican trader Juan Rodriguez. The landing of the Pilgrims at Plimoth (Plymouth) Plantation in what became known as the Massachusetts Bay Colony, comprising much of present-day New England, became an important settlement in the North.

Distant from their homelands, these and other settlers traded mainly under the barter system, exchanging goods and commodities. In time, coins were also used, but they were in short supply. The coins used included silver and gold coins from the Spanish-American colonies in Mexico and Central and South America and European coins of Great Britain, the Netherlands, France, and other countries. Gold coins from many lands were also used, and various copper coins and tokens were prevalent for small transactions.

Landing of the Pilgrims in America, 1620.

Credit was also common, especially in rural areas. Merchants in the colonies kept ledgers for their local and regional customers, adding credits and subtracting debits. Depending on the demand for them, hand-made clothing and shoes, poultry and eggs, produce, and other goods would be traded. Outside of cities, such commerce lasted for a long time—well into the early decades of the 19th century in some areas. Labor also had an exchange value. In New Hampshire in the early 1800s a day's labor might be valued at 50¢ in trade, and a day's labor of a man and a workhorse at $1. A popular term for produce and crafted goods in lieu of coins was "country pay." Other examples of bartered commodities included musket balls, nails, and hides.

Some international trade in commodities was conducted by barter. Lumber, fish, flour, and other products were shipped from New England to the Caribbean and related islands and traded for sugar and molasses, which were then used in the British colonies to make rum. This *ardent liquor*—strong alcoholic spirits—was "money" in its own right, easily transported, and used to buy other products. Most notably, rum was used in the *triangular trade*, wherein the products of the New World, such as rum or tobacco, were traded in Europe, where traders acquired more goods, which they used to buy slaves from African traders, who were then taken to the New World to labor in the making of rum and the processing of natural resources such as sugar and cotton. Tragically, the slaves themselves were considered trade goods, and slavers stuffed and overcrowded slaves into the holds of their ships—where many suffocated, were smothered, or otherwise died due to conditions—in order to ensure delivery of the maximum possible amount of these humans who were being used as a trade commodity in New World markets.

MONEY OF THE COLONIES

VIRGINIA AND MARYLAND

In the Mid-Atlantic region Virginia and Maryland were under British rule, and each often shared monetary and commercial rules. The primary agricultural product of both was tobacco, which also served as currency, both in everyday commerce, and for the payment of taxes. In 1618 the Virginia Company placed a value on tobacco of 3 shillings per pound. At the time, for a tobacco grower working for the Virginia Company to transport a young woman brought from England and take her as a wife cost 100 pounds of tobacco. This was not a slave sale, as the women volunteered to come to the New World and marry. The Virginia Company even gave credit to men who could not pay to transport a woman because they believed marriage was so essential to a worker's happiness, and therefore an exploitable resource. At one time, women wishing to journey to the colonies became scarce, and the price for a wife was raised to 150 pounds. This was "under protection of a law that debt for wives should take precedence of all other debts and be the first recoverable."[1]

Furs obtained from trade with Native Americans also had exchange value, as did gunpowder and lead shot. For purposes of evaluation and comparison with each other, these items often had their worth stated in the British system of pounds, shillings, and pence, although such coins were scarce. It took 12 pennies (or pence: the penny was abbreviated as "d," after the Roman *denarius*) to make a shilling (designated "s" after the Roman *solidus*), and 20 shillings to make a pound (£). Twenty-one shillings equaled a gold guinea.

Grain, country pay, or corn were general term for corn, peas, and wheat, which were often traded in bushel units. These forms of pay often had specified exchange values as did fish, oatmeal, and cattle in some colonies north and south along the Eastern Seaboard. In the Carolinas pine tar and rice often served as currency.

Over time values changed, depending on the quantities of each trade good available. In Virginia in 1640, 21 years after tobacco was mandated as the official currency, overproduction resulted in a sharp decrease in value. All tobacco considered to be of "bad" quality and half of that considered "good" was ordered to be burned. Payments were to be made in coins after that time, but coins were not plentiful. Most in circulation were British copper and silver and Spanish and Spanish-American silver and gold. These coins tended to be exported to buy goods in distant lands, particularly in Europe. At various times cross-currency exchange standards were set, as in Virginia in 1645 when the legislature established the value of the Spanish 8 reales (also known as the *piece of eight* or *dollar*) as equal to 6 British shillings.

On June 3, 1642, Virginia and Maryland levied a tax of 2 shillings on every *tithable* person—a member of the productive labor force—in the colonies. To facilitate payment it was decreed that corn would be valued at 10 shillings per barrel, wheat at 4 shillings per bushel, beef at 3-1/2 pence per pound, "good hens" at 12 pence, capons at 1 shilling 6 pence, six-week-old calves at 25 shillings, butter at 8 pence per pound, "good weather" goats at 20 shillings, "pigs to roast at three weeks old" at 3 shillings, cheese at 6 pence per pound, and geese, turkeys, and kids at 5 shilling each. These were to be collected at appointed places in the colonies and sent by boat to Jamestown to be deposited in the governor's treasury.[2]

Tobacco continued in general use and was subject to regulations as to value and the maximum amount that a recipient was obliged to take. In 1666 Virginia and Maryland banned the planting of tobacco in order to preserve its value, an action that caused great distress for those raising the crop.

Cattle and other livestock were used as money in certain colonies.

Picking tobacco—a popular medium of exchange in Maryland,
Virginia, and certain other areas during the colonial era.

THE NORTHEAST

In New York beaver pelts were actively traded, and in New Hampshire farm products, lumber, and salted fish had value. Certain crops and cattle had to be cared for or stored with protection to preserve their monetary value. On the plus side, such exchange goods were immune from counterfeiting, and when quantities increased or decreased they did slowly, with owners being aware of the change in value.

A list of commodities valued as money by the Massachusetts General Court in 1727 includes pork, beef, wheat, barley, Indian corn, oats, peas, hemp, flax, beeswax, butter, hides, leather, dried codfish, mackerel, oil, bayberry wax, turpentine, bar iron, cast iron pots and kettles, well cured tobacco, and good tallow, among other things.[3] On March 4, 1635, the Massachusetts General Court adopted legislation which provided that musket balls of full bore were to trade at a farthing (one fourth of a penny), but no recipient was compelled to take more than a shilling (24) of them at a time. Values were adjusted from time to time.

Indian wampum (also known as *wampampeage* and other variations) made from shells was popular in trade, particularly in the Northeast, and was subject to specified exchange rates in British money.

In 1630 wages and prices were strictly regulated in an effort to control the economy. A common laborer could receive no more than 12 pence per day, a skilled carpenter no more than 16. Despite the rigid regulations, by 1640 England and Ireland had drained the colonies of nearly all cash money, and coins became so scarce and coveted that they were worth far more than their normal exchange value in goods. The value of wampum rose to four white or two purple per penny.

At Harvard College, founded in 1636, the tuition of £1 6s 8d could be paid several ways, including by about 1,900 beads of white and blue (purple) quahog wampum. In time, mass-produced glass beads became plentiful as did counterfeit shell beads, leading to value depreciation, and the use of wampum declined.

North America in 1733. (H. Popple, map)

Rural New England in the 18th century. Most commerce was done by barter.

OTHER COLONIES

As trade with the Dutch in New Amsterdam (New York) was extensive, the Massachusetts General Court on September 27, 1642, specified that a Holland "ducatour being worth 3 gilders" would be worth 6 shillings, and a *rix dollar* of 3-1/2 guilders, worth 5 shillings. Massachusetts was especially important in maritime trade. The large silver

Dutch *leeuwendaalder* (Lyon or Lion dollar) and French *écu* saw extensive circulation, as did the gold coins of Europe and their New World colonies. Some New York bills of 1709 were even denominated in Lyon dollars.

The Province of Carolina (later separated into North Carolina and South Carolina to become part of the 13 original colonies) and Georgia had their own rules. Their commerce was primarily based on agriculture. In time, cotton and tobacco became important. No mints were established in those districts in pre-federal times, and thus they are not discussed in detail here.

FOREIGN COINS IN AMERICA

Spanish and Spanish-American Coins

Silver and gold coins of Spain and the Spanish colonies in America were important in American commerce. After the first several generations of British settlements, the Spanish coins became more significant in trade than English issues. All the while, Spanish silver and gold coins were dominant in worldwide maritime trade. Sailing vessels out of the ports of Boston, Salem, New York, Baltimore, Charleston, New Orleans, and other cities typically carried thousands of silver 8-reales coins, known as Spanish dollars. These were readily exchangeable for goods at all foreign ports.[4] No other *coins* in world monetary history, since the ancient days of the famous "owls" of Athens, ever enjoyed such a high status, although later currencies such as the American dollar did.

From the 17th century onward, Spanish mints in Mexico, Chile, Peru, Columbia, and elsewhere produced large quantities of coins, bearing the portraits of Spain's monarchs. These were denominated in reales (singular: *real*), with eight reales equaling a Spanish dollar. The dollar was divided into eighths or "bits," the real worth of each bit being 12-1/2¢. From this, the term "two bits" for a 25¢ coin passed into the modern idiom.

1757 Spanish 1-real coin or "bit," worth 12-1/2 cents in America.

1759 8 reales or "piece of eight" struck at the mint in Mexico City, with the Mo mintmark to the right of the date. Such coins were standard worldwide in maritime trade.

In the 1850s the Meschutt family, who operated the Metropolitan Coffee Room at 433 Broadway, New York City, advertised their business by counterstamping coins, such as on this well-worn 1785 2-reales coin that seems to have been in circulation for many years. Many foreign coins were legal tender in the United States until August 21, 1859.

Gold coins were reckoned in escudos, with 8 escudos equaling a gold doubloon, worth about 16 Spanish dollars. In early times Bolivia's Potosí mint was the dominant coiner, and it was eventually succeeded by the mint in Mexico City. Over time, 14 different mints were active in Mexico. The doubloons found their place in legend and lore relating to pirates.

Deeds, contracts, and other documents involving money were usually drawn in English or Spanish money. In 1775, when the Continental Congress began issuing paper money in the prelude to American independence, and continuing several years later when the United States of America was formed, the bills were denominated in Spanish milled dollars.

Almanacs and navigation manuals often listed exchange rates, as did currency-exchange booklets called *cambists*. In these the Spanish dollar was usually featured more than any other monetary unit.

1724 8 escudos, or doubloon, struck at the mint in Lima, Peru, from early and somewhat rustic dies.

1797 8 escudos struck in Bogotá, Columbia, depicting King Charles IIII (IV) of Spain.

UNITED STATES OF AMERICA.

Accounts are kept here in different ways, but chiefly in Dollars, which are *Monies of Account.* divided into 10 Dimes, 100 Cents, or 1000 Mills. This is called Federal Money, to distinguish it from the various currencies which were formerly the monies of the United States, and which are still partially retained in domestic traffic; but in foreign commerce and all Government concerns Federal Money is used, which was established by an Act of Congress in 1789, and in which the Dollar is valued at 4s. 6d. sterling.

Accounts were originally kept, in all the British Colonies of America, as in England; and British coins circulated, as well as Spanish Dollars; but in consequence of excessive issues of paper, various depreciations took place, which were liquidated at different rates of composition. Hence arose the different currencies, which were established by raising the nominal value of the Pound sterling and of the Spanish Dollar, so as to correspond with the depreciations respectively. The following are the different ratios of sterling to currency, and also the proportional values of the Spanish Dollar:

1. In New Hampshire, Massachusetts, Connecticut, Rhode Island, Virginia, *Currencies.* Kentucky, Ohio, Tennessee, Indiana, and Mississipi, the ratio is as 3 to 4. Thus £1 sterling = £1 6s. 8d. currency, and £1 currency = 15s. sterling. Hence also the Dollar is worth 6s. currency.

2. In New York and North Carolina, the ratio is as 9 to 16; and hence £1 sterling = £1 15s. 6⅔d. currency, and £1 currency = 11s. 3d. sterling. The Dollar here is therefore worth 8s. currency.

3. In New Jersey, Pensylvania, Delaware, and Maryland, the ratio of currency to sterling is as 3 to 5. Hence £1 sterling = £1 13s. 4d. currency, and £1 currency = 12s. sterling. The Dollar is therefore worth 7s. 6d. currency.

4. In South Carolina and Georgia, the ratio is as 27 to 28; and hence £1 sterling = £1 0s. 8⅓d. currency, or £1 currency = 19s. 3⅓d. sterling. The Dollar is therefore worth 4s. 8d. of this currency.

The above currencies may be reduced into each other by the proportional *Currencies Reduced.* values of the Dollar. Thus, to reduce the currency of New Hampshire to that of New York, multiply by 8 and divide by 6; or, add ⅓.

VOL. I.   Z Z

A page from *The Universal Cambist and Commercial Instructor*, by P. Kelly, LL.D., second edition, London, 1821, reflecting the omnipresent use of the Spanish dollar in America and conversion rates into British money. The rates varied from colony to colony.

British Coins

The British colonies in America used pence (d), shillings (s), and pounds (£) in official documents and regulations, often with conversion tables to show equivalent values in reales and escudos. Copper coins in circulation in the colonies were predominantly English halfpence, about the size of a later (1793 onward) United States' large cent. Halfpence circulated by the millions and were the mainstay of small transactions. As was popular practice with most other countries, British coins featured the image of the monarch currently on the throne. Scattered other foreign coppers were seen on occasion, and if they were approximately the size of a halfpenny, they were valued as such.

1754 Great Britain copper halfpenny with the portrait of King George II. These small-denomination coins were very plentiful in America and served for change and small purchases.

The shilling was the most commonly mentioned British coin in colonial America. This 1703 shilling depicts Queen Anne and is inscribed VIGO. It was made from silver captured by the British Navy from ships in the harbor of Vigo, Spain, and commemorating that victory.

A shilling was worth 12 pence. As there were no coins of the penny denomination in wide circulation until 1797, it took 24 halfpence to exchange for a shilling.[5] The shilling was ideal for most transactions by citizens buying goods or services, and many if not most transactions were reckoned in multiples of that coin. Even well into the 19th century in America many goods and services, such as admissions to events, were priced at such values as 12-1/2 cents, 25 cents, and 37-1/2 cents, these being tendered in silver shilling coins.

Five shillings equaled a crown, roughly equivalent to a Spanish-American 8-reales or a French 5-franc piece. Twenty British shillings equaled a pound, which for a time was equal to a gold guinea. After 1717 a guinea was equivalent to 21 shillings instead of 20. Guineas were struck until 1814. However, in Great Britain certain goods such as luxury items continued to be priced in guineas instead of pounds, until British money was decimalized in the 1970s—tradition, you know!

British pounds, shillings, and pence continued to be used in American ledgers and commercial accounts into the 1820s and 1830s, especially in small towns in the Northeast.

Other Foreign Coins in Colonial Times

Coins in the colonies came in many other forms as well, particularly silver and gold issues from Holland, Portugal, various German states, Venice, and elsewhere. After the American Revolution, some of these were made legal tender by federal law. Such rules were revised from time to time. Newspapers in the larger cities often had "Prices Current" and related columns giving the going rates for certain commodities and also exchange equivalents for coins. These rates could vary from colony to colony and were often expressed in "New York money," "Pennsylvania money," or similar. Travel of citizens between the colonies was small; most people, excepting those in the military or mercantile pursuits, were apt to be born, live, and die without going far from home. Accordingly, rules of commerce and rates of exchange were often regional, and none encompassed all of the colonies. There were, however, stage roads connecting cities in the East that were used for mail and commerce.

In the 18th century, as part of the flood of European coins brought to America, such low-denomination coins as Mark Newby and Wood's Hibernia coppers were used here, although they were not made specifically for this side of the Atlantic, nor did they bear any inscriptions relating to the colonies. Mark Newby coppers were legal tender for a time in West Jersey (largely today's New Jersey), but no such mantle was ever placed on Hibernia coinage. Irish coins of the Voce Populi type have no known connection with America. These European coppers have been adopted into American collections by numismatists, although there is no logic for doing so. Collectors also cherish the brass pieces produced in England circa 1616 for the Sommer Islands (Bermuda). At the time these were under the management of the Virginia Company, but that division

1767 advertisement for a stage line connecting Philadelphia and New York City. The price for a through trip was 20 shillings.

was separate from the colony on the American mainland. The authorization provided that these little coins, each depicting a wild hog, only be used on those islands. The main "currency" of the Sommer Islands during that time continued to be tobacco.

FOREIGN COINS IN LATER TIMES

It may come as a surprise to learn that Spanish and Spanish-America coins continued to be legal tender for many years. Anyone going shopping in, say, Charleston in 1825, Baltimore in 1840, or New York City in 1852, would see many Spanish-American silver coins and but few coins bearing the imprint of the United States of America. Spanish-American gold coins dominated commerce in the states until the 1840s.

A time capsule of how plentiful foreign coins were in America in the 1840s is furnished by coins recovered from the SS *New York*, a sidewheel steamer bound out of Galveston, Texas, headed for New Orleans, that was lost in a hurricane in the Gulf of Mexico on Monday, September 7, 1846. Gold included coins from small denominations to large from Chile, Columbia, Denmark, France, German states, Great Britain, Italy, Mexico, the Netherlands, and Spain, with those of France and the German states being especially plentiful. The silver carried on board was less diverse, but it included many foreign issues. The Act of February 21, 1857, ended the legal-tender status of foreign gold and silver coins, effective two years from the time of declaration, although the deadline was later extended by six months. It was not until the mid-1850s that there were enough federal (minted by the United States) silver coins in circulation to be dominant.

The SS *New York*.

1844 Denmark 2–Christian d'or,
Altona mint.

1811-A France 20-francs.

1845 8-reales, Guanajuato Mint, Mexico, from
the SS *New York*, lost at sea on the third day
out of Galveston, Texas, headed to New Orleans.

1843 Great Britain sovereign.

1795 gold Saxony 10-talers.

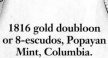

1824 gold doubloon
or 8-escudos, Chile.

1816 gold doubloon
or 8-escudos, Popayan
Mint, Columbia.

In a letter dated February 6, 1857, Mint Director James Ross Snowden advised Secretary of the Treasury James Guthrie that although earlier laws had made Spanish and other dollars legal tender at the Mint, the same was not true of fractional denominations. Notwithstanding this, the minor coins have "held a far larger place in our currency than [dollar-sized coins] named in those acts." Snowden said that many fractional coins had been received at the Mint for recoining into United States silver denominations. The average buying price for coins at the Mint was as follows:

> Spanish pillar dollars $1.053. "There is a class of dollars which is generally mistaken for the true Spanish pillar dollar, but is decidedly inferior, being worth only $1.01 on the average, and withal very irregular. They were coined by the royal party in the revolutionary times of Mexico, and may be known by their defaced appearance, which is not due to wear, but to blows of the hammer, by which they were coined, whence they are called hammered dollars."
>
> Dollars of Mexico, mixed $1.063.
>
> Dollars of Peru, prior to recent changes $1.065.
>
> Dollars of Bolivia $1.065.
>
> Dollars of Central America $1.027.
>
> Five francs of France, mixed $0.99.
>
> Worn Spanish quarters $0.235.[6]
>
> Worn Spanish eighths $0.109.
>
> Worn Spanish sixteenths $.052.

EARLY PAPER MONEY

PAPER MONEY IN COLONIAL TIMES

From time to time paper currency issues were produced by the various colonies, the earliest being the issue of Massachusetts dated December 10, 1690.

From this point the 13 colonies issued their own paper money in many forms, extending into the Revolutionary War era and beyond, by which time they had become states of the Union. Vermont, admitted as the 14th state in 1791, also issued its own currency. Most of these notes were *bills of credit* or, in today's parlance, IOUs. Currency was issued and circulated, often in payment for obligations of the state, but had a redemption date of a later time, such as three years hence.

Paper issues printed for colonial monies of account usually were reckoned in either English pounds, shillings, or pence, or in Spanish dollars. For example, an early note of Pennsylvania bore the inscription:

> This bill shall pass current for five shillings within the Province of Pennsylvania according to an Act of Assembly made in the 31st year of the reign of King George II. Dated May 20, 1758.

A note of Delaware was inscribed:

> This indented bill shall pass current for Fifteen Shillings within the Government of the Counties of New Castle, Kent, Sussex on Delaware, according to an Act of Assembly of the said Government made in the 32nd Year of the Reign of our Sovereign Lord King George II. Dated the 1st Day of June, 1759.

Paper money issued by the colonies tended to stay within those particular colonies and was of little value or interest elsewhere. A bill of Massachusetts, if presented in South Carolina, would be accepted only at a discount, if at all. The British Crown controlled the colonies on the coast, and did not specifically authorize paper money. However, its use was tolerated as an expedient for commerce or financing projects such as military activities. Generally, bills were receivable for taxes, but they often expired or became invalid after a specific length of time.

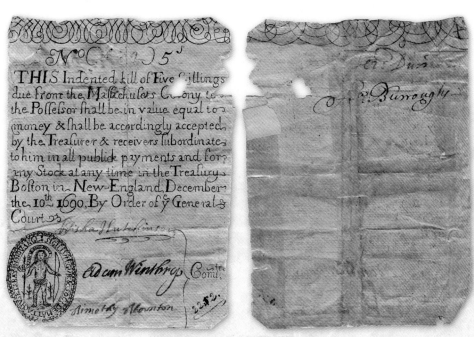

December 10, 1690, Massachusetts 5-shilling note—
the first paper-money issue of an American colony.

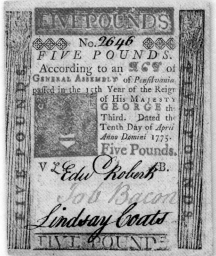

Pennsylvania note for 5 pounds, May 10, 1775.

Counterfeiting was endemic in the colonies, despite warnings printed on many notes that it was punishable by death to do so. The prevalence of bad bills in commerce led the authorities to change the designs of official currency from time to time. In some instances it was difficult to distinguish the good from the false, as when Robert Towle, formerly an authorized printer for certain New Hampshire notes, made some for himself.

Continental Currency

On May 10, 1775 the Continental Congress authorized the issuance of paper money. By this time each of the colonies had produced its own bills, making the subject of currency very familiar to the delegates. Although independence was coming, these entities were still under the British crown. A typical note of the first issue included this printing:

> This Bill entitles the Bearer to receive EIGHT Spanish milled DOLLARS, or the Value thereof in Gold or Silver, according to the Resolutions of the CONGRESS, held at Philadelphia the 10th of May, 1775.

Under the name of the United Colonies, the first issue was followed, through 1779, by others, all denominated, as above, in "Spanish milled dollars, or the value thereof in gold or silver," although the well-intentioned colonies had no formal government or any silver or gold coins to fulfill the promise. Notes valued from 1/6 of a dollar up to $80 were issued in what is known as the Continental Currency series.

$30 note issued by the United Colonies, May 10, 1775, a year before the Declaration of Independence. These and other notes were payable in Spanish milled dollars (8 reales), but in reality there was little coinage available to redeem them.

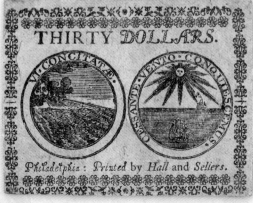

All of these bills were printed by Hall and Sellers, of Philadelphia, well known for producing state currency as well. Certain of the backs featured the already popular "nature printing" devised by Benjamin Franklin, in which plant leaves were used as part of the design.

During the time they were issued the bills constantly depreciated in value, giving rise to the saying, "Not worth a continental."[7] By 1790 the value of Continental bills had diminished to the point at which a silver dollar coin could buy several hundred dollars or more in such currency. Then came the act of Congress of August 1, 1790, which provided that old Continental bills would be receivable at the United States Treasury at the rate of $1 in coin for $100 in paper, an action which is said to have yielded a fortune to certain speculators holding large quantities of such money. This offer expired on September 1, 1791. Then the act of May 8, 1792, revived the policy, which was extended until March 7, 1793. After that time the bills had no official exchange value.

$8 Continental Currency note of May 11, 1778. By this time newly issued notes traded at a sharp discount from face value. The back plate had real leaves that imprinted their design, in a technique known as "nature printing."

Lord Timothy Dexter, as he styled himself, lived in a mansion on High Street in Newburyport, Massachusetts, with a yard decorated with life-size statues of figures in history. Although facts are scarce, Dexter is said to have made a fortune speculating in Continental bills and also in maritime trade with the West Indies.

EARLY MINTS IN BRITISH NORTH AMERICA

MINTS IN THE STATES

The colonies in what would later become the United States were mostly under the control of the British Crown, New Amsterdam excepted, however it became British by capture in 1664 and was renamed New York.[1] Rules for coinage were created by Parliament and by royal decree, and the colonists had little say, though they were allowed to set the exchange rates of foreign coins in circulation.

Establishing a local mint would have helped alleviate the nearly continuous shortage of coins in circulation, which is exactly what the Massachusetts Bay Colony did, although not with the blessing of Great Britain. Under the auspices of Lord Baltimore a mint was contemplated for Maryland in the late 1650s. It did not materialize, but many Maryland-themed silver coins were struck at the Royal Mint in London and circulated in the colony.

After the American Revolution the situation changed, and several mints were set up under state management (in Massachusetts) or by contract with private individuals.

In the meantime, from the early days onward a number of jewelers, assayers, and other small enterprises—a list which includes Samuel Higley in Connecticut, John Chalmers in Annapolis, and Ephraim Brasher in New York, among others—privately produced unofficial coins. Each of these is discussed in this chapter.

Colonial and early American coins have been numismatic favorites for a long time. Reflective of this, the entry below for Vermont copper coins gives details of the 12 different varieties struck at a small mint operated by Reuben Harmon Jr. in Rupert, Vermont, between 1785 and 1786. For expanded information on all issues see the *Whitman Encyclopedia of Colonial and Early American Coins*, 2008.

EUROPEAN COINS WITH AMERICAN INSCRIPTIONS

Outside the scope of mints in America, but with varying degrees of importance in the monetary system, were coins made in Europe, primarily in Great Britain, intended to be circulated on this side of the Atlantic and with specific inscriptions relating to America.

Cecil Calvert, the second Lord Baltimore and living in England, was in charge of the Maryland colony and as such assumed that he had the right to coin money for use in

Maryland. Denominations including the copper penny and silver fourpence (groat), sixpence, and shilling were made in London and shipped to his brother Philip Calvert, who circulated them in commerce. They were familiar sights for many years afterward.

In 1688 Richard Holt received a license or patent from the British Crown to make coins for circulation in the American plantations. These were denominated at 1/24 of a real, or about a farthing, one-half of a halfpenny.

A Maryland sixpence of 1659, undated, struck in London and sent to the Maryland colony.

William Wood, a British entrepreneur, under a patent from King George I (said to have been arranged through some sort of liaison with the Duchess of Kendal), minted Rosa Americana coins in denominations of halfpenny, penny, and twopence, mostly in 1722, 1723, and 1724. These were made of a special alloy called *bath metal* and had a yellowish cast. The inscription ROSA AMER. UTILE DULCI translates to "the American rose, the useful with the sweet." Many of these circulated in the British Isles as well.

An undated (1688) token struck in tin for circulation in the American plantations. These were denominated at 1/24 of a real.

In 1773 the Royal Mint struck a large quantity of copper halfpence with inscriptions relating to Virginia. These circulated widely at the cusp of the Revolutionary War and remained in commerce for years afterward. Several thousand Mint State coins survived in a hoard later owned by Colonel Mendes I. Cohen—these account for most of the high-grade pieces in existence today.

A Rosa America halfpenny of 1722 struck in bath metal. Coins of this type were made by William Wood in Great Britain and were intended for circulation in America.

In the 1780s large quantities of half-penny-size copper tokens were struck in Birmingham, England, a center for the private minting of coins and tokens. Imported in quantity to the United States they circulated widely here.

In 1796 half dollar–size silver coins or medalets were struck in Paris for use in Castorland, a district in Upstate New York

A 1773 Virginia copper halfpenny struck in London for circulation in the Virginia colony.

A 1785 Nova Constellatio halfpenny-size copper coin, part of a series of such pieces dated 1783 and 1785.

A 1796 Castorland medalet or coin from a French settlement in Upstate New York. *Castor* is the Latin word for beaver, one of which is shown at the bottom of the reverse.

settled by the French. It is not known if these circulated in regional trade or if they were considered to be awards of some kind.

France struck many coins for the New World, but none with inscriptions relating to lands that later became part of the United States. *A Guide Book of United States Coins* lists many varieties.

MINTING IN COLONIAL AMERICA

Massachusetts Bay Colony
Boston, Massachusetts (1652–1682)

The Massachusetts Bay Colony traces its formation to the Puritan settlements of Salem (1628) and Boston (1629), a few years after the Pilgrims established the Plymouth Plantation in 1620. Under the new charter of 1691, these two colonies, together with Maine and Nova Scotia, were combined into a single political entity. In early times commerce was conducted by the barter system and also by assigning monetary values to certain goods and commodities. Coins, primarily British, Spanish, and Spanish-American, were scarce and were often exported.

On May 27, 1652, the Massachusetts General Court passed an act that provided that a mint be established in the colony. John Hull was appointed as mint master, and his associate, Robert Sanderson, was a partner. Silver threepence, sixpence, and shillings, denominated III, IV, and XII, were to be coined. As compensation Hull received 1 shilling 3 pence from every 20 shillings coined. This arrangement was renegotiated in 1667. In due course it was determined that a mint house "16 feet square, 10 feet high, and substantially wrought" be established in Boston on land belonging to Hull.

The initial Massachusetts silver coins produced were of simple design and consisted of a silver disc stamped on one side with NE, for New England, and on the other side and at the other end of the planchet (so that the NE impression would not be flattened), the denomination III, VI, or XII. Most coinage was of the shilling denomination, with many fewer being made of the sixpence. Only a few threepence were made, as evidenced by one known to exist today. These were the first coins struck in what would become the United States.

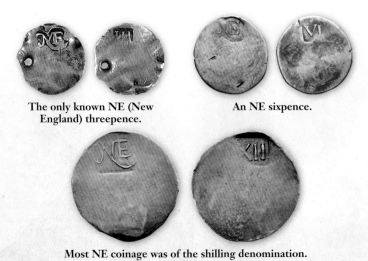

The only known NE (New England) threepence.

An NE sixpence.

Most NE coinage was of the shilling denomination.

It was soon thought that the simple design made it easy to clip silver slivers from the edge. On October 19, 1652, the General Court determined that:

> Henceforth all pieces of money coined as aforesaid shall have a double ring on either side, with the inscription MASSACHUSETTS and a tree in the center on one side, and NEW ENGLAND in the year of our Lord on the other side . . .

The change necessitated new procedures. Two steel dies were now needed—one for the obverse and the other for the reverse. Each was hand-engraved, and rather crudely, thus the letters N and E as in NEW ENGLAND would each be different. Each of the three denominations—again the III, IV, and XII—bore the date 1652, referring to the year that Massachusetts silver coinage was first authorized. This year, with one exception, was used on all later coinage, which extended to 1682.

W. Elliot Woodward in his Sixth Sale, 1865, suggested the design resembled a "palmetto tree." The earliest use of *willow* found by Sydney P. Noe, a student of the series, was in Woodward's 1867 sale of the Joseph Mickley Collection. The Willow Tree pieces, as they are universally designated by numismatists today, were apparently produced by crude means, for all known specimens show evidence of multiple strikings. Although details are not known, it is probable that the blank silver planchet discs were placed between two loosely fastened dies and hit with a sledgehammer. The Willow Tree coins are thought to have first been made in 1654 and, according to current thought, continued until 1660. If so, the production must have been sporadic, for all such coins are rare today.

Following the Willow Tree coinage, the design was changed to the new so-called Oak Tree style. Certain of these were likely made by a rocker-type press, as used on the Willow Tree issues. Oak Tree coinage was effected in several different denominations, including the customary threepence, sixpence, and shilling, all with the 1652 date. A new denomination was introduced, the twopence which bore the date 1662, thus standing alone among all Massachusetts silver coins in this regard. This is the year that the General Court ordered the mint master "to coin twopence pieces of silver, in proportion according to the just value and alloy of the monies, allowed here, to answer the occasions of the country for exchange."

Only three examples of the 1652 Willow Tree coinage are known today.

A 1652 Willow Tree sixpence.

Most 1652 Willow Tree coinage was of the shilling denomination.

Oak Tree coinage continued in production until the type was replaced by a new design displaying on the obverse a clear representation of a pine tree.

The new motif depicts the tree shown on the flag of the Massachusetts Bay Colony. Denominations in the Pine Tree coinage are the threepence, sixpence, and shilling. Shillings were made in two formats, earlier styles (until about 1675) on thin, broad planchets and using a rocker press, and later issues (1675–1682) on smaller, thicker planchets struck on a screw press. Both styles were made in large numbers. The number of specimens existing today suggests that the small-planchet Pine Tree shillings may have been made in the largest quantities of all.

Massachusetts silver coins remained in circulation in the Northeast until well into the 1700s. They were also familiar sights in Canada to the north and the Middle-Atlantic states to the south.

The Oak Tree twopence was authorized in 1662 and examples of this, the lowest of the Massachusetts denominations, are unique in bearing that date instead of 1652.

A 1652 Oak Tree threepence.

A 1652 Oak Tree sixpence.

Most 1652 Oak Tree coinage was of the shilling denomination, following the general rule of popularity throughout the Massachusetts silver issues.

A 1652 Pine Tree threepence.

A 1652 Pine Tree sixpence.

A 1652 Pine Tree shilling with large planchet.

A 1652 Pine Tree shilling with small planchet.

Higley Coinage
Simsbury, Connecticut (1737–1739)

Among the most interesting and rare of all early American issues are the copper coins or tokens struck circa 1737 by Dr. Samuel Higley, of Granby, Connecticut, or by persons of his acquaintance. Higley, a medical doctor with a degree from Yale College, also practiced blacksmithing and made many experiments in metallurgy. In 1727 he devised a practical method of producing steel.

In 1728 Higley purchased property on a hill in Simsbury (the area later known as East Granby) which furnished the site for many copper mines, the most famous being the extensive mine corridors and shafts which were later used as the Newgate Prison. Mines on the hill were worked extensively during the early and middle 18th century.

Higley conducted a small mining business that extracted and refined high-quality copper ore. It is thought that most of the metal was exported to England. In or around 1737 Higley, or someone associated with him, is thought to have made copper coins for local and regional use. Traditional lore, such as in S.S. Crosby's *Early Coins of America*, 1875, is rich, but facts and contemporary information are scarce.

While on a voyage to England in May 1737, on a ship loaded with copper from his own mine, Samuel Higley died. His brother, John Higley Jr., together with Rev. Timothy Woodbridge and William Cradock, probably engraved and struck the issues of 1739 and possibly those of 1737 as well. Numismatic historian David Gladfelter has noted:

> The problem in ascribing the 1737-dated tokens to Samuel Higley is the lateness of their dates. The date 1737 would not have been used in Connecticut until March 25, 1737 (then New Year's Day under the old-style calendar) and thereafter. Samuel Higley lost his life at sea in May 1737, so he would have had only a few weeks to prepare dies and strike tokens dated 1737.[2]

The obverse of a popular variety depicted a standing deer with the legend THE VALUE OF THREE-PENCE. The reverse showed three crowned hammers (derived from the arms of the English black-smiths' guild) with the surrounding legend, CONNECTICUT, and the date 1737.

A 1737 Higley copper with standing deer and three crowned hammers.

Ruins of the old Newgate Prison atop the old copper-mine tunnels in East Granby, Connecticut.

A 1739 Higley copper with standing deer, broad axe, and I CUT MY WAY THROUGH.

A 1739 Higley copper. THE WHEELE GOES ROUND combined with the broad axe die.

Legend has it that drinks in the local tavern sold at the time for threepence each, and Higley was in the habit of paying his bar bill with his own coinage. There was a public cry against this, for the Higley copper threepence was of a diameter no larger than the contemporary English halfpence that circulated in the area—coins that had a value of just 1/6 of that stated on the Higley coin. Accordingly, the inscription was changed so that the obverse legend read VALUE ME AS YOU PLEASE. The pieces still bore an indication of value, the Roman numeral III below the standing deer.

Two new reverses were designed, one of which pictured three hammers with the inscription I AM GOOD COPPER. The other reverse, picturing a broad axe, had the legend I CUT MY WAY THROUGH.

The third obverse design, of which only a single specimen is known, depicted a wagon wheel with the legend THE WHEELE GOES ROUND.[3]

Apparently the quantity of original Higley coinage was small, and circulation was effected mainly in Granby and its environs. The average diameter of a Higley copper is about 28.6 mm. All have a plain edge.

MINTAGES OF 1776

New Hampshire Coinage

On March 13, 1776, the New Hampshire House of Representatives voted that a committee be established to consider the production of copper coinage. The members reported that it would be beneficial to produce copper issues as the Continental Currency and other paper-money issues in circulation were too large in denomination for small transactions.

A 1776 New Hampshire copper equal in value to 1/108 of a Spanish dollar.

A design was submitted and adopted, showing on the obverse a tree and the words AMERICAN LIBERTY, emblematic of the revolutionary spirit prevailing at the time. The reverse was to depict a harp and the date 1776.

William Moulton, an accomplished silversmith, was recommended for the franchise to produce up to 100 pounds of coppers to be submitted to the General Assembly prior to circulating the pieces. It was further recommended that 108 of these pieces be equal to one Spanish milled dollar, with the weight of each individual coin being equal to the current English halfpenny. These were not made in quantity, and only a few exist today.

Massachusetts Patterns

While silver coins were produced by the Massachusetts Bay Colony beginning in 1652 (and extending until 1682), it was not until nearly a century later in 1776 that copper pieces were made. In the latter year at least three varieties of coins, all presumably patterns, were struck. Very little is known today concerning their origin or under what circumstances they were made.

A 1776 Massachusetts copper "penny" bears on the obverse the representation of a pine tree with MASSACHUSETTS STATE. The reverse shows a woman with a staff (inspired by British coppers) seated on a globe, with LIBERTY AND VIRTUE / 1776. The only known example is in the Massachusetts Historical Society.

A 1776 Massachusetts Pine Tree copper with a seated figure on the reverse.

Another 1776 Massachusetts copper is of a different design and features on the obverse the standing figure of an Indian. The reverse depicts a goddess seated. This is also unique and is in the cabinet of the American Numismatic Society.

A 1776 Massachusetts copper with a standing Indian on the obverse and a seated figure on the reverse.

The third 1776 pattern issue of Massachusetts, also unique, has been known as the "Janus" copper halfpenny and on the obverse has three heads in one. Of the coin, once a part of the Garrett Collection, much has been speculated in the hobby's literature.

A 1776 Massachusetts "Janus" copper halfpenny.

MINTING AFTER THE REVOLUTION

John Chalmers
Annapolis, Maryland (1783)

John Chalmers, an Annapolis, Maryland, goldsmith and silversmith, produced in 1783 a series of silver coins of his own design. Values were of threepence, sixpence, and shilling, and several varieties were made.

The building in which these pieces were coined stood at Fleet and Cornhill streets. Although the coins had no official status and were privately issued, apparently the government took no exception to them. Some, perhaps all, dies were engraved by Thomas Sparrow, who also made plates for Maryland paper money, though some could have

A Chalmers threepence. The obverse has two clasped hands at the center, and the reverse is with an inscription enclosing a branch surrounded by a wreath. This variety is very scarce.

A Chalmers sixpence. The obverse has a star surrounded by a wreath, and the reverse has an ornamented cross. Several die varieties of this coin exist.

A Chalmers shilling. The obverse has two clasped hands encircled by a wreath, and the reverse has a beaded circle at the center. An adjacent circular line encloses a scene with a bar or fence across, a serpent above, and below, two birds competing for the same worm. Two die varieties exist: Short Worm (illustrated) and Long Worm.

A Chalmers shilling with rings on the reverse. The obverse has three lines of script at the center: *Equal / to One Shi*. With an ornament above and clasped hands below. The reverse has 12 linked rings around border, 11 of which enclose a five-pointed star. At the center, a ring from which a liberty pole and cap extend upward with a five-pointed star to each side of the cap. Five are known to exist.

been engraved by Chalmers, who certainly had the talent. Coinage apparently was quite extensive, for several hundred examples are known today. Most show evidence of considerable use in commerce.

On all, inscriptions are around the borders. Diameters vary but average for the threepence is 13 mm, sixpence 17.5 mm, and shilling 22 mm. Their edges were reeded by hand.

COPPER COINAGE OF 1785–1788

Republic of Vermont
Rupert, Vermont (1785–1786)

The brief series of 12 copper coins attributed to the mint in Rupert, Vermont, provides the opportunity to discuss an early American specialty at length, perhaps conveying the reason why such series have attracted much interest over the years. More extensive series, such as Massachusetts silver coins of 1652 to 1682 and Connecticut copper coins of 1785 to 1788, if discussed here in the same detail, would necessitate arranging the present study into several volumes.[4]

COINAGE AUTHORIZED

Of the several states that issued copper coins during the 1780s the earliest was Vermont, an area which was not formally a state until its admittance to the Union in 1791. Its coins were issued by the Republic of Vermont, an independent entity at the time, although it considered itself to be the 14th state.

On June 10, 1785, Reuben Harmon Jr., a storekeeper and entrepreneur of Rupert, Bennington County, petitioned the House of Representatives for permission to produce coinage for Vermont. A committee consisting of Messrs. Tichenor, Strong, and

Williams, with the addition of Ira Allen from the Legislative Council, was formed to consider the proposal. On June 15, 1785, a bill authorizing Harmon to coin copper pieces was sent to the governor and Council for consideration and possible amendment:

Whereas Reuben Harmon Jr., Esq., of Rupert in the county of Bennington, by his petition has represented that he has purchased a quantity of copper suitable for coining, and praying this legislature to grant him a right to coin copper, under such regulations as this assembly shall think meet; and this assembly being willing to encourage an undertaking that promises so much public utility, therefore:

Be it enacted and it is hereby enacted . . . that there be and hereby is granted to the said Reuben Harmon Jr., Esq., the exclusive right of coining copper within this state for the term of two years from the first day of July, in the present year of our Lord one thousand seven hundred and eighty-five; and all coppers by him coined shall be in pieces of one third of an ounce troy each, with such devices and mottoes as shall be agreed on by the committee appointed for that purpose by this assembly.

And be it further enacted . . . that the said Reuben Harmon before he enter on the business of coining, or take any benefit of this act, shall enter into a bond of five thousand pounds, to the treasurer of the state, with two or more good and sufficient sureties, freeholders of this state, conditioned that all the copper by him coined as aforesaid, shall be a full weight as specified in this act, and that the same shall be made of good and genuine metal.

The bill was passed on the same day. On June 16 the required financial bond was obtained. Harmon found that his coins, regulated to be the weight of one-third-ounce each, would be too heavy and would weigh more than contemporary pieces of half-penny size circulating at that time throughout the United States. Accordingly, an amendment was passed on October 27, 1785:

Be it enacted and it is hereby enacted . . . that all coppers coined by said Reuben Harmon Jr., Esq., shall be of genuine copper in pieces weighing not less than four pennyweight fifteen grains each, and so much of the aforesaid act that regulates the weight of said coins is hereby repealed.

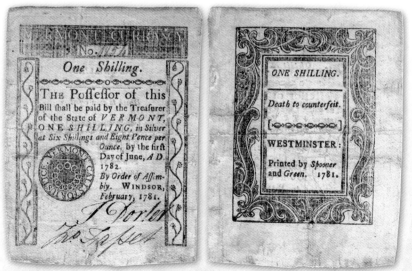

A Vermont 1-shilling note of 1781, issued
several years before the official coinage.

HARMON AND THE RUPERT MINT

A letter from historian Benjamin H. Hall of Troy, New York, on March 3, 1855, gives this account.[5]

> Reuben Harmon Jr. came from Suffield, Connecticut, in company with his father, Reuben Harmon Sr., about the year 1768, and settled in the northeast part of Rupert, Vermont. He was a man of some note and influence while there. At a meeting of the inhabitants of the New Hampshire Grants held at Dorset, September 25, 1776, initiatory to their Declaration of Independence, Mr. Reuben Harmon [probably Jr. was one of the representatives from Rupert]. He was representative in the Vermont Legislature from Rupert in 1780, was justice of the peace from 1780 to 1790, and held several minor offices. In the year 1790 or thereabouts he left Rupert, for that part of the State of Ohio called New Connecticut, and there died long since.
>
> His Mint House was located near the northeast corner of Rupert, a little east of the main road leading from Dorset to Pawlet, on a small stream of water called Millbrook, which empties into Pawlet River. It was a small building, about 16 by 18 feet, made of rough materials; sided with unplaned and unpainted boards. It is still standing, but its location and uses are entirely different from what they were originally. Its situation at present is on the border of the adjoining town of Pawlet whither it was long since removed, and what was once a coin house is now a corn house.
>
> Col. William Cooley [Coley], who had worked at the goldsmith's trade in the city of New York, and who afterwards moved to Rupert, made the dies and assisted in striking the coin.

THE MINT AND COINAGE

The same correspondent wrote to Charles I. Bushnell on July 18, 1855, with additional information:

> The sundial, or "Mind Your Business" copper coin, common in New England at the close of the last and at the commencement of the present century, was first manufactured by Abel Buel [Buell] at New Haven, Connecticut, the original dies having been designed and cut by himself. Not long after this, his son William Buel, removed from the manufactory to the town of Rupert, Bennington County, Vermont, and in connection with a Mr. Harmon, established the mint-house on what is now known as Millbrook.
>
> William had taken with him the original dies used by his father at New Haven, and continued at Rupert the coinage of the coppers above referred to, until the coin had depreciated so much in value as to be worthless or nearly so, for circulation. The remains of the dam which rendered the waters of Millbrook eligible are still to be seen, and pieces of copper and specimens of the old coin are still occasionally picked up on the site of the old mill and in the brook below.

It is doubted today that 1787-dated Fugio coppers, referred to above as the "Mind Your Business copper coin," were actually struck in Vermont, for the act which authorized Reuben Harmon Jr. to strike coins specifically delineated the type of coins to be produced and provided strong penalties if there were deviations from those authorized. Eric P. Newman gives the opinion that Buell brought with him punches and tools but not coining dies.[6] These would have been Abel Buell's punches that were used to create the 1786 and 1787 Mailed Bust Left dies, similar in style to those used on Connecticut coppers. However, the workmanship of the lettering placement and style on the obverses and reverses is amateurish in comparison to the Connecticut dies believed to have been completely done (punches plus lettering) by Abel Buell.

It was related that his son, William Buell, had become involved in an altercation with Indians who accused the younger of killing one of their friends. It seems that Buell had obtained from a druggist a quantity of nitric acid and upon returning to his residence with a jug of this substance he was

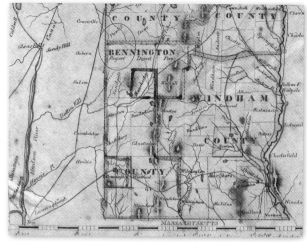

A map showing the lower part of Vermont. Near the New York State border at the upper left is Rupert and, to the north, Pawlet. (Zadock Thompson, *Map of the State of Vermont*, 1842)

approached by some Indians who wanted to drink what they thought was rum. Buell told them the jug contained acid and would poison them, but the Indians did not believe it. Taking the jug, one of them swallowed a substantial portion and died soon thereafter from the effects. Buell was then accused of killing the Indian, and in accordance with tradition, the dead man's companions sought revenge in kind. To escape the situation Buell is said to have fled to Rupert, Vermont, conveniently distant from Connecticut.

Julian Harmon, the grandson of Reuben Harmon Jr., shed some further light on the early Vermont coiners in a letter dated June 4, 1855. He believed that Reuben Jr. came to Vermont in about 1760, having moved from *Sandisfield*, Massachusetts. He verified that William Buell of New Haven cut the dies and stated:

> The mint house stood on Pawlet River, three rods from [Reuben Harmon Jr.'s] house, a story and a half house, not painted, a furnace in one end for melting copper and rolling the bars, and in the other [west] end machinery for stamping, and in the center that for cutting. The stamping was done by means of an iron screw attached to heavy timbers above and moved by hand through the aid of ropes. Sixty per minute *could* be stamped, although 30 per minute was the usual number. William Buel assisted in striking the coins. Three persons were required for the purpose, one to place the copper, and two to swing the stamp. At first, the coins passed two for a penny, then four, then eight, when it ceased to pay expenses. The English imported so many of the "Bungtown coppers," which were of a much lighter color.

It was further stated that the coiner moved to Ohio in about 1800 and engaged in the business of making salt at the Salt Spring Tract, in Wethersfield Township, Trumbull County, which he continued until his death on October 29, 1806, in his 56th year.

B.H. Hall, in a letter to Charles I. Bushnell, dated June 14, 1856, gave another view of the mint:

> On the north side of Millbrook the "old copper house" was first erected. . . . From this location, in the town of Rupert, the "Mint House" was afterwards removed to and placed on the eastern bank of Pawlet River in the same town. Here it was also used for minting purposes. When the manufacture of coins was abolished, it was

allowed to remain on Pawlet River for several years, but we could not learn to what uses it was put. Its third removal was to a spot north of the house of John Harwood, Esq., in the town of Rupert, on the east side of the main road. While here it was occupied as a residence by a family named Goff. It was again removed from its third location to a site nearly opposite, where it remained until its final journey which took place many years ago. This placed it on the farm of William Phelps about a mile north of John Harwood's residence in "the edge" of the town of Pawlet. Here it stood until last winter, when it was blown down.

Z. Humphrey, in *The History of Dorset*, 1924, states that William Coley, a New York City goldsmith (and silversmith), did not come to Rupert until after the legislature extended the coinage franchise in 1786, nor did William Buell arrive with "dies" until after that time.[7] We do know that Coley was in Rupert by 1787 (see Machin's Mills account on page 30). The federal census taken on July 6, 1790, listed Coley as residing in Rupert and the head of a household there (including, in addition to Coley, three males under age 15 and two females with ages not specified).

VERMONT LANDSCAPE TYPE COINS

The first coins produced by Harmon, probably mostly if not entirely at the first mint location on Millbrook, have a scenic motif and portray on the obverse a typical Vermont rocky mountain ridge forested with pine trees. The dies were probably by Coley, or by another craftsman at the partnership of Van Voorhis and Coley at 27 Hanover Square, New York City (Coley later moved to Rupert). Today, numismatists often refer to these as the "Landscape" type. To the right a sun peeping over the ridge is seen and below is a plow. Around the border is the legend VERMONTIS RES PUBLICA and the date 1785.

VERMONTIS was the Latinized of "Vermont." This translation of "Vermont" was not standardized, so in 1785 and continuing into 1786, such other variations as VERMONTS and the lengthy VERMONTENSIUM (creating "Republic of the Vermonters" in literal translation) were used.

The reverse depicted an all-seeing eye from which emanated 13 short rays, with a star above each, and 13 long rays. The legend STELLA QUARTA DECIMA ("the 14th star," a reference to Vermont's ambition to become the 14th state) surrounded. The orientation of the eye (with the eyebrow above the eye) varies with regard to the surrounding inscription. The central motif was copied from the reverse of a 1783-dated Nova Constellatio "Blunt Rays" copper, a coin then in general circulation (see earlier discussion of Nova Constellatio copper coinage). Later issues are of the "Pointed Rays" style.

The following Vermont coppers are attributed to Bressett numbers (which give numbers for each obverse die and letters for each reverse) as delineated by Kenneth Bressett in "The Vermont Copper Coinage," part of *Studies on Money in Early America*, American Numismatic Society, 1976; RR or Ryder-Richardson numbers by Hillyer C. Ryder, 1920, modified by John M. Richardson, *The Numismatist*, May 1947; and W numbers from the *Whitman Encyclopedia of Colonial and Early American Coins*, 2008.

VERMONT LANDSCAPE VARIETIES

1785, VERMONTS (Bressett 1-A, Ryder Richardson-2, Whitman-2005):

Obverse: A ray of sun points to the period after RES. *Reverse:* The period after STELLA is above the left element of the ray, the tail of the Q is above the double

points of the ray, and the star points to right leg of the second A in QUARTA. ***Notes:*** Landscape obverses of 1785 are probably from a different hand than those of 1786. The letter punches are different and are heavier, the rocky outcrop is hand-cut and differs considerably on each of the three obverses, and the overall appearance is more rustic than those of the next year. The reverse of this and other 1785 issues of the landscape type are loosely copied after the 1783 "Blunt Rays" Nova Constellatio copper then in circulation. This is the first and most available of the 1785-dated landscape types. Planchet fissures of one degree or another are the rule, but exceptions occur. They are often seen with areas of light striking. On late states there is a large die crack extending upward from the 8 in 1785. It is not known whether this or W-2010 was the first struck, but there is no doubt both were from the mint building in its first location on Millbrook. ***Rarity:*** 500 to 999 are known.

1785, VERMONTS (Bressett 2-B, RR-3, W-2010): ***Obverse:*** A ray of sun points to the right of the period after RES. ***Reverse:*** This issue has blunt rays, a period after STELLA above the star, the tail of the Q is between the star and the tip of a ray, and the star points to the period after QUARTA.

Notes: This is the scarcest of the 1785 landscape varieties and planchet flaws are common. They are sometimes seen on a wide planchet with extensive dentils visible on both sides, but these are in the minority. According to Tony Carlotto, this is "probably the most intriguing coin in the landscape series. This is because of the striking peculiarities and planchet oddities." There is one struck over a 1785 Connecticut (Miller 4.1-F-4). ***Rarity:*** 33 to 64 are known.

1785, VERMONTIS (Bressett 3-C, RR-4, W-2015): ***Obverse:*** No lettering is below the date. ***Reverse:*** These coins have blunt rays, a period after STELLA is above the ray, the tail of the Q is between the star and the tip of a ray, and the star points to the left leg of the second A in QUARTA. The all-seeing eye on the reverse at the center is *inverted*, a detail first noted by Tony Terranova. This die is not used elsewhere. ***Notes:*** Planchet defects are common. In its late state the obverse die bulged considerably at the center. ***Rarity:*** 65 to 124 are known.

1786, VERMONTENSIUM (Bressett 4-D, RR-6, W-2020): ***Obverse:*** A ray of the sun points to the upright of the R in RES, the fourth tree from the left is in midair, and the U in PUBLICA is repunched. ***Reverse:*** These coins have pointed rays and a ray points at the tail of the Q. ***Notes:*** The landscape coppers of 1786 are probably from a different hand than are those of 1785. On three 1786 dies the letter E is broken at the bottom, but was sometimes strengthened

in the die, and the top of the R is very weak or broken. The rocky cliff is quite similar on all three and was probably done with a punch, then adjusted in the die. There are two reverse dies for 1786: Bressett D and E. Each is copied after the 1783 and 1785 Nova Constellatio coppers with pointed rays. These are usually seen with planchet flaws and rifts and lightly struck in areas. This is the most available 1786 landscape type and is more often seen than any of the 1785 issues. *Rarity:* 500 to 999 are known.

1786, VERMONTENSIUM (Bressett 5-E, RR-7, W-2025): *Obverse:* A ray of sun points to the right foot of the R and the tree touches lower left of the N. *Reverse:* This issue has pointed rays and a ray points at the U. *Notes:* These are often seen on nice planchets, though usually with some lightness of strike in areas. It is the first of the Vermont varieties with normal die rotation (180° difference between the obverse and reverse), although variations from rotated dies occur throughout the later series. Roy Bonjour considers this to be the most elusive of the VERMONTENSIUM varieties. *Rarity:* 65 to 124 are known.

1786, VERMONTENSIUM (Bressett 6-E, RR-8, W-2030): *Obverse:* A ray of sun points to the right foot of the R, the tree is distant from the N, and the 1 in 1786 is sharply double punched. *Reverse:* These used the same die as preceding. *Notes:* This is the scarcest of the three 1786 landscape varieties. They are often seen well struck (except for the central part of the rocky ledge, which on this variety is not as well defined as on the other two VERMONTENSIUM dies), and planchet flaws are common. *Rarity:* 125 to 249 are known.

VERMONT PORTRAIT COINAGE

It is likely that in 1786 it was decided that the Vermont coppers, being of a unique obverse design not familiar to the citizens of Vermont or surrounding states, did not circulate as well as they would have if the design had been a more standard motif. Accordingly, new obverses and reverses were adopted, imitating the familiar coinage of England, with a portrait on the obverse of King George II and on the reverse a seated figure similar to Britannia on the halfpence.

A different style, known today as the *Baby Head* variety, depicts a boyish bust facing right with the legend AUCTORI VERMON: ("by the authority of Vermont") surrounding. The reverse shows the seated figure of a woman with INDE ET LIB (an abbreviation for "independence and liberty") at the border. The date 1786 is below.

The initial coining franchise was granted to Harmon for a period of two years. Before the expiration, Harmon, on October 23, 1786, petitioned the General Assembly to extend the term on the grounds that it was too short to indemnify him for the great expense of erecting a factory, acquiring machinery, and otherwise beginning the coinage of copper. On October 24, 1786, the franchise was granted an extension. The design was officially modified to copy the English style, and following an initial period, Harmon was to pay a royalty for the coining privilege:

Be it enacted by the General Assembly in the State of Vermont that there be and hereby is granted and confirmed to the said Reuben Harmon Jr., Esq., the exclusive right of coining copper within the state, for a further term of eight years from the first day of July, 1787; and that all copper by him coined shall be in pieces weighing not less than four pennyweight, fifteen grains each; and that the device for all coppers by him hereafter coined shall be, on the one side, a head with the motto "Auctoritate Vermontensium" abridged, and on the reverse a woman with the lettering INDE: ET LIB:, for "Independence and Liberty."

And be it further enacted by the authority aforesaid that the said Reuben shall have and enjoy the aforesaid privilege of coining coppers within the state free from any duty to this state as a compensation therefore, for the full term of three years from the first day of July, 1787; and from after the expiration of said three the said Reuben shall pay for the use of this state two and one-half percent of all the copper he shall coin for and during the remainder of aforesaid term of eight years.

It was further specified that appropriate bonds and guarantees were to be provided. Presumably, the "landscape" coins were discontinued around this time, and the "head" type with the George II portrait was instituted.

PORTRAIT VARIETIES

1786, Baby Head (Bressett 7-F, RR-9, W-2040): *Obverse:* An infantile portrait facing right is on the obverse. This is the only variety with AUCTORI: at left and VERMON: to the right. The top of the R punch is broken. *Reverse:* INDE: [group of dots] ET: LIB: at the border. The promi-

nent cloth or sash from the seated figure's shoulder near the branch hand curves downward, goes under the pole, and ends in a curl. There are four wheat sheaves on the shield, reflective of Vermont agriculture. *Notes:* These are nearly always seen on imperfect planchets with fissures and usually in lower grades. They are usually dark brown or black. The wheat sheaves are usually mostly or completely missing. The inspiration for the youthful portrait has been a matter of conjecture, and it is not punched like any other Rupert die. It is one of the most mysterious issues among copper coinages of the 1780s. For many years this has been one of the most popular and sought-after Vermont varieties. *Rarity:* 250 to 499 are known.

1786, Bust Left (Bressett 8-G, RR-10, W-2045): *Obverse:* A portrait of King George II, copied from English coinage, is on the obverse with VERMON: / AUCTORI: [colon after *both* words] at the border. The V is distant from bust. *Reverse:* The branch hand is opposite the space past the E. There

are four wheat sheaves on the shield, a distinctive feature that is visible on only a few specimens. *Notes:* The planchet is usually dark and with laminations or rifts. The obverse dies of this and the next two are by an unknown diesinker using punches attributed to Abel Buell, who was a principal in Connecticut of the Company for Coining Coppers. However, the portraits differ slightly from those used on Connecticut coppers, obvious differences being the epaulet strips are shorter and the neck treatment different on the Vermont dies. The lettering on the reverses is from different punches than used on Connecticut issues. *Rarity:* 125 to 249 are known.

1786, Bust Left (Bressett 9-H, RR-11, W-2050): *Obverse:* A portrait of King George II, copied from English coinage, is on the obverse. The V is closer to the bust than on the preceding, but the C is too small and is placed high. *Reverse:* The branch hand is opposite the upright of the E. *Notes:*
These are usually seen with planchet defects and dark surfaces. The obverse die was used to strike some examples of RR-11, then RR-15, then more of RR-11. *Rarity:* 65 to 124 are known.

1787, Bust Left (Bressett 9-I, RR-15, W-2060): *Obverse:* The obverse is the same as Bressett 9-H. *Reverse:* All examples seen have a massive die crack obscuring much of the date. The rarity of this issue suggests that the die had a very short life. *Notes:* These are usually seen with planchet
defects and dark surfaces. The obverse die was first used to strike W-2050, then some of W-2060 as seen here, then additional specimens of W-2050. While this is a rare variety, its market value goes beyond that consideration as it is actively collected as the only year in which this type was made, and is listed in *A Guide Book of United States Coins.* An example is known struck over a Bressett 3-C (RR-4, W-2015) and George III halfpence, the latter most curious. *Rarity:* 17 to 32 are known.

In 1787 Reuben Harmon Jr. entered into an arrangement with a number of other individuals involved in coinage. Ownership interest and connection was formed between the Vermont coining enterprises and Machin's Mills, a private mint (see below), located on the shores of Orange County near Newburgh, New York. At the new location coinage of Vermont coppers continued, but of different designs.

Machin's Mills Mint
Newburgh, New York (1787–1789)

The private mint known to numismatists as Machin's Mills was probably the most prolific North American coiner during the years from 1786 to 1789. With the exception of Vermont coppers, which will be studied first in the listings to follow, each and every coin was a counterfeit. Other than the authority from Vermont that Reuben Harmon Jr. brought from Rupert, Vermont, in 1787, none of the Machin's Mills coins had any legal status. This made little difference at the time. Legal tender was ill-defined, and the millions of British coppers in circulation at the time had been made in London by the country from which America had declared independence. Few Americans would have cared if anyone counterfeited them, more important, any copper coin more or less the size of a British halfpenny readily circulated in American commerce.

All Machin's Mills coppers were the numismatic equivalent of cowbirds—they appeared under different names, and never was the source identified. Beginning with Vermont historian Benjamin H. Hall in the 1850s and continuing to include Charles I. Bushnell, Sylvester S. Crosby, and C. Wyllys Betts in the 19th century, Machin's Mills became an object of mystery, fascination, and study. Today it remains so, but not to the same extent, as the study of coins and history has unraveled many mysteries. Still, much is theoretical, providing a rich field for research.

AN IMPRESSIVE MINT PLANNED

In 1787 Reuben Harmon Jr., holder of the franchise to coin Vermont coppers, entered into a partnership with a number of other individuals involved in minting, as previously. Ownership interests were formed between Harmon and Machin's Mills, a private mint located on the shore of Orange Pond near Newburgh, New York.

Machin's Mills was established by an agreement dated April 18, 1787, which united the interests of Samuel Atlee, James F. Atlee, David Brooks, James Grier, and James Giles, all of New York City, with Thomas Machin of Ulster County, New York.

Captain Machin was of English birth, and prior to the Revolution had served as an officer with the British forces. During the war he entered the American Army as an engineer and in 1777 was employed by Congress to erect fortifications along the Hudson River and to stretch an iron chain across the river at West Point to prevent the passage of British ships. Following the war, Machin relocated near Newburgh, where he erected buildings subsequently used for the coinage venture.

The agreement provided that the profits from the coining enterprise should be split into six shares valued at £50 each, directly in proportion to the original stock which consisted of £300 of capital. It was stated that Samuel Atlee and James F. Atlee "being possessed of certain implements for carrying on said trade, do agree to lend them to the parties to these presents for and during the continuance of their co-partnership without any fee or reward for the same." What these "certain implements" were is not known today, but punches and dies were probably included. It is popular conventional numismatic wisdom that James was a skilled engraver, but modern research has furnished no evidence of this.[8] He and Samuel were brewers by trade.

In the Machin's Mills agreement it was further provided that "Thomas Machin, being possessed of certain mills, doth hereby agree to let the parties have free use of them for and during the continuation of their co-partnership." Brooks, Grier, and Giles agreed to pay an additional £10 each toward readying the Machin's Mills facility to make it suitable for coinage. The management was to be by James F. Atlee and Thomas Machin.

On June 7, 1787, another agreement was drawn up. Ten partners participated, including the original six involved in the Machin's Mills enterprise, plus four others: Reuben Harmon Jr. of Rupert, Vermont; William Coley, also of Rupert; Elias Jackson of Litchfield County, Connecticut; and Daniel Van Voorhis, a New York City goldsmith. The latter had worked as a silversmith in Princeton, New Jersey, circa 1782–1784, then in New York City with Coley, Simeon Bayley, and Albion Cox in the same trade in the firm of Van Voorhis, Bayley, Coley & Cox, which dissolved in April 1785. It was succeeded by Van Voorhis, Bayley & Coley, which ended in July of the same year, after which the business was continued as Van Voorhis & Coley at 27 Hanover Square. This arrangement continued until Coley moved to Rupert. Later, Albion Cox became involved in the coinage of coppers for the state of New Jersey. It was noted in the agreement that Harmon, after obtaining the coinage privilege from the legislature of Vermont, took in Coley, Jackson, and Van Voorhis as equal partners.

It was proposed that coinage be conducted in two locations: at Machin's Mills, then in the process of readying for coinage, and at Rupert, Vermont, in the existing facilities there. Duties were divided amongst the various partners, with provisions being made for audits, settling accounts, and other business necessities.

Whether much if any coinage under the new agreement was done at Rupert is questionable, in the view of the author, although the 1786-dated Baby Head Vermont copper is of a different style than any others now attributed to Rupert. The Machin partners paid no attention to dates and, for example, some of its counterfeits were dated 1785. Nearly all of the Vermont coinage in Rupert was on very rough planchet stock. Nearly all of the Machin's Mills coinage attributed to Newburgh is on fairly good stock.

It is not known who engraved dies for the enterprise. Dies and punches could have been ordered easily enough from jewelers and others. Abel Buell advertised that he sold such supplies. No account has been found of a known skilled engraver residing in Newburgh in this time frame.

The Machin enterprise is believed to have wound down in 1789, at which time the coppers panic made further production of coppers unprofitable.

DESCRIPTION OF MACHIN'S MILLS

The coin factory later called Machin's Mills was located on Orange Pond, which at one time was also called Machin's Pond. A new outlet, which provided water to a large extent of Chambers Creek, was tapped and the minting structure was erected at the outlet. Originally, this outlet was an overflow for times of high water, the natural outlet being further west at a place called Pine Point.

The business of Machin's Mills was conducted with secrecy. Tradition holds that a guard with a hideous mask was employed to frighten away the curious. In *An Outline History of Orange County, New York*, 1846, Samuel W. Eager noted that "operations there, as they were conducted in secret, were looked upon at that time with suspicion, as illegal and wrong."

To aid in their acceptance into commercial channels the counterfeit English halfpence were struck from dies deliberately made to produce coins which looked weak, as if they had been in circulation for a long time. It was related that in the year 1789, 1,000 pounds of copper coins saw production. This relatively small amount may have represented the tail end of the operation. Although no figures survive, it is likely that the total mintage ran into the hundreds of thousands of coins, if not more. On October 14, 1790, James F. Atlee wrote to Thomas Machin to request that the partners dissolve the enterprise on suitable terms so as to avoid a tedious and expensive lawsuit.

MACHIN'S MILLS COINAGE

Numismatists today believe that Machin's Mills coined a wide variety of coppers, probably anything that they thought could be circulated at a profit. Included were numerous counterfeits of contemporary English halfpence. As noted by Crosby, these bore on the obverse the portrait of King George III with GEORGIUS III REX surrounding. The reverse depicted the seated figure of Britannia. At the same time it is virtually certain that pieces bearing legends relating to Connecticut, New York, and possibly other coinages were made as well. As stated, Vermont coins were perfectly legal, as partner Harmon held the contract. Other coinages were counterfeit. In some instances the dies were mixed, probably inadvertently, resulting in illogical combinations, such as a 1787 copper coin with a Vermont inscription on the obverse and with a reverse showing Britannia, a style intended for use on a counterfeit English halfpenny.

Many 1787 Vermont coppers and probably all of those dated 1788 were struck at Machin's Mills. Nearly all of these are lighter in weight than the earlier 1785 and 1786 Rupert issues, are less carefully engraved, and are often carelessly struck. Among counterfeit issues, those with Connecticut inscriptions are nearly always lightweight and on high-quality planchets. Features are often indistinct, as made, to make them appear worn and encourage their circulation. Coins with IMMUNE COLUMBIA and various New York–related inscriptions were made. It is likely that counterfeit silver coins were struck there as well, but this probability has had no more than casual study.

Machin was not the only counterfeiter in America. Others remain to be chronicled, perhaps including the makers of the 1786, Baby Head Vermont and the most curious 1787, "Muttonhead" Connecticut copper.

NUMISMATIC ASPECTS

Machin's Mills counterfeits of Connecticut and Massachusetts, and New York–related copies, are in great demand, and rarities command high prices. Counterfeits of British regal halfpence of King George II and, more numerous, George III *specifically* attributable to Machin's Mills are in great demand today. These are identifiable by having letter, numeral, or device punches connected with Vermont coppers and counterfeits of Connecticut and other state coinages, although with unrelated inscriptions. Certain of these are quite distinctive in their appearance: George III almost appears as if wearing lipstick, from some sort of an "apostrophe" punch used to make the lips. As a general guideline, this includes many English halfpence with a date of 1778, 1787, or 1788. No regal (authentic, made in England) halfpence were struck after 1775. For counterfeits dated from 1747 to 1778, guidelines are more challenging. To be attributed to Machin's Mills, a coin must have sawtooth denticles, the 1 in the date must not be a "J," there will be no berries in the obverse wreath, the solid shield lines will have no outlines, and the lips will be distinctive. *Caveat:* There are many counterfeit English halfpence offered as "Machin's Mills" coins on the Internet, but which were not made in America. Note that counterfeit Connecticut, Massachusetts, and New York coins are listed under those specific states.

VERMONT COPPERS OF MACHIN'S MILLS

After the move of operations to Machin's Mills in Newburgh, New York, coins were made lighter in weight, but usually on far better planchets. Today these later issues are found in varying shades of brown color. Nearly all have light striking at the centers with few details. Likely, the dies were made this way so that a freshly minted coin, if darkened, could be passed readily in circulation, giving the impression it had gone from hand to hand for a long time.

MACHIN'S MILLS COUNTERFEIT ENGLISH HALFPENCE

Varieties of counterfeit British halfpence with the king's portrait on the obverse and a seated figure on the reverse can be attributed to numbers created by Robert A. Vlack (see Bibliography) and are also listed in *Walter Breen's Complete Encyclopedia of Colonial and Early American Coins* (1988), and in the *Whitman Encyclopedia of Colonial and Early American Coins* (2008), to which one may refer for details. The dates on the coins bear no particular relevance to the years in which they were counterfeited. A brief synopsis is included here, but new discoveries are often made.[9]

A sample Vermont copper of Machin's Mills:
the 1787 Mailed Bust Right variety, with a
portrait of King George III, and is struck
over a Nova Constellatio copper with
prominent traces of the undertype remaining.

This Machin's Mills copper combined
a Vermont obverse with the reverse
of a counterfeit British halfpenny.

The Machin's Mills coiners combined a
crudely engraved Vermont obverse die with
the IMMUNE COLUMBIA reverse die.

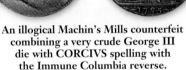

A Machin's Mills counterfeit English
halfpenny, 1747, with a GEORGIVS II
head facing left (the only obverse
with the head in this direction).

An illogical Machin's Mills counterfeit
combining a very crude George III
die with CORCIVS spelling with
the Immune Columbia reverse.

Connecticut Coinage
(1785–1788)

On October 18, 1785, Samuel Bishop, James Hillhouse, Joseph Hopkins, and John Goodrich petitioned the General Assembly of Connecticut to authorize the production of copper coins to alleviate a chronic shortage of small change. Much of the change then in circulation was counterfeit. On the 20th the authorization was given to coin no more than £10,000 worth of lawful money of the standard of English half-pence, to weigh six pennyweight (equal to 144 grains) each, and to bear a design of a man's head on one side with the letters AUCTORI: CONNEC: ("By the Authority of Connecticut"). The reverse was to depict the emblem of Liberty with an olive branch in her hand and with the inscription INDE: ET. LIB: 1785, the lettering being an abbreviation of the Latin words for "Independence and Liberty." Similar wording was also used on Vermont coppers of 1786 and 1787 coined in Rupert (see pages 22–30). One provision for these coins stated: "Nothing in this act shall be construed to make such coppers a legal tender in payment of any debt, except for the purpose of making even change, for any sum not exceeding three shillings."

THE COMPANY FOR COINING COPPERS

Facilities were set up, and coinage began soon thereafter. The state did little if anything to audit contractors. In 1787 a committee was appointed to review the coinage, but no report has been found. In May 1789 another committee gave an overall review to the Assembly. It was related that on April 7, 1789, a meeting of the parties involved at the inception of minting was held at a private home in New Haven. Attending were Samuel Bishop, James Hillhouse, Mark Leavenworth, and John Goodrich. The history of the coinage was revealed.

On November 12, 1785, the four original persons named in the coinage act entered into an agreement with Pierpoint Edwards, Jonathan Ingersoll, Abel Buell, and Elias Shipman to form the Company for Coining Coppers. Buell was an experienced plate and die engraver, and brought his expertise to the enterprise. The others were said to be citizens of high integrity and repute, men who were important in the state.

CHANGES AND NEW AGREEMENTS

This arrangement continued until February 1786, when Ingersoll and Edwards sold one sixteenth part of the company each to Goodrich. In March 1786 Hopkins sold one sixteenth part to Goodrich, and in April 1786 Edwards, Shipman, and Ingersoll each sold one sixteenth part to James Jarvis. The latter continued production of Connecticut coppers until sometime in the summer of 1786, when the supply of copper metal was depleted.

On September 10, 1786, the company leased its minting apparatus to Mark Leavenworth, Isaac Baldwin, and William Leavenworth for a period of six weeks. Likely, the latter individuals had access to copper. Additional transfers of interest took place, so that in June 1787 the ownership stood as follows: James Jarvis, 9/16 part; James Hillhouse, 2/16 part; Mark Leavenworth, 2/16 part; Abel Buell, 2/16 part; and John Goodrich, 1/16 part. In early summer 1787 the Company for Coining Coppers wound down, and the enterprise became known as James Jarvis & Co.

Inspectors from the legislature found out that 28,944 pounds' weight of coppers was produced by the Company for Coining Coppers, with one twentieth part, or 1,477 pounds and a few ounces, being transmitted to Connecticut as a royalty. At the time the coppers passed in circulation at 18 pieces to a shilling. As a shilling was comprised of 12 pence, the coppers were worth three-quarters of a penny.

No records have been located as to the extent of the Jarvis operation, or if its coinage was officially sanctioned by the state. Its main business was the production of 1787-dated Fugio coppers, and emphasis was on trying to make as much profit as possible.

The committee further learned that Major Eli Leavenworth, apparently a relative of the other two Leavenworths earlier associated with the venture, made blank coppers in autumn 1788 and had them stamped in New York with various impressions. Some of these apparently had inscriptions relating to Connecticut, but others were of different designs.

TWILIGHT OF THE COINAGE

It was disclosed that Abel Buell had gone to Europe. Before leaving he gave Benjamin Buell, believed to be a son, the right to produce coppers. As of the committee report in 1789, Benjamin had just begun to issue pieces of undetermined design, which would have been dated 1788 or earlier.

In the meantime, Captain Thomas Machin and his partners in Newburgh, New York, operated Machin's Mills. Many different varieties of lightweight counterfeit Connecticut coppers were made there. It can be argued that as the State of Connecticut legislated that the coppers were not legal tender and were simply a convenience for small change, they were fair game for anyone to make!

On June 20, 1789, the official right for anyone to coin coppers was suspended. Thus concluded the official Connecticut production, although it is likely that most coinage after the Company for Coining Coppers dissolved was not reported to the state. About this time the "coppers panic" (public distrust at the prevalence of counterfeit and lightweight coppers in circulation) sharply devalued Connecticut and other coins, making further production unprofitable, even for counterfeiters. Machin's Mills was shuttered about this time. As a further end to such business, the U.S. Constitution became effective on March 9, 1789, after which the federal government reserved the exclusive right to produce coins.[10]

COMMENTARY

The dies for coins struck by the Company for Coining Coppers are thought to have been made by Abel Buell, one of the firm's partners. Others are thought to have produced many other dies of different styles used by counterfeiters, notably Machin's Mills, to which coinage of many pieces dated from 1786 to 1788 are attributed.

As a general rule, the weight of these coppers declined steadily, with those of the first year, 1785, being much heavier than those of the last, 1788. During this interim more than 350 die combinations were made, by far the largest number of any state copper coinage.

Connecticut coppers, which circulated widely, were considered to be fair game for other coiners as the pieces had no legal-tender status. In addition to coinage at Machin's Mills, other pieces were struck outside of the state, and some were made in New York City. If a letter punch on a Connecticut copper is identical to one used on a New Jersey copper, it does not logically follow that a mint in New Jersey was busy making Connecticut coins. It could have been that various dies for Connecticut and New Jersey were made in the same shop, not necessarily in either state.

A 1785, Mailed Bust Right. M 2-A.1, W-2305
as illustrated and many other varieties.

A 1785, Mailed Bust Left. M 7.1-D, W-2440
as illustrated and several other varieties.

A 1785, African Head. M 4.1-F.4, W-2355
as illustrated and several other varieties.

A 1786, Mailed Bust Right—a type
with several portrait variations. M 2.1-A,
W-2465. This is a Machin's Mills issue with
a distinctive portrait and on the reverse the
legend is transposed as ET LIB / INDE.

A 1786, Mailed Bust Left. M 4.1-G, W-2525.
It is one of many varieties of the type.

A 1786, Mailed Bust Left. M 5.3-B.2,
W-2565. "Hercules Head" obverse;
a die also used in 1787.

A 1786, Draped Bust Left. M 7-K, W-2695.
This is one of two varieties of this type, the
other being 6-K with the same reverse.

A 1787, Mailed Bust Right. M 1.1-A, W-2700.
Small head on obverse with reverse legend
transposed as ET LIB / INDE. It is one of a
number of Mailed Bust Right varieties.

A 1787, Mailed Bust Right. M 1.2-C,
W-2720. The "Muttonhead" variety
with a topless figure on the reverse. It is a
contemporary counterfeit in great demand.

A 1787, Mailed Bust Left. M 3-G.1, W-2805.
It is one of many varieties of the type.

A 1787, Draped Bust Left. M 16.2-NN.2,
W-3005. This is one of more than 100
varieties of the obverse type.

A 1788, Mailed Bust Right. M 2-D, W-4405.
It is one of many varieties of the obverse type.

A 1788, Mailed Bust Left. M 7-E, W-4480. It
is one of many varieties of the obverse type.

A 1788, Draped Bust Left. M 17-Q, W-4640.
It is one of many varieties of the obverse type.

As Connecticut coppers circulated at a higher value than certain of their contemporaries, many Nova Constellatio and some other pieces were overstruck with Connecticut designs, probably at Machin's Mills.

These coins have been popular with collectors for a long time. Authorized issues and counterfeits are collected with great fervor, the counterfeits often being rarer and in greater demand due in some instances to their rustic die work. Attribution is to "M numbers" devised by Henry C. Miller and published in *The State Coinages of New England*, published serially (then in book form) by the American Numismatic Society in 1920. Since then many new varieties have been discovered and added to the original listing. A comprehensive listing with Miller numbers as well as Whitman numbers can be found in the *Whitman Encyclopedia of Colonial and Early American Coins* (2008), which gives an easy-finding guide to aid in the attribution of varieties.

New York Related Coinage
New York, New York (1785–1787)

Although no official state authorization relating to a native New York coinage is known, a number of issues were made with legends relating to that state. A news dispatch from Worcester, Massachusetts, on March 16, 1786, was reprinted widely, including in the New York *Daily Advertiser*, on March 25, 1786:

> A large quantity of counterfeit copper coin manufactured in this and the neighboring states has lately been in circulation in this commonwealth, but it is now generally refused a currency by the trading part of the community: nearly one half of the copper coin in this country for twenty or thirty years past has been of a base kind, manufactured at Birmingham in England: however, it crept into circulation, and did, until the late additional quantity above-mentioned made its appearance, pass for the same value as those which were genuine.

The above raises some questions that numismatists have yet to answer. Nothing is known about the specific nature of counterfeit copper made in Massachusetts and nearby states, seemingly anterior to "twenty or thirty years ago" when British counterfeits were imported in quantity. These mysterious earlier coins would seem to have been American-made imitations of British halfpence prior to the Revolution. Continuing from the same account:

> New York, Connecticut, and Vermont have authorized a person in each of those states to coin coppers: numbers of them are now in circulation: they are in general well made, and of good copper, those of New York in particular. [Were a person to be] authorized in this state for the same purpose it would undoubtedly prevent the manufacture of those made in base metal.

The "made in base metal" term seems to refer to a base metal other than copper, perhaps lead (which was easy to cast).

PETITIONS FOR NEW YORK COINAGE

In petitions said to have been dated February 11, 1787, Ephraim Brasher and John Bailey appealed to the New York State Assembly for the right to produce copper coins. The *Assembly Journal* records that "The several petitions of John Bailey and

Ephraim Brasher, relative to the coinage of copper within the state were read and referred (to committee)." The original petition can no longer be traced. It is not known whether separate petitions were presented by Brasher and Bailey or whether they combined their efforts. Brasher did not always work alone, and at one time John Bailey was associated with him, as were his brother Abraham and George Alexander at various times. In 1787 the city directory lists Brasher's address as 77 Queen Street and Bailey's residence as 22 Queen Street, which indicates the possibility of a close relationship at the time.

On March 3, 1787, Captain Thomas Machin presented his own petition to the assembly. Neither proposal was acted upon favorably. In the same year the legislature studied money in circulation and endeavored to regulate existing coinage, which came from many different sources domestic and foreign. It was reported at the time that the principal copper coins circulating within the state of New York were composed of:

> A few genuine British half-pence of George the Second, and some of an earlier date, the impressions of which are generally defaced. A number of Irish half-pence, with a bust on the one side, and a harp on the other. A very great number of pieces in imitation of British half-pence, but much lighter, of inferior copper, and badly executed. These are generally called by the name of "Birmingham Coppers," as it is pretty well known that they are made there, and imported in casks, under the name of Hard Ware, or wrought copper.
>
> There has lately been introduced into circulation a very considerable number of coppers of the kind that are made in the State of New Jersey. Many of these are below the proper weight of the Jersey coppers and seem as if designed as a catchpenny for this market.

It was further noted that it took 48 genuine British halfpence, when new, to weigh one pound *avoirdupois*, but of the imitation or counterfeit pieces believed to be made in Birmingham, it took 60 to compose one pound weight. The genuine New Jersey coppers at six pennyweights, six grains each, equaled 46-2/5 pieces to the pound, or a slightly heavier standard than the British. It was noted at the time that all of these coppers passed without discrimination at 14 pieces to a shilling, or slightly less than the value of a penny, a shilling being composed of 12 pence. It was noted that at this rate a 57-percent profit would go to coiners of genuine British halfpence, a 96-percent profit to makers of imitation "Birmingham" halfpence, and a 65-percent profit to those who made New Jersey coppers.

To remedy these inequities, on April 20, 1787, an act was passed to provide that after August 1:

> No copper shall pass current in this State except such as are of the standard weight of one third part of an ounce avoirdupois, of pure copper, which copper shall pass current at the rate of 20 to a shilling of the lawful current money of this state, and not otherwise. . . .

An extensive issue of 1787-dated copper coins, each with a bust on the obverse surrounded by NOVA EBORAC ("New York") appeared. The reverse of these issues depicted a seated goddess with a sprig in one hand and a liberty cap on a pole in the other hand, with the legend VIRT. ET. LIB. ("Virtue and Liberty") surrounding, with the date 1787 below. The letter punches used on this issue are identical to those used on the Brasher doubloon die. It seems likely that John Bailey and Ephraim Brasher operated a minting shop in New York City and produced these and possibly other issues.

1787 NOVA EBORAC COPPERS

The Nova Eborac (New York) coppers were produced in quantity. Today they are widely collected. The *Whitman Encyclopedia of Colonial and Early American Coins* (2008), and *Walter Breen's Complete Encyclopedia of Colonial and Early American Coins* (1988), can be used to attribute the varieties, selected examples of which are given below.

Among other copper coins which have been associated with New York, directly or through die linkages, are certain issues bearing the figure of Liberty seated on a globe with the legend IMMUNIS COLUMBIA surrounding. Another issue depicts George Clinton on the obverse and the New York State seal on the reverse. Still another illustrates the standing figure of an Indian.

Another issue features on the obverse a crude bust of George Washington with the legend NON VI VIRTUTE VICI ("Not by force but by virtue have we won"). The reverse contains a Latinized New York, NEO EBORACENSIS, surrounding a seated figure of Liberty holding the scales of Justice and a Liberty cap, with the date 1786 below.

With the exception of Nova Eborac coppers, which are scarce, most copper pieces bearing legends relating to New York and all die combinations of such pieces range from rare to very rare today.

A 1787, Nova Eborac Copper, Large Bust (Whitman-5750, Breen-985).

A 1787, Nova Eborac Copper, Medium Bust, Seated Figure Facing Left (W-5755, B-986).

A 1787, Nova Eborac Copper, Medium Bust, Seated Figure Facing Right (W-5760, B-987).

A 1786, Non Vi Virtute Vici Copper, Small head / Neo-Eboracensis (W-5730, B-977).

A 1787, Heraldic Eagle with E PLURIBUS UNUM, Excelsior, Eagle Facing Left / Arrows at Right (W-5785, B-980).

A 1787, Liber Natus Libertatem Defendo, Indian / Excelsior (W-5795, B-990).

1786 AND 1787 BRASHER GOLD COINS

One of the most famous of all American coin issues is the 1787 gold doubloon issued by Ephraim Brasher (pronounced as "bray-zher"), a New York goldsmith, silversmith, and jeweler who may have been associated with New York copper coinage as well. In the few instances that specimens have appeared in auction catalogs during the past century, great acclaim and publicity has been given to them.

The Bank of New York, established in 1784, distributed a list of various foreign gold coins, their weights, and the accepted value at which they would be received in payment. In stores and counting houses it was considered unwise to accept a coin until it was pronounced genuine. Confusion arising from a wide variety of denominations, designs, and countries of origin aided counterfeiters. To this was added the crimes of clipping (deliberate removal of small amounts of metal from a coin's edge) and sweating (washing a coin with nitric acid to remove a small layer of metal).

It is believed that Ephraim Brasher was called upon to regulate foreign gold coins as well as certain silver issues. This was done by clipping or plugging gold coins in circulation to bring them up or down to the standard of gold coins published by the Bank of New York in 1784, after which appropriate hallmarks would be added. Other goldsmiths active in this trade included Standish Barry, John Burger, Jacob Boelen III, Fletcher & Gardiner, John Letellier, and Thomas Underhill.[11] For Brasher a counterstamp, usually in the form of the letters EB in an oval, was impressed upon each coin as a permanent identification. Specimens of EB-marked foreign gold coins are known today.[12]

During this era, Brasher himself minted two styles of gold doubloons, his coins being valued at $15. The first, antedated 1742 and bearing the date 1786 partially visible in the border, is of the Lima style, so called because it loosely copied a much earlier Spanish doubloon design used at the mint in Lima, Peru. These are signed BRASHER on the obverse. Apparently, it was thought that this familiar motif would encourage the circulation of the coins. Facts are scarce and theories are abound. Next came the 1787-dated doubloon showing a sun rising over a mountain on one side and an eagle on the other, also signed with Brasher's full surname. Both have the counterstamp EB. Inscriptions include the state motto of New York. A lighter-weight half-doubloon was made from the same 1787 die pair.

A 1742 (1786) Lima-Style Gold
Doubloon (W-5820, B-984).

A 1787 Brasher Doubloon,
Hallmark on Breast (W-5845, B-982).

Immune Columbia, Immunis Columbia, Confederatio, and Related Coppers
(1785–1787)

Stylistically related in design to the Nova Constellatio coppers are certain pieces with an obverse from the IMMUNE COLUMBIA die combined with several different reverses. It has been suggested that the legend refers to America (Columbia) being immune to the problems of distant Europe. Three of these reverses are of the Nova Constellatio design: two with pointed rays and one with blunt rays. IMMUNIS COLUMBIA coins are related.

The CONFEDERATIO legend appears on an interesting series of coins, all extremely rare, dated in the 1780s. There have been many theories about the makers of these dies and places of coinage, but facts are scarce. Some may be related to New York coiners in New York City, or to Machin's Mills. It has been suggested that these are mostly patterns. The average condition of most examples would seem to belie this, for well-worn coins are the norm, suggesting they circulated as money instead of being preserved as samples. Whatever their origin, the original quantities issued must have been small for the majority of varieties. Of all early American coppers, coins in this series are among the most enigmatic.

Combinations with the CONFEDERATIO dies were made in different formats, including with the standing figure of an Indian surrounded by the legend INIMICA TYRANNIS AMERICANA, a die featuring George Washington, a die with the seated figure of Columbia and with the legend IMMUNIS COLUMBIA, and others. Several examples are listed below.

A 1785, Immune Columbia / Nova
Constellatio with Star in Border.
Copper (Whitman-1960, Breen-1117).

A 1785, Inimica Tyrannis Americana /
Confederatio, Small Stars
(W-5635, B-1124).

A 1787, Immunis Columbia /
Large Eagle. Plain Edge
(W-5680, Breen–1136 and 1137).

New Jersey Coinage
(1786–1788)

MINTS AND MINTING

Coins bearing the imprint NOVA CAESAREA (New Jersey latinized) were produced under contract. On June 1, 1786, the General Assembly awarded the privilege to three partners: Thomas Goadsby, Albion Cox, and Walter Mould. The coppers, which bore no denomination, were to weigh 150 grains each. The coinage was to be completed by June 1788. As compensation the coiners were to keep one-tenth of the pieces produced. Matthias Ogden, prominent in state politics, was also involved.

The partnership arrangements changed over a period of time, there were disagreements. Most production took part at official facilities in Rahway and Elizabethtown, supplemented by addition production in New York City and some counterfeit issues by Machin's Mills. The study and collection of New Jersey coppers is a fascinating specialty. The standard reference is *New Jersey State Coppers: History, Description, Collecting*, by Roger Siboni, John Howes, and A. Buell Ish, 2013.

DESCRIPTION OF THE NEW JERSEY COPPERS

Although there are variations, the basic New Jersey copper depicts on the obverse the head of a horse facing left (from the State Seal), below which is a horse-drawn plow with two handles. Around the border is NOVA CÆSAREA (with AE as a ligature). At the bottom border is the date: 1786, 1787, or 1788. Certain coppers of 1788 have the horse's head facing left.

The reverse has a spade-shaped shield at the center with E PLURIBUS UNUM around the border. The dies were all made by hand and differ slightly from one another in letter spacing and other features.

TYPES OF NEW JERSEY COPPERS

As noted above, the coppers of New Jersey are of one basic type, with variations.

1786, Date Under Plow Beam: There are three varieties: Maris 7-C, 7-E, and 8-F. These are all rare and are the crème de la crème of the New Jersey coppers. In any grade the offering of an example at auction creates much interest.

1786, Coulterless: There are 14 varieties from 7 different obverse dies: Maris 8.5-C, 9-G, 10-G, 10-h, 10-gg, 10-oo, 10.5-C, 11-G, 11-H, 11-hh, 11.5-G, 12-G, 12-I, and 22-P. They are very popular, but are somewhat scarce as a type.

1786: There are 35 varieties from 15 different obverse dies: Maris 13-J, 14-J, 15-J, 15-L, 15-T, 15-U, 16-J, 16-L (Protruding Tongue variety), 16-S, 16-d, 17-J, 17-K, 17-b (PLUKIBUS reverse, defective R), 17-J (Bridle variety), 18-L, 18-M, 18-N, 19-M, 20-N (Drunken Diecutter obverse: illustrated), 21-N, 21-O, 21-P, 21-R, 21.5-R, 23-P, 23-R, 23.5-R, 24-I, 24-M, 24-P, 24-Q, 24-R[13], 25-S, 26-S, and 26-d.

1787: There are 70 varieties from 49 different obverse dies: Maris 6-C, 6-D, 27-j, 28-L, 28-S, 29-L, 29.5-L, 30-L, 32-T, 33-U, 34-J, 34-V, 35-J (1787 Over 1887), 35-W (overdate), 36-J, 37-J (Goiter variety with lump under chin as are others from this obverse die), 37-X, 37-Y, 37-f, 38-L (Small Head as are others from this obverse die), 38-Y, 38-Z, 38-a, 38-b, 38-c, 39-a, 40-b, 41-c, 42-c, 42.5-c, 43-Y, 43-d, 44-c, 44-d, 45-d, 45-e, 46-e, 47-e, 47.5-e, 48-X, 48-f, 48-g (Outline to Shield), 52-I, 53-j (period after CAESAREA.), 54-k (Serpent Head), 55-l (PLURIRUS reverse), 55-m, 56-n (The Camel Head variety is known to have been struck over many other coins including Connecticut, Machin's Mills, Vermont, Nova Eborac, Nova Constellatio, English, and Irish coppers—a wider variety of undertypes than found on *any* other state copper coin: illustrated, over Clinton copper), 58-n (Star Below Plow Handles[14]), 59-o (Sawtooth variety for denticles on the obverse), 59-mm (Sawtooth obverse), 60-p (PLURIBS reverse), 61-p (PLURIBS), 62-q (WM initials on some), 62-r, 62.5-r (WM in high relief), 63-q, 63-r, 63-s, 64-t (large and small planchets), 64-u, 68-w, 69-w[15], 70-x, 71-y, 72-z, 73-aa, 73.5-jj, 81-ll (contemporary counterfeit), 83-ii (contemporary counterfeit), and 84-kk (contemporary counterfeit).

1788, Horse Head Left: There are three varieties from three different obverse dies: Maris 49-f[16], 50-f, and 51-g. These are scarce and are a very popular type.

1788, Horse Head Right: There are 11 varieties from 10 different obverse dies: Maris 64.5-r, 65-u, 66-u, 66-v, 67-v, 74-bb (Running Fox reverse), 75-bb (Running Fox: illustrated), 76-cc (Running Fox), 77-cc (Running Fox), 78-d, and 79-ee (Frightened Head[17]).

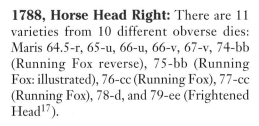

Massachusetts State Coinage
Boston, Massachusetts (1787–1788)

DESIRE FOR A COPPER COINAGE

In 1786 the Massachusetts government gave serious consideration to the production of copper coins. Two proposals were received, one from Seth Reed and the other from James Swan. A committee was formed and recommended that the state itself should coin coppers. It was estimated that in order to coin £20,000 in coppers it would require a cost of £8,250 worth of copper metal plus £1,950 for the workhouses, presses, plating (strip-producing) mill, and other apparatus together with the fuel, wages of the workmen, and other costs, for a total expenditure of £10,200, returning to the Commonwealth of Massachusetts a profit of £9,800. Denominations of penny, halfpenny, and farthing were envisioned.

On October 17, 1786, legislation was passed which provided:

> ... that there be a mint erected within this Commonwealth, for the coining of gold, silver, and copper: and that all the coin that shall be struck therein shall be of the same weight, alloy, and value, and each piece bear the same name. ...
>
> There shall be a quantity of copper coins struck, equal to the amount of 70,000 dollars, in pieces of the two different denominations mentioned in the said resolve, and in convenient proportions: one of which to have the name Cent stamped in the center thereof, and the other Half Cent, with such inscriptions or devices as the governor with the advice of Counsel may think proper: and the said coin, when struck, shall be received in all payments in this Commonwealth.

It was further provided that a mint would be established and that an assayer, workmen, supplies, and other accessories would be obtained for the purpose of coining. On May 2, 1787, the coinage committee reported that subsequent investigation found it necessary to erect a furnace made of special fire brick, the clay for which could not be obtained until that spring. Conversations were held with Captain Joshua Witherle (as he signed his name: also given as Wetherle, Wetherly, and Wetherlee in early accounts), and the committee found that he was suited to superintend the setting up and conducting of the business. During the same month copper suitable for coining was located, including 3,434 pounds weight of copper and 650 pounds of "sprews," belonging to the Commonwealth, as well as several mortars and cannons suitable for melting down for use in coinage. On June 27, 1787, the designs were established:

> [The] device on the copper coin to be emitted in this Commonwealth to be the figure of an Indian with a bow and arrow and a star on one side, with the word COMMONWEALTH, and on the reverse a spread eagle with the words OF MASSACHUSETTS AD 1787.

When coinage did materialize, "OF" and "AD" were omitted.

THE MINT IN OPERATION

In a speech on October 18, 1787, the governor of Massachusetts noted that, "in consequence of an act made October 1786, a mint has been erected for coining cents, and a very considerable quantity of copper will soon be ready for circulation." The building used as a mint house was a wooden one owned by Witherle. It measured one story in height, of high stud, about 20 feet wide by 40 feet in length. It seems that some coins had been struck by this time, as the manifest dated September 2, 1787, for the consort sloop *Lady Washington*, departing Boston on a voyage of exploration to the Pacific Northwest coast, included "300 medals, 500 cents, and 500 half cents," and an unspecified quantity of the same items aboard the flagship *Columbia*. In the legislative halls there was much interest and discussion that autumn of making silver coins as well, but nothing came of the proposals.

The *New-Haven Gazette, and the Connecticut Magazine*, Thursday, November 1787, included this item from Massachusetts:

> Springfield, November 13. We are informed, on authority not to be doubted that the copper coin of this Commonwealth, struck by order of the General Court, will, in a very few days, be in circulation, when, after that time, the coinage of any state, with the base metal imported from Birmingham, will not be a currency among us.

Problems developed with producing the half cents and cents. On January 16, 1788, the government expressed its concern with the slowness in mintage operations and ordered an investigation. Witherle reported that certain of the equipment did not work properly, that he had difficulties obtaining dies, and that there were other problems.

Most of the dies for Massachusetts half cents and cents were made by Joseph Callender, a Boston engraver, who charged £1 4s each. His invoices reflected the making of 39 new dies and the repair of three used dies. This expense was considered excessive, so Jacob Perkins, a young silversmith of Newburyport, was engaged to produce additional dies at a payment of one percent of the value of the coins to be struck from the dies. It is believed that the shape of the letter S on the coins was distinctive with each engraver. In all issues of 1787 and in some of 1788, the S is open at the top and bottom, these being attributed to Callender. In both half cent die varieties of 1788 as well as several of the cents of the same year, the S is narrow and the serifs at the upper and lower part are close to the curves, somewhat resembling a figure eight, these considered to be the work of Perkins. A later review disclosed that during the operation of the mint £3 18s 10d had been paid to Perkins for dies, as compared to, under a different arrangement, £48 12s paid to Callender. In addition to the foregoing, some contemporary counterfeit Massachusetts cents are attributed to Machin's Mills.

On June 17, 1788, Joshua Witherle reported that many other unfortunate circumstances, including particularly harsh weather the preceding winter, caused additional difficulties and delays. It was stated that $2,500 in coins had already been struck and had been deposited to the account of the Commonwealth. As of that time £2,136 5s 7d had been expended in the operation of the mint, for which just £939 value of copper coins had been struck, leaving a loss of £1,197 5s 7d. On November 22, 1788, realizing that each half cent and cent cost more than twice its face value to produce, the Commonwealth began plans to close the mint. The accounts were closed in January 1789 and employees were dismissed.

Witherle had probably occupied the building previous to the period of the state coinage. He was a coppersmith by trade, and was commonly known among the boys of Boston as "the cent maker."

Crosby stated:

> The copper for the coins of Massachusetts (excepting a small quantity which was drawn at Newton), after being cast into ingots at the mint in Boston, was carted to the mill at Dedham, where it was drawn under a triphammer, and rolled into sheets, when it was returned to the mint, where the blanks were prepared and the coin stamped.

After the mint closed, Massachusetts reviewed other proposals to make coins for the state on a contractual or royalty basis, but none were adopted. As was true of other state coinages, likely the coppers panic of 1789 dimmed the possibility of any profits being made from such a venture. As evidenced by wear, these coins circulated for many years afterward.

DESCRIPTIONS OF THE COINS

Half cents and cents were struck bearing the dates 1787 and 1788. As most of the coinage materialized after late 1787, most of the 1787-dated pieces were probably struck in 1788. The Massachusetts copper coins were well received in the channels of commerce and saw active circulation for several decades following their issue.

Coined in large quantities, Massachusetts half cents and cents of 1787 and 1788 are fairly plentiful today. Numerous die varieties exist and all have a plain edge.

The obverse of each half cent and cent depicts a standing Indian, facing to the viewer's left, with a bow in his right hand and an arrow in his left. Between the top of the bow and the Indian's face is a five-pointed star. To the left and right are COMMON / WEALTH. The relative positions of the bow, arrow, and star are keys to identifying die varieties.

The reverse of each half cent and cent displays a heraldic eagle with an olive branch to the viewer's left and a bundle of arrows to the right. On its breast is a shield, with HALF / CENT in incuse letters on the smaller denomination, and CENT on the larger. An exception is Ryder 2-F, the most desired collectible die variety in the series, which has the arrows and branch transposed left to right, and CENT in raised letters. Around the top and side border is MASSACHUSETTS. The date 1787 or 1788 is under a horizontal line below the eagle. Sometimes the line (called a *dash* by Crosby) is single, other times it is doubled.

These copper coins represent the first appearance of the word CENT on a coin made within the United States. The denomination was intended to refer to a "cent" or one hundredth part of a Spanish milled dollar, the same money of account used on Continental currency bills issued from 1775 to 1779.

Attributions are to numbers assigned by Hillyer C. Ryder, and were issued in 1920 by the American Numismatic Society as part of *The State Coinages of New England.* Some later discoveries have since been added.

MASSACHUSETTS HALF CENTS

1787: There are nine varieties from five obverse dies: Ryder 1-D, 2-A, 3-A, 4-B, 4-C, 4, 5-A, 6-A, and 6-D.

1788: There are two varieties from one obverse die: Ryder 1-A and 1-B.

MASSACHUSETTS CENTS

1787: There are 14 varieties from 8 obverse dies: Ryder 1-B (Aged Face—Machin's Mills), 2-A (Horned Eagle die break)[18], 2-C, 2-E, 2-F (Transposed Arrows: arrows on left side, and CENT incuse), 2-G, 3-G, 4-C (Bowed Head[19]: same die used for other 1878 cents and for 1788 17-I), 4-D, 4-J, 5-I (Machin's Mills), 6-G (Stout Indian[20]), 7-H (Stout Indian—Machin's Mills), and 8-G.

1788: There are 25 varieties from 18 obverse dies: Ryder 1-D, 2-B, 3-A, 3-E, 4-G, 6-N (No Period after Massachusetts), 7-M, 8-C, 9-M, 10-L, 11-C, 11-E, 11-F, 12-H (Stout Indian), 12-I (Stout Indian), 12-K (Stout Indian), 12-M (Stout Indian), 12-O (Stout Indian), 13-I, 13-N (No Period after Massachusetts), 13-P, 14-J (Stout Indian—Machin's Mills), 15-M, 16-M, 17-I, and 18-M.

LATER PRIVATE COINAGE

Standish Barry
Baltimore, Maryland (1790)

Standish Barry, a Baltimore silversmith, struck a distinctive silver threepence token in 1790. Barry, 27-years old at the time, may have intended the piece to commemorate the anniversary of American independence, or perhaps some special celebration was held in Baltimore which occasioned its issue. The piece bears a male portrait on the obverse—long thought to possibly be George Washington or even Barry himself, but today theorized to be James Calhoun, a judge (and first mayor of Baltimore) active in the 1790s. The specific day July 4 in addition to the year (expressed as 90) is a distinctive feature. The coinage production apparently was extremely limited, for his threepence are exceedingly rare today.

About this time Barry is believed to have made imitation Spanish-American gold doubloons hallmarked SB, per the research of John Kraljevich in the description of a coin in the Eliasberg Collection sale, 2005, lot 3012—a scenario similar to Ephraim Brasher's imitation 1742 "Lima" gold doubloon, an important addition to early American numismatics, described earlier.

1790 Standish Barry threepence (W-8510, B-1019).

1735 Standish Barry Imitation
Lima doubloon (W-8515).

The Campbell "Mint"
Bergen County, New Jersey
(Early 1800s)

Of all the privately owned mints in America, the one operated by the Campbell family of Bergen County, New Jersey, was easily the most colorful, unique and long lasting.[21] Although some may maintain that their factory was not exactly a mint, it cannot be denied that the wampum they manufactured for use in early America was indeed a special form of money and a substitute for coinage.

The term wampum is derived from the Algonquian word *wampomeag* referring to strings of white *(wab)* beadwork *(umpe)* arranged in the form of symbols and patterns that tell the history, religion, and culture of people without a formal written language. Strings and belts made from wampum beads were often used as gifts and as documents or treaties, and to record important events. The tiny beads that made up these important artifacts were fashioned with great care from the colorful blue and white portions of quahog shells that were indigenous to New England.

The specific role of Native American wampum and its use as a trade item or money in Colonial America is often misunderstood because history has most frequently recorded its use solely as money. In Colonial New England it became a necessity for trade

Wampum from the Campbell "Mint."

between the European settlers and native Indians who valued it as a customary adjunct to their transactions. In time it became so essential to trade that it was widely accepted and officially valued at six white beads, or three purple beads, to one English penny.

Roger Williams (1603–1683), an early advocate of fair dealings with Native Americans and founder of the colony of Providence Plantation, on October 7, 1640, recorded that "The want of coin enhances the rate of wampum, and it is ordered that white wampum shall pass at four a penny and blue at two a penny and not above 12d at a time except the receiver desire more." Massachusetts Court records indicate that on May 22, 1661, the law authorizing the use of wampum as legal tender was repealed. The introduction of silver coins made in Boston eventually drove the use of beads out of circulation.

This turn of events, however, did not totally eliminate the use of trade beads for transactions with the natives. By then many different kinds of beads, made from glass, bone and other media, had begun to flow into the area from European sources, and were in great demand for decorative use and personal adornment. There were several places in New Jersey where traditional wampum was still being made; Cape May and Pascack being among the best. It was into this setting in 1775 that the Campbell family set out to make a living supplying a product that would accommodate the commercial needs. An early backer in the venture was allegedly John Jacob Astor, who used Campbell wampum to purchase furs which were then sold in Europe.

The Campbell "mint" operated for a time in the family home but with growth was transferred to a special building where water power was used to operate the grinding and polishing equipment. After 1930 much of their efforts went into making other decorative items of interest for more western tribes. The Campbell factory closed sometime around 1850 with the beginning of increased production of United States coins, and the end of an era.

THE FIRST FEDERAL COINAGE

CONTINENTAL DOLLAR COINS

One of the most significant early American issues is the 1776 Continental dollar, which was the first federal coinage, although no specific documentation has been found concerning the 1776 Continental dollars. The pieces are believed to have official status because of their inscriptions. Bearing devices and inscriptions

A 1776 Continental Currency dollar in pewter with CURENCY spelling.

taken from Continental currency paper money (of the authorization of February 17, 1776), these bear on the obverse the date 1776 and, surrounding, the inscription CONTINENTAL CURRENCY. Within is a sundial, below which is MIND YOUR BUSINESS, with FUGIO ("I fly," a reference to the passage of time) to the left. The reverse displays 13 intertwined circles, each with the name of a state, joined to form a linked chain border. Within the center is the inscription AMERICAN CONGRESS and WE ARE ONE.

No specific documentation has been found regarding the origin of the Continental dollar. The resolution of February 1776, pertaining to the issuance of paper money, resulted in the production of different denominations from the 1/6 dollar through $8, including the $1 denomination. The resolution of May 9, 1776, provided for the various denominations from the $1 through $8. However, the resolution of July 22, 1776, omitted the $1 and contained denominations from $2 through $30. Likewise, the final resolution of that year, November 2, 1776, omitted the $1 note and began with the $2. It is likely that it was intended that the pewter Continental dollar coin serve in the place of the $1 note during the latter part of 1776. The reason for striking brass impressions is not known. A few silver examples are known, of full intrinsic worth at the time, and may represent the intention for all of the coinage. The Continental Congress was short on funds, and no quantity of silver was on hand for production.

Certain varieties have the inscription E.G. FECIT meaning "E.G. made it." Eric P. Newman, who has studied the series extensively, identified the Continental dollar dies to be the work of Elisha Gallaudet of Freehold, New Jersey, who also engraved plates for printing Continental Currency paper money. The same engraver cut all the dies for this series, as evidenced by the craftsmanship.

Several different die varieties were made. *Currency* was spelled three ways: CURRENCY, CURENCY, and CURRENCEY, the latter being imitative of an error found on the 1/6 dollar note of February 17, 1776, indicating that the engraver may have copied the specific legends on this particular design while making the dies. The typical pewter strike is about 41 mm in diameter and has the edge ornamented with twin olive leaves.

A 1776 Continental Currency dollar in pewter with CURRENCY spelling.

A 1776 Continental Currency dollar in pewter with CURRENCY spelling and E.G. FECIT.

A 1776 Continental Currency dollar in pewter with CURRENCEY spelling.

A 1776 Continental Currency dollar with CURRENCEY spelling corrected by changing the second E to a Y and putting an ornament over the Y originally in place.

PROPOSED MINT OF NORTH AMERICA

NEW COINAGE SYSTEM

In 1783 a remarkable group of pattern coins made its appearance: the "bit" of 100 units, the "quint" of 500 units, and the "mark" of 1,000 units. Sylvester S. Crosby said: "These are undoubtedly the first patterns for coinage of the United States and command an interest exceeding that of any others of this class." To these can be added a five-unit piece in copper, unknown to Crosby. These were intended to illustrate coinage to be produced in the proposed Mint of North America, operated under government auspices.

Governor Morris, who had been assistant financier of the Confederation, proposed these pieces. Jared Sparks, in *The Life of Gouverneur Morris* (1832), quotes Morris, a dramatic exposition of the omnipresence of the Spanish dollar and how its values varied:

> The various coins which have circulated in America have undergone different changes in their value so that there is hardly any which can be considered of the general standard, unless it be Spanish dollars. These pass in Georgia at five shillings, in North Carolina and New York at eight shillings, in Virginia and the four eastern states at six shillings, and in all other states, excepting South Carolina, at seven shillings and sixpence, and in South Carolina at thirty-two shillings and sixpence. The money unit of a new coin to agree, without a fraction, with all these different values of a dollar, excepting the last, will be the fourteen hundred and fortieth part of a dollar, equal to the sixteenth hundredth part of a crown. Of these units, twenty-four will be a penny of Georgia, fifteen will be a penny of North Carolina or New York, twenty will be a penny of Virginia and the four eastern states, sixteen will be a penny of all the other states, excepting South Carolina, and forty-eight will be thirteen pence of South Carolina.
>
> It has already been observed that to have the money unit very small is advantageous to commerce: but there is no necessity that this money unit be exactly represented in coin: it is sufficient that its value be precisely known. On the present occasion, two copper coins will be proper, the one of eight units, and the other of five. These may be called an Eight and a Five. Two of the former will make a penny proclamation, or Pennsylvania money, and three a penny Georgia money. Of the latter, three will make a penny New York money and four a penny lawful, or Virginia money. The money unit will be equal to a quarter of a grain of fine silver in coined money. Proceeding thence in a decimal ratio, one hundred would be the lowest silver coin, and might be called a Cent. It would contain twenty-five grains of fine silver, to which may be added two grains of copper, and the whole would weigh one pennyweight and three grains. Five of these would make a Quint or five hundred units, weighing five pennyweight and fifteen grains: and ten would make a Mark, or one thousand units weighing eleven pennyweight and six grains.

Some of the circumstances surrounding the issue of 1783 pattern coins were recorded by Robert R. Morris, the financier. His diary noted that on April 2, 1783, "I sent for Mr. [Benjamin] Dudley who delivered me a piece of silver coin, being the first that has been struck as an American coin."[1]

MINT OF NORTH AMERICA

On April 16 it was noted that he "sent for Mr. Dudley and urged him to produce the coin to lay before the Congress to establish a mint." On the following day, the 17, he "sent for Mr. Dudley to urge the preparing of coins, etc. for establishing a mint."

Morris reported on April 22 that "Mr. Dudley sent in several pieces of money as patterns of the intended American coins." On July 5 he noted that "Mr. Benjamin Dudley also informs of a minting press being in New York for sale, and urges me to purchase for the use of the American Mint."

On August 19 he reported as follows:

> I sent for Mr. Benjamin Dudley, informed him of my doubts about the establishment of a mint and desired him to think of some employment in private service, in which I am willing to assist him in all my power. I told him to make out an account for the services he had performed for the public and submit at the Treasury office for inspection and settlement.

On August 30 it was reported that "Mr. Dudley brought the dies for coining in the American Mint."

NOVA CONSTELLATIO PATTERNS

The dies for the bit and quint denominations were cut by hand. The largest value, the mark, was made by the use of prepared punches. The pedigree of two pieces, the mark and the quint, obtained by John W. Haseltine in 1872 and years later sold into the Garrett Collection, was attested to by the seller, Rathmell Wilson, who on May 28, 1872, stated:

> The history of the two coins which you obtained from me, viz. Nova Constellatio 1783, U.S. 1,000, Nova Constellatio, 1783 U.S. 500 is as follows.
>
> They were the property of Hon. Charles Thomson, secretary of the first Congress. At his death the property was left by will to his nephew, John Thomson, of Newark, state of Delaware. These two coins were found in the desk of the said deceased Charles Thomson, and preserved by his nephew during his life: at his death they came into the possession of his son Samuel E. Thomson of Newark, Delaware, from whom I received them. So you will perceive that their genuineness cannot be questioned: as they were never out of the possession of the Thomson family, until I received them.

Among the accounts of the United States, under the category of "Expenditure for Contingencies," between January and July 1783, the following entries appear in relation to coinage:

> February 8. Jacob Eckfeldt, for dies for the Mint of North America, $5 and 18/90ths.
>
> March 21. Benjamin Dudley employed in preparing a mint. $75 and 24/90ths.
>
> April 17. John Swanwick, for dies for the public mint $22 and 42/90ths.
>
> May 5. A. DuBois, for sinking, casehardening, etc., for pairs of dies for the public mint $72.
>
> June 30. Benjamin Dudley employed in preparing a mint $77 and 60/90ths.

The Nova Constellatio patterns were created with care and with the intention of initiating a federal coinage. Their time had not come, and had it not been for the fortuitous saving of the silver pieces by Thomson, we might not know of them today. The copper five-unit piece is a later discovery found in Europe in 1977. Most of the different ex-Thomson 1783 Nova Constellatio issues were acquired at the Garrett Collection sale by John J. Ford Jr., and later went into another fine private cabinet, along with the five-unit coin.

A Nova Constellatio
pattern five units in copper.

A 100-unit coin, or a bit,
in silver, the metal used
for large denominations.
These coins typically have
an ornamented edge, but a
plain-edge variety also exists.

A Nova Constellatio pattern 500 units,
or quint. The obverse has the all-seeing
eye with rays surrounding and 13 stars
between the longest rays. NOVA
CONSTELLATIO is around the border.

A Nova Constellatio pattern quint.
The obverse has the all-seeing eye
with 13 closely spaced rays surrounding
and 13 stars between the highest tips
of the rays. There is no inscription.

A Nova Constellatio pattern 1000 units, or mark.

The Mint of North America did not come to pass. The primary 1783 pattern designs were later used in a private venture to coin and distribute Nova Constellatio copper coins mostly dated 1783 and 1785 struck by a private mint in Birmingham, England.

Fugio Copper Coinage
New Haven, Connecticut (1787)

BACKGROUND OF THE COINAGE

Fugio coppers, often referred to as "cents" in numismatic circles, were authorized by the American Congress. Produced on contract, they were the first widely circulated coin of the federal government. The obverse depicts a sundial with the inscription FUGIO ("I Fly," a reference to the passage of time: a motif borrowed from the earlier Continental Currency fractional notes dated February 17, 1776, also used on the 1776 Continental dollar coins), and MIND YOUR BUSINESS, aphorisms suggested by Benjamin Franklin. On the reverse a circle of links represents the colonies, with the inscription UNITED STATES / WE ARE ONE at the center.

Minting apparently occurred in several different locations, including New Haven, Connecticut, and possibly New York City. Many of the hubs used in the die-making process were engraved by Abel Buell, who was associated with other coinages, most notably Connecticut, but whose punches were used on some later coins made in Rupert, Vermont. While Abel Buell is credited with making the hubs, inspection of the coins reveals that workmanship on the details of the coining dies varied widely. Some are done skillfully with the letters well aligned, while others are amateurish.

MINTING AND DISTRIBUTION

The initial production apparently commenced in New Haven, Connecticut, under the supervision of James Jarvis. By June 1, 1788, coinage was halted. Abel Buell went to Europe leaving the dies and other apparatus to others, who various accounts refer to as his sons William and Benjamin. It is believed that Samuel Broome and Jeremiah Platt (Jarvis's father-in-law), associated with copper coinage for the state of Connecticut, continued to make Fugio coppers on a subcontract basis at mints located near New Haven at Westville and Morris Cove.

On September 30, 1788, it was reported that:

> There are two contracts made by the Board of Treasury with James Jarvis, the one for coining 300 tons of copper of the federal standard, to be loaned to the United States, together with an additional quantity of 45 tons, which he was to pay as a premium to the United States for the privilege of coining: no part of the contract has been fulfilled. A particular statement of this business, so far as relates to the 300 tons, has lately been reported to Congress. It does not appear to your committee that the Board was authorized to contract for the privilege of coining 45 tons as a premium, exclusive of the 300 mentioned in the Act for Congress.
>
> The other contract with said Jarvis is for the sale of a quantity of copper amounting, as per account, to 71,174 pounds: this the said Jarvis has received at the stipulated price of 11 pence farthing, sterling, per pound, which he contracted to pay in copper coin, the federal standard, on or before the last day of August 1788, now past: of which but a small part has been received. The remainder it is presumed, the Board of Treasury will take effectual measures to recover as soon as possible.

Apparently somewhat fewer than 400,000 Fugio coppers were actually struck, instead of several million. An entrepreneur, Royal Flint, took over the enterprise, but seems to have accomplished very little.

NUMISMATIC ASPECTS

A large quantity of original 1787 Fugio coppers, estimated at several thousand pieces, was for many years in the vault of the Bank of New York. Tony Terranova reported that 712 remained as of 1998. Most Mint State pieces known today trace their origin to this particular group.

The typical Fugio copper has some areas of lightness due to casual striking. This is often seen at the bottom of the obverse and on the reverse toward the outside of the rings. Denticles are usually incomplete. Planchet rifts or fissures are not uncommon. More often than not, these problems are overlooked in descriptions, and certified holders don't mention them. These defects are to be expected and add to the "personality" of each coin. Grading interpretations have been relaxed over time, and Fine coins of the 1950s are often graded Very Fine or Extremely Fine today. While collecting by Newman varieties appeals to specialists, most numismatists seek either a single representative example or the various major styles listed in *A Guide Book of United States Coins*.

1787 Fugio, Cross (Quatrefoil) After Date, No Cinquefoils: There are four varieties from one obverse die: Newman 1-B, 1-L, 1-Z (Raised Rims on Center Label: illustrated), and 1-CC (AMERICAN CONGRESS on Label; extremely rare).

1787 Fugio, Club Rays with Concave Ends, Cinquefoils: There are six varieties from five obverse dies (two with FUCIO): 2-C (FUCIO Misspelling: illustrated), 5-F, 5-HH, 23-ZZ (FUCIO), 24-MM, and 25-PP. Nearly all of these coins are in lower grades.

1787 Fugio, Club Rays with Convex Ends, Cinquefoils: There are two varieties from two obverse dies: 3-D and 4-E.

1787 Fugio, 1 Over Horizontal 1 in Date: There are three varieties from one obverse die: 10-G, 10-T, and 10-OO. These are usually seen in lower grades.

1787 Fugio, Pointed Rays, Cinquefoils: There are 45 varieties from 17 obverse dies: 6-W, 7-T, 8-B (many in the Bank of New York Hoard), 8-X (many in the BNY Hoard), 9-P, 9-Q, 9-S, 9-T, 11-A (UNITED above, STATES below), 11-B (dozens in the BNY Hoard), 11-X (many in the BNY Hoard: one of the varieties most often seen today), 12-M, 12-S, 12-U, 12-X (many in the BNY Hoard), 12-Z (Raised Rims on Center Label), 12-KK, 12-LL, 13-N, 13-R, 13-X (726 remaining in the BNY in 1948, fewer later: planchet and striking problems are typical: illustrated), 13-JJ, 13-KK, 14-H, 14-O, 14-X, 15-H, 15-K, 15-V, 16-H, 16-N, 17-I, 17-S, 17-T, 17-WW, 18-H, 18-U, 18-X, 19-M, 19-Z (Raised Rims on Center Label), 19-SS, 20-R, 20-X, 21-I, and 22-M.

1787 Fugio, Narrow Rings on Reverse, Copy Dies: There is only one variety, 104-FF. Struck circa 1859 from copy dies made to the order of Horatio N. Rust, they are sometimes called "New Haven restrikes." Hundreds exist in brass (sometimes called "copper"), a dozen or so in silver, and two in gold.

PHILADELPHIA MINT, 1792 TO DATE

FIRST PHILADELPHIA MINT (1792–1832)

PROVIDING FOR A FEDERAL MINT

While the Fugio coppers provided useful small change for circulation, the issue hardly represented the sovereignty of the new nation. President George Washington in his first annual message to Congress, January 8, 1790, said, "Uniformity in the Currency, Weights and Measures of the United States is an object of great importance, and will, I am persuaded, be duly attended to."

A portrait of
Thomas Jefferson
by Gilbert Stuart.

In the same year Thomas Jefferson initiated a contact with Jean Pierre Droz, a Swiss engraver, trained in France. Droz (spelled as Drost in certain correspondence) discussed a contract, but never signed it, to come to America in the spring of 1792 to make coining machinery and engrave dies at the proposed national mint, but on October 13 advised William Short in Paris that he had decided against it.[1]

Washington's address to Congress, October 25, 1791, included this:

> The disorders in the existing currency, and especially the scarcity of small change, a scarcity so peculiarly distressing to the poorer classes, strongly recommend the carrying into immediate effect the resolution already entered into concerning the establishment of a mint. Measures have been taken pursuant to that resolution for procuring some of the most necessary artists, together with the requisite apparatus.

A portrait of
Alexander Hamilton
by John Trumbull.

Plans for a national mint continued, and many discussions were held in Congress and elsewhere.

A lengthy paper by Secretary of the Treasury Alexander Hamilton titled "On the Establishment of a Mint," communicated to the House of Representatives, January 28, 1791, outlined the need for a national mint and the challenges it presented, including:

1st. What ought to be the nature of the money unit of the United States?

2d. What the proportion between gold and silver, if coins of both metals are to be established.

3d. What the proportion and composition of alloy in each kind?

4th. Whether the expense of coinage shall be defrayed by the government, or out of the material itself?

5th. What shall be the number, denominations, sizes, and devices of the coins?

6th. Whether foreign coins shall be permitted to be current or not; if the former, at what rate and for what period?

Hamilton, giving extensive details, proposed that the dollar and its decimal divisions be used. At the time, commerce was flooded with countless British copper halfpence of proper weight, many more imitations of these privately made in Birmingham, and domestic copper coins of 1785 to 1788 from Vermont, Connecticut, New Jersey, Massachusetts, and counterfeits from Machin's Mills. Silver and gold coins were mostly of Spanish inscriptions, as had been the case for generations, with many British, French, Dutch, and other issues as well.

He recommended that one unit of gold in coinage be equal in pure weight to 15 units of silver. These metals should be alloyed as copper and other metals occur naturally with gold and silver, and such added strength that reduced wear. He suggested that ideal coins would be:

One gold piece equal in weight and value to 10 units or dollars.

One gold piece equal to a tenth part of the former, and which shall be a unit or dollar.

One silver piece which shall also be a unit or dollar.

One silver piece which shall be in weight and value a tenth part of the silver unit or dollar.

One copper piece which shall be the value of a hundredth part of a dollar.

One other copper piece which shall be half the value of the former.

As no federal silver or gold coins were yet in circulation, Hamilton's document expounded extensively on the practices of other countries.

Congress, through the resolution approved on March 3, 1791, authorized President George Washington to establish a federal mint and to obtain the needed personnel and equipment. This was followed on April 2, 1792, by the "Mint Act," full title, "An Act Establishing a Mint, and Regulating the Coins of the United States." These denominations were provided:

Act of April 2, 1792, Authorized Coins

Eagle. $10. 247-4/8 grain (16.0 g) pure or 270 grain (17.5 g) standard gold.

Half eagle. $5. 123-6/8 grain (8.02 g) pure or 135 grain (8.75 g) standard gold.

Quarter eagle. $2.50. 61-7/8 grain (4.01 g) pure or 67 4/8 grain (4.37 g) standard gold.

Dollar or unit. 371-4/16 grain (24.1 g) pure or 416 grain (27.0 g) standard silver.

Half dollar. 185-10/16 grain (12.0 g) pure or 208 grain (13.5 g) standard silver.

Quarter dollar. $0.25. 92-13/16 grain (6.01 g) pure or 104 grains (6.74 g) standard silver.

Disme. $0.10. 37-2/16 grain (2.41 g) pure or 41 3/5 grain (2.70 g) standard silver.

Half disme. $0.05. 18-9/16 grain (1.20 g) pure or 20 4/5 grain (1.35 g) standard silver.

Cent. $0.01. 11 pennyweights (17.1 g) of copper.

Half cent. $0.005. 5-1/2 pennyweights (8.55 g) of copper.

The legislation further provided that one side was to have an impression emblematic of liberty, with the inscription "Liberty" and the year of the coinage. The reverse of each of the gold and silver coins was to have an eagle and UNITED STATES OF AMERICA. The reverse of copper coins was to have an inscription giving the denomination.

Silver and gold were to be valued at the ratio of 15 to 1. The ratio calculation proved to be incorrect in terms of market conditions, the ratio of silver was fractionally higher than the 15 set, and gold coins became "cheap" in terms of silver, eventually causing many gold coins to be melted or exported. Other provisions gave further details.

There were opposing voices. Private coiners in England made copper cents on speculation in 1791, these featuring the portrait of Washington.

In 1792 Peter Getz, an engraver in Lancaster, Pennsylvania, prepared a pattern half dollar with Washington on the obverse and the required eagle on the reverse. Other proposals could be mentioned.

George Washington depicted on an obverse die by Samuel Brooks, a Boston artist and engraver. This portrait was taken from life, possibly during the president's visit to Harvard College in October 1789, and is the earliest medal with a true portrait. Examples in brass, as here, were sold for $2 in February 1790 by Jacques Manly.

The most-familiar portrait of Washington is by Gilbert Stuart. Not used on coins or medals, this depiction became famous on paper money.

A British proposal for an American cent, 1791.

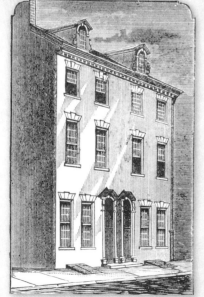

An engraving from the late 1800s of the joined front buildings of the first Mint.

A proposal for a half dollar by Peter Getz, an engraver in Lancaster, Pennsylvania.

Coinage of Half Dismes

Philadelphia, the nation's capital at the time (until 1800), was the natural choice for the mint's location. President George Washington lived a couple blocks away, and other federal officials were nearby. The Bank of the United States, authorized under a 20-year charter by Congress in 1791, was a convenient nearby institution. In time, drafts on the Bank of the United States were accepted by some foreign mercantile interests in lieu of coins, especially in countries in which goods were imported from America and these could be used in payment.

On April 14, 1792, David Rittenhouse, a well-known scientist, inventor, and scholar, was appointed to be Mint director. Announced by President Thomas Jefferson:

David Rittenhouse, oil on canvas by Charles Willson Peale, 1796.

> Know Ye, That reposing special Trust and Confidence in the Integrity and Abilities of David Rittenhouse of Pennsylvania. I have nominated and by and with Advice and Consent of the Senate do appoint him Director of the Mint of the said United States, and do authorize and empower him to execute and fulfill the duties of that Office according to law, and to have and to hold the said Office with all the Powers, Privileges, and emoluments to the same of right appertaining during the pleasure of the President of the United States for the time being.

> In Testimony whereof I have caused these letters to be made Patent, and the Seal of the United States to be hereunto affixed. Given under my Hand, at the City of Philadelphia, the fourteenth day of April in the year of our Lord one thousands seven hundred and ninety two, and of the Independence of the United States of America the Sixteenth.

Rittenhouse recommended that the property owned by Frederick Hailer at 27 and 29 North Seventh Street (after 1855 known as 37 and 39 North Seventh Street) and the contiguous lot at 631 Filbert Street be purchased for the site of the new Mint. The buildings had been used formerly by Michael Shubert as a distillery. On July 9 President Washington wrote to the director, the letter beginning:

> Having had under consideration the letter of the Director of the Mint of this day's date, I hereby declare my approbation of the purchase he has made of the house and lot for the Mint, of the employment of Mr. Voigt as coiner; of the procuring fifteen tons of copper, and proceeding to coin the cents and half cents of copper, and dismes and half dismes of silver . . .

By early summer a coining press and other equipment was stored in the cellar of a house owned by John Harper on North Sixth Street and Cherry Street. On July 11 Thomas Jefferson took $75 in silver to Harper's cellar, the metal likely having come from a withdrawal of $100 from the Bank of the United States the previous day. It was converted into 1,500 tiny five-cent pieces, each with the inscription HALF / DISME, which Jefferson received on July 13.[2]

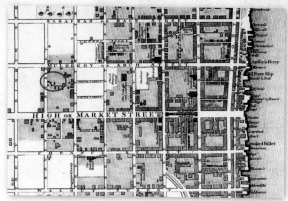

**Philadelphia streets with the location of the future
(1792) Mint circled in red. (B. Eastburn, 1776)**

On July 18 the real estate transaction was consummated, and the deed to the property was obtained from Frederick Haller. The cost was $4,276.67.[3] On the next day the destruction of unwanted structures was begun by John Keyser, John Bitting, Matthias Sumers, John Maul, John Christian Glouse, and Nicholas Sinderling. On July 30 Baltis Clymer delivered the first five loads of mortar sand. At 10 o'clock in the morning on July 31 a foundation stone for the southern half of the front building (No. 27 North Seventh) was laid by Director Rittenhouse, this being the second structure to be erected for the government (the first was a lighthouse in Virginia). The two front buildings were joined and appeared to be a single edifice.

On August 25 the foundation and walls were complete, and the roof was raised. On September 7 the structure was completed. As Rittenhouse was in ill health, it was left to Voigt to begin the Mint's operation. As early as September 3, Voigt ran a notice for copper in *Dunlap's Daily Advertiser*: "The highest price will be given for old copper at the Mint, No. 29 North Seventh Street." Space was also purchased in *Bache's General Advertiser*. On September 11 the Mint made its first purchase of copper, six pounds for 1s 3d per pound. In early October there were three presses in the Mint, one of which was employed to strike additional half dismes.[4] The dies had become slightly rusted by that time. It is thought that several hundred or more were coined.

Frank H. Stewart in his Mint history described the facilities:

> Facing on Seventh Street was the so-called Office Building [the two joined front buildings]. Behind it was a yard 18 feet wide, with an alley to one side. The Coinage Building [a misnomer, although some coining was accomplished there, it was mainly used to melting and refining[5]] measured 30 by 35 feet and was in the rear. The first floor of the Coinage Building measured 28 feet, 11 inches in width, 32 feet, 3 inches in depth, and 11 feet, one inch in height. All inside measurements. The second floor was 29 feet, 2 inches in width, 31 feet, 10 inches in depth, 12 feet, 1 inch high on the front and 9 feet, 10 inches high on the back. One vault in the Coinage Building was 6 feet, 5 inches in length, 5 feet, 7 inches in width, and 6 feet high.[6]

THE MINT BEGINS PRODUCTION

Following the coinage of half dismes for circulation, over a half-dozen varieties of patterns were made, including cents from dies by Robert Birch and a pattern presumed to be a quarter dollar by Joseph Wright.

In his account book, on December 17, Henry Voigt wrote, "Struck off a few pieces of copper coins."

On the next day Jefferson wrote to Washington:

> Th. Jefferson has the honor to send the president 2 cents made on Voigt's plan by putting a silver plug worth three-fourths of a cent into a copper worth one-fourth of a cent.
>
> Mr. Rittenhouse is about to make a few by mixing the same plug by fusion with the same quantity of copper. He will then make of copper alone of the same size and lastly he will make the real cent as ordered by Congress four times as big. Specimens of these several ways of making the cent may now be delivered to the committee of Congress now having the subject before them.

The price of copper began to rise. It became necessary for Rittenhouse to petition Congress to reduce the amount of copper metal required for each cent and half cent. As a result, the Act of January 14, 1793, was passed, changing the weight of the proposed copper coinage. It stated that "every cent shall contain two hundred and eight grains of copper, and every half cent shall contain one hundred and four grains of copper" and that the earlier act with regard to the weight of copper coinage should be repealed. The January act therefore reduced the copper in the cent to 8-1/2 pennyweights from the original specified weight of 11.

A pattern silver-center cent by Henry Voigt, 1792. Judd-1.

Most duties at the Mint required strong body and temperament. Work began at six in the morning in the warmer months and at seven in the winter, the start of an 11-hour stint, this for six days a week. Rum was provided, although drunkenness was not allowed. A coining-press operator earned about 90 cents to one dollar per day. These presses were operated by human muscle. Horses provided the power to operate the roller mill which converted copper ingots into flat strips. Nero, a dog purchased for $3, helped guard the Mint while a watchman remained through the night.[7] Lighting was first provided by candles and whale-oil lamps, later augmented by lanterns.

Copper one-cent pieces were first made in quantity for circulation in February 1793 and were delivered by the coiner on March 1. The earliest versions had a head of Miss Liberty on the obverse and a chain of 15 links on the reverse, one for each state (by this time Vermont had joined the Union in 1791, and Kentucky joined in 1792). Apparently, the design was less than pleasing. One newspaper account described Miss Liberty as being "in a fright," and the chain on the reverse as an ill omen for a land of freedom. The motif was changed to a different face of Miss Liberty, and, on the reverse, a wreath. That summer, copper half cents were made for the first time. Some early dies were made by Joseph Wright (who died of yellow fever that September), and likely by Henry Voigt, coiner on the Mint staff. Formerly a clockmaker, Voigt had engraving skills. Likely, Wright made the 1793 Liberty Cap cent dies, which have an artistry and consistency not found on the cents of 1794.

Robert Scot, a local maker of copper plates for bank notes, maps, and illustrations, was hired as engraver on November 23, 1793, and would continue in the post until his death on November 3, 1823. His work included, but was not limited to, engraving and sinking original dies, creating punches such as for portraits, leaves, and the like, and making hubs and dies. A reverse die for a half dime took about six days on average. Although policies varied over time, letter punches were often obtained from trade sources, such as those handing jewelers' supplies. There seems to be little consistency as to the life of dies. Some cracked or failed while early on the press, and others lasted for thousands of impressions.

In 1793 copper for coining half cents and cents was obtained from many sources, including by advertising to the public in local papers. In London, diplomat Thomas Pinckney purchased 30 "cases" of copper to be shipped to New York, then to Philadelphia. On another occasion 10 casks of copper nails were imported. Coinage this year was limited to copper cents and half cents. No silver or gold coins were made, as surety bonds of $10,000 each were required for chief coiner Henry Voigt and assayer Albion Cox. Neither man could raise such a sum, so that if gold or silver coins had been desired to be struck, it could not have been accomplished.

In 1793 and continuing for a long time afterward there were no numismatists in America who systematically sought and saved the new federal coinage. Their interest was mainly limited to ancient coins, European coins, and medals. In England, however, there was a great interest in modern coins and tokens. Some numismatists acquired U.S. issues as part of world coinage. America, of course, had been a British colony a generation earlier.

COINAGE OF SILVER AND GOLD

In autumn 1794 the first silver coins were struck at the Philadelphia Mint, consisting of half dollars and dollars of the Flowing Hair design. The press was not strong enough to coin the dollars properly, with the result that many had to be discarded from this run, and only 1,758 were deemed worthy to release. Even these were weakly struck at the lower left of the obverse and the corresponding part of the reverse. Dies dated 1794 for the half dime were made, but were not used until 1795. By this time a congressional action of March 3, 1794, had reduced the surety bond requirements to $5,000 for Voigt and just $1,000 for Cox. Mint Director David Rittenhouse, Frederick Augustus Muhlenberg, Peter Muhlenberg, Henry Kammerer, and Nicholas Lutz guaranteed the bond for Voigt, while Charles Gilchrist stood as guarantor for Cox. Gold coinage could now begin, but none was accomplished in this year. A new, large press weighing 3,232 pounds was delivered on May 6, 1795, and was suitable for coining silver dollars. The cost was $937.17 payable to Samuel Howell, Jr. & Co.

In his 1861 book, *Washington and National Medals*, James Ross Snowden wrote:

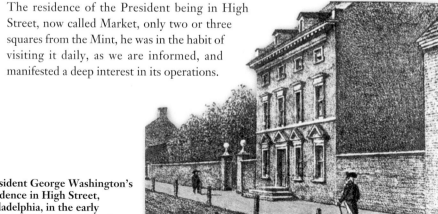

The residence of the President being in High Street, now called Market, only two or three squares from the Mint, he was in the habit of visiting it daily, as we are informed, and manifested a deep interest in its operations.

President George Washington's residence in High Street, Philadelphia, in the early 1790s. (J.F. Watson, *Annals of Philadelphia*, 1830)

Snowden drew heavily upon Mint "tradition," passed along by the Eckfeldt family, and information gathered by assistant assayer and Mint Cabinet curator William E. DuBois. Unfortunately, no daily diary entries for George Washington have been located that mention daily Mint visits or, for that matter, details of *any* visit. An ambiguous record suggests that Washington may have personally witnessed an assay of foreign coinage at the Mint on December 28, 1792, but this is the lone instance in which any documentation supports the notion of a visit from Washington.[8] If they occurred, such visits in the aggregate would have occupied a significant amount of his time, and no doubt he would have expressed many opinions and ideas.

In this year the Mint opened an account with the First Bank of the United States (chartered in 1791). The first deposit was made on June 17, 1794, to the amount of $440.07. For years thereafter much business would be done with the First Bank and its successor, the Second Bank of the United States (chartered in 1816). The Bank of North America and the Bank of Pennsylvania also received Mint deposits in 1794.

On December 27, 1794, James Davy, an Englishman who sought commercial involvement in coinage, sent these observations to the secretary of state:

> I shall offer my opinions on the Mint of the United States of America . . . I find the supply of copper has not been regular, that the power now applied is not adequate, nor are many parts of the machinery adapted for performing the work to the best advantage.
>
> Other observations might have been made, & Improvements suggested, if I had seen the whole of the machinery at work, or had the opportunity of a more minute inspection of the Mint; There is no doubt but that the coinage may be much improved, by rendering the cents brighter, and clearer from rough black streaks which much disfigure many of them, the expenses considerably diminished & a greater quantity of coins produced; besides laying a foundation for other important advantages to this country, by adopting the plan herein offer'd.—
>
> If it should appear that my services will be of use, I shall be happy to engage my attention to the execution of the improvements suggested, & in procuring materials, & the necessary workmen in addition to those already engaged, some of whom appear to possess considerable merit.

Davy's expertise, such as it may have been, was not tapped.

The Bank of the United States.

THE MINT IN 1795

In 1795 the House of Representatives appointed a committee to investigate the state and progress of the Mint. A report dated February 9, 1795, by Congressman Elias Boudinot (later to become Mint Director), included this:

> The works consist of two rolling machines, one for hot and the other for cold metal, worked by four horses, and require five hands constantly to attend them, while in operation. There is a third, nearly completed, to be appropriated to the smaller coinage. A drawing machine for the purpose of equalizing the strips for cutting the planchets, and are worked by the same hands as are last mentioned. Three cutting presses for the planchets of larger and smaller coins, which are worked by one man each. A milling machine, which is intended to be worked by the horse mill, but, at present, requires one hand.
>
> Three coining presses, with the improvement for supplying and discharging themselves by machinery. Six hands will attend three, if in one room. A fourth, for dollars and medals, in particular, will be finished in about three months. Two turning lathes for dies, and a boring machine for making holes in the large frames, screws for presses, stakes, rollers, and an infinite variety of instruments and tools, necessary to carry on the coinage. There are, besides three annealing and one boiling furnace, with two forges, the assay, melting and refining furnaces.
>
> The net produce of these works, from the establishment of the mint to this time, consists of one million and eighty-seven thousand five hundred cents, paid into the treasury of the United States, equal to ten thousand eight hundred and seventy-five dollars; in silver coins delivered, thirty-five thousand one hundred and sixty-five dollars. The future produce, it is said, will be about two hundred thousand cents per month.[9]

Boudinot also told of the duties of Mint officials:

> The Director, whose duties, besides those contained in the act instituting the Mint, are the general superintendence of the whole business, in all its various departments, the making or approving of all contracts and purchases, relative to the institution, determining on the expediency of all improvements, buildings, machines, and whatever may be thought necessary for promoting the utility of the Mint, and lastly, to inspect all receipts and issues of the Mint, with the accounts of the expenditure, and to draw warrants for the same.
>
> The Assayer. He assays all metals brought to the Mint, and reports their respective qualities to the treasurer, for his direction. He attends and inspects the melter and refiner, and has charge of the bullion, jointly with the treasurer and chief coiner. He has, hitherto, also had the care of the melting and refining all the precious metals that have come to the Mint.
>
> The Chief Coiner prepares all the necessary machines, belonging to the different branches of coining the several metals directed by law; works all the ingots, received from the melter and refiner, into a proper state for coining, and, when completed, delivers them over to the treasurer, and lastly, oversees all the different workmen employed in the coinage, and keeps them in their duty.
>
> The Engraver, whose actual duties are the raising and furnishing all punches that are requisite for the completion of the dies, the engraving and sinking all original dies, and raising all hubs that are struck out of them. He has an assistant, occasionally, as the business is urgent.

The Treasurer, whose duty it is to take charge of all bullion, received or deposited in the Mint, for coinage. He delivers it out, as wanted, for working, and receives, in return, all the coins, as they are completed, from the chief coiner. He registers all the qualities of the metals, as reported by the Assayer, and pays out all the coins, when completed, on the warrants of the Secretary of the Treasury and Director, makes all payments on account of the Mint, and renders his account every three months to the treasury of the United States.

The Clerks. At present, there are three clerks, one of them performs all writings, relative to the Mint, required of him by the director and assayer, makes out all orders of the Director, and keeps regular entries of the same. He keeps an account of all bullion, received and delivered by the assayer, acting in the capacity of refiner, and does such out of door business, for the use of the Mint, as is required of him by director or assayer. He also keeps an account of the workmen employed, pays them their wages, and procures the necessary materials.

One other is clerk to the chief coiner, whose duty it is to keep an account how much, and what, metal is received by the chief coiner, from the treasurer, and of the sum returned in coin. He weighs out, daily, the several metals to the proper workmen, and receives it back at night, by weight of which, he keeps the proper entries. He makes out the pay rolls, for the Director's inspection, on which warrants on the treasurer are issued. He is present at the payment of the workmen, and takes their receipts. He also keeps the accounts of the contingent expenses attending the institution.

The remaining clerk belongs to the treasurer and keeps all accounts relative to his receipts, issues, and expenditures, whether of bullion, coin, or payments. He also weighs the bullion, in the first instance, registers the quantity of alloy, and counts the coin issued from the Mint. Each clerk receives a salary of five hundred dollars per annum, except the treasurer's, who receives from the Director, out of his own salary, an addition of two hundred dollars per annum.

On June 30, 1795, David Rittenhouse resigned as Mint director. On July 8 Henry William DeSaussure was appointed in his place, but served only until October 27. Elias Boudinot then took the post from October 28, 1795, to July 1805. The first gold coins, $5 half eagles, were delivered in July 1795, followed by $10 eagles in the autumn. On October 10, 1795, Adam Eckfeldt, who had done work as a mechanic and blacksmith from time to time since 1792, was given full-time employment as a "die sinker and turner" for $500 per year. Soon afterward, he was named as assistant coiner at $700. His duties were to "personally forge, turn, and harden all dies and hubs and to do, or direct to be done everything necessary thereto, except the engraving and polishing."[10]

Mint Director Henry William DeSaussure.

Eckfeldt, who remained at the Mint for many years, had a deep interest in numismatics. In time he furnished desired coins as an accommodation to interested collectors and others. Sometime around the mid-1820s he began to save unusual or special coins he encountered, and when the Mint Cabinet was officially launched in 1838 he supplied many specimens.

On October 10, 1795, the officers and certain staff members and their annual salaries were:

Henry William DeSaussure, director—$2,000.

Nicholas Way, treasurer—$1,200.

Henry Voigt, chief coiner—$1,500.

Albion Cox, assayer—$1,500.

Robert Scot, engraver—$1,200.

David Ott, melter and refiner *pro tem*—$1,200.

Nathaniel Thomas, clerk to the treasurer—$700.

Isaac Hough, clerk to the director and assayer—$500.

Lodewick Sharp, clerk to the chief coiner—$500.

John S. Gardner, assistant engraver—$936.[11]

Adam Eckfeldt, die forger and turner—$500.

An engraver preparing a coin die. (Marcia Davis 1982 for the John W. Adams catalog of 1794 cents)

In 1796 the quarter dollars and quarter eagles were coined. This was the first year in which all denominations authorized by the Mint Act of April 2, 1792, were struck.[12] During this era John Smith Gardner worked as an assistant engraver from time to time and made punches, hubs, and dies.

DIFFICULTIES WITH COIN CIRCULATION

Designs changed and evolved over the years. It was general practice to have copper coins bear one common design, silver coins another standard motif, and gold coins to share still another, although sometimes there was overlapping. The Draped Bust obverse was first used on silver coins of 1795, then copper coins of 1796, while gold coins depicted Miss Liberty with Draped Bust and wearing a conical cloth cap.

In the 1790s there were no significant domestic mines or natural sources for silver and gold. Most metal was obtained by melting foreign coins. Copper was obtained from many domestic sources as well as by importing. On the London market, the worldwide arbiter of values, the ratio of gold to silver hovered around 1 to 15.5 during this era, making it profitable to export American gold coins. From their first coinage in 1795, countless pieces met this fate.

It was the desire of Congress that silver and gold coins of the Mint would become common in the channels of trade, and that the authorization permitting certain foreign coins to become legal tender would expire relatively soon. During the 1790s and extending into the next century, many silver and gold coins were exported with the result that they were a minority in circulation. Many investigations and reports were made concerning this. As an example, on December 11, 1797, Representative Abraham B. Venable of the congressional committee charged with reporting on coinage, stated in part:

It appears, from the best information they can obtain, that very little of the silver coin of the United States has circulated at any considerable distance from the mint, especially in the interior parts of the country.

That, by the operation of the law, which provided that, at the expiration of three years after the coinage of gold and silver should commence at the mint, all foreign silver coins, except Spanish milled dollars, and the parts of such dollars, should cease to be a legal tender, considerable embarrassments have been produced, and many losses sustained, as a very considerable quantity of foreign silver coins, other than Spanish milled dollars, and the parts of such dollars, was, at that time, in circulation. Your committee also find, that, by the operation of the said act, all foreign gold coins will cease to be a legal tender, after the thirty-first day of July next; that a great quantity of it is now in circulation, and must necessarily continue so, until that period arrives, as it will be scarcely possible for the mint, on its present establishment, to coin a sufficient quantity to replace it.

Your committee are, therefore, of opinion, that provision ought to be made by law, authorizing and requiring the collectors of the revenue to receive, in discharge of all demands of the United States, foreign silver coins, other than Spanish milled dollars, and the parts of such dollars, at the rates, and under the regulations, by which they were receivable before the fifteenth day of October last; that this regulation should continue for two years, and until the end of the next session of Congress thereafter; and that so much of the said act, as relates to the circulation of foreign gold coins, be suspended for the like time.

All the while the Bank of the United States, as well as other banks, merchants, and others in commerce, received and paid out large quantities of Spanish-American gold and silver coins and other foreign issues with not enough federal coins in circulation to make much difference.

Continuing Activity

Half cents gained very little traction in commerce, and from 1793 onward they formed only a small percentage of the copper mintage, exceptions being 1804 and 1809 (when relatively few cents were made). The most important reason was efficiency. Whereas silver and gold coins were made to the order of the depositors of these metals who specified what denominations they desired (resulting in erratic production of some), copper coins were produced on the Mint's own order. The difference between face value and the cost of metal and striking represented a profit. It was easier and cheaper to strike one cent instead of two half cents. Great efforts were made by the Mint to acquire copper in the early dates. Later, finished planchets were supplied by Matthew Boulton in Birmingham and various companies in the United States. Production of cents was nearly continuous during the life of the first Philadelphia Mint, the only exception being in 1815.

In 1797, as in some other years in the 1790s and in 1803, yellow fever swept Philadelphia, forcing the Mint to close from late August to late November and driving many citizens to refuge in the countryside. The source of the scourge was not known. The doorkeeper at the Mint remained in residence for security purposes, and Adam Eckfeldt, who had been employed by the Mint since 1792, spent the next few weeks working on a coining press. The dies were removed from the presses and put into a vault in the Bank of the United States. Such periodic closing of the Mint resulted in some scrambling of obverse and reverse die combinations in the first decade of Mint operation, this affecting various denominations. In the same year, deposits of silver at the Mint reached a low, affecting coinage of that metal, but silver deposits resumed with vigor in 1798, mostly coined into dollars. Many dollars were exported, never to return. The Spanish-American eight reales coins remained dominant in domestic as well as international commerce.

On January 3, 1799, Mint Director Elias Boudinot included this in his *Annual Report* for 1798:

> With pleasure he refers the President to the enclosed returns of issues of the several species of coin from the Mint, since the first of January, 1798; during which time, the coinage has been stopped near three months, occasioned by the late calamitous fever, and the decay of some of the machinery.
>
> Yet, by these returns, it will appear that the coinage of gold amounts, in value, to 205,610 dollars; that of silver, to 330,291 dollars; and that of copper, to 9,797 dollars; in the whole, amounting to 545,698 dollars; exceeding, in value, nearly double what has ever been coined at the mint in any one preceding year, and increases the whole amount of the coinage, since the commencement of the business, in October, 1794, to 483,245 dollars, in gold; 792,643 dollars, 75 cents, in silver; and 41,004 dollars, 74 cents, in copper; amounting, in the whole, to 1,316,893 dollars 49 cents.
>
> From information the Director has received, he has no doubt but there will be a full supply of silver bullion for the ensuing year, at the present establishment of the Mint; and the frequent deposits of gold, give him encouragement to suppose a proportionate supply of that precious metal will be kept up. The present arrangement, with regard to copper coin, will enable the Director, during the course of the next summer, to supply any demand that is likely to be made for cents, and at present there are a considerable number on hand.
>
> The Director cannot, with propriety, close this report, without mentioning, that, during the last summer, a scheme was discovered for robbing the Mint, by persons out of it, in concert with one person employed in the Mint; and although the offenders have been detected, prosecuted and punished, yet it fully justifies the observations heretofore offered to the President, on the unprotected state of the Mint, to which the Director begs leave to refer.

In 1799 yellow fever spread once again, and the Mint was closed from August 24 to October 23. On December 14, 1799, George Washington died at his Mount Vernon home. He was widely mourned. In time, coins, tokens, and medals with his portrait would become a specialty in numismatics.

Elias Boudinot engraving by J.W. Paradise after a painting by Waldo & Jewett. (*National Portrait Gallery of Distinguished Americans*, 1836)

On January 8, 1800, the *Annual Report* submitted to President John Adams by Mint Director Elias Boudinot included this regarding 1799:

> The Mint has been regularly supplied with bullion, both gold and silver, so as to keep it in constant operation, on the present establishment, during the year past year, excepting two months, in which, the works were totally stopped, on account of the then prevailing fever; and there is a rational prospect that the supply will be continued for the present year.
>
> From the late arrangements with regard to supplies of copper planchettes, for the coinage of cents, there is no doubt but that one press, equal to the coining of 14,000 per day, may be kept in constant operation.

Copper coins continued to be the mainstay of the Mint, as the profit from them supported other expenses.

During the summer of 1800 in the last official year of the 18th century, the capital of the United States was moved from Philadelphia to what was first called the Federal City, then Washington City, then simply Washington. The transfer was done by 123 federal clerks and other works. This precipitated a call that the Mint be moved, and various correspondences addressed this, such as in a report of a Senate committee on March 14, 1800, that included:

> In that the existing law requires the removal of the Mint to the permanent seat of government; that such removal would, in many respects, be inconvenient; but the policy of keeping up that establishment, in a situation where its operations will not be under the immediate superintendency and direction of the principal officers of government, is questionable.

The Executive Mansion, later known as the White House, was occupied by President John Adams, who had been elected in 1796. There was no plumbing yet, and water needed to be carried from a distance. The Library of Congress was established by a congressional appropriation of $5,000 for the purchase of approximately 900 books, including 740 from London booksellers Cadell & Davies. The second decennial census recorded a population of 5,308,483, including 896,849 slaves.

Although federal half cents and, in particular, cents had become common in circulation, countless copper coins of the 1780s minted under authority of the states, counterfeit as well as regal British halfpence, and other coins were still in circulation in quantity. The *Annual Report* for 1800 included this suggestion:

> The cents issued from the mint, amounting, now, to the sum of seventy-nine thousand three hundred and ninety dollars and eighty-two cents, the proclamation required by law ought to be issued, by which all other copper coin will be put out of circulation . . .
>
> The great rise in the price of copper, in Great Britain, has prevented so large an importation, and of course so large an issue of cents from the mint, as would, otherwise, have been done, the public being now tolerably well supplied with that species of small change.

INTO THE 19TH CENTURY

Thomas Jefferson was inaugurated on March 4, 1801, following his controversial election finalized on the 36th ballot by the House of Representatives on February 17. He abolished lavish social events at the White House and dressed casually to receive visitors.

With the nation's capital in Washington, many legislators and others continued to desire to have the Mint moved there as well. Amid controversy, on March 3, 1801, Congress authorized the Mint to remain in Philadelphia for a further two years. Despite other efforts to move it, the Philadelphia Mint in its various iterations remained in the city of its founding.[13] From 1800 to 1803 many attacks were directed at the Mint, most of which were out of its control—such as the amount of gold and silver deposited for coinage and the flight of many such coins to foreign countries. Time and again during the early years of the Mint various proposals for coinage by contract were advanced, but no changes were made.

On March 22, 1802, Mint Director Elias Boudinot prepared an inventory that included:

> Five striking presses with machinery. Three cutting presses. One milling machine. Five pairs of rollers, great and small. One drawing machine. A large number of hubs and dies on hand, of different denominations. Engraver's tools.

Boudinot's *Annual Report* dated January 1, 1803, included this:

> The Director of the Mint of the United States, begs leave respectfully to make his annual report on the issues of the state of the Mint.
>
> He is happy to inform the President that the bullion deposited in the Mint during the past year has far exceeded what was expected at the beginning of it, notwithstanding the considerable check given to deposits for some time by frequent reports from the seat of government during the last session of Congress that the Mint would be abolished.
>
> Since the first day of January, 1802, there has been issued from the Mint a sum, amounting in the whole, to five hundred and sixteen thousand one hundred and fifteen dollars, eighty-three cents . . . which been added to the current coin of the Union. Of this sum, one hundred and twenty-nine thousand seven hundred and thirty dollars, and ninety-one cents, in value, in gold, have been coined from bullion and gold dust imported into the United States, and collected to the Mint, as a center, from the different parts of the Union. The balance of the gold coinage has been coined from clipped, plugged, and otherwise spoiled foreign coins, which have been sent to the Mint as bullion. Had not this whole sum been coined in the United States, it must have been remitted to the European markets; in which case the freights, insurance, and commissions, with the profits on the cents, would have amounted to a sum nearly equal to the current expenditures of the Mint. . . .
>
> The current expenses of the Mint for the past year, have amounted to seventeen thousand four hundred and sixty-two dollars and sixty-five cents, as will appear from Schedule, No. III, from which the profits on the copper coinage, amounting to five thousand six hundred and forty-four dollars and thirty-three cents, should be deducted.
>
> Besides the cents on hand, we have near twenty-four tons of copper planchets ready for striking; the coinage of which are in daily operation, at the rate of fifteen thousand cents a day.
>
> It is a duty incumbent on the Director of the Mint, respectfully to call the President's attention to the expiration of the law of the United States for continuing the Mint at Philadelphia, on the fourth of March next, by its own limitation. It therefore becomes absolutely necessary that the subject should be brought before Congress so early that provisions may be made for the contingency. If Congress should

rise without doing anything therein, the Mint could not be continued in Philadelphia with propriety; neither could it be removed to the seat of government, for want of a law to authorize it.

It is but doing justice to merit to say, that the officers of the Mint, concerned in the coinage, and the workmen, have greatly increased in their professional knowledge, and have acquitted themselves with strict integrity, and particular attention to their several departments, for many years past;—so that not a dollar has been lost, except in one solitary instance, when the culprit was detected by their assiduity and care, prosecuted, and punished; and it was by their exertions that the Mint was kept open during the late distress of the city, by the fever of last summer.

If the Mint should remain in its present situation, there will be a necessity of at least two additional horses, and some repairs to the machinery, part of it having been repaired the past year, from necessity. At least five hundred dollars will be necessary in that case, to be added to the usual estimate, to be appropriated for the purchase of horses, and further repairs to the present machinery.

There was a shortage of copper, and the Mint arranged with Matthew Boulton of Birmingham, England, to ship every spring and autumn 20 to 25 tons of prepared planchets.

Moving or Abolishing the Mint

The following two letters shed light on the uncertain state of the Mint in 1802 and also provides a description of the Mint's equipment.

Amidst the continuing discussions in Congress of moving the Mint to Washington as well as terminating the institution and having coinage done by private contractors, engraver Robert Scot, in the post since autumn 1793, wrote to the Secretary of the Treasury Albert Gallatin on March 2, 1802:

> The probability of the abolition of the Mint establishment, induces me, thus early, to state to you, that, if the Legislature should not be disposed altogether to abandon the copper coinage, or might be willing, after repealing the laws establishing the mint, to allow of a copper coinage, provided it may be done without any expense to the public, I would solicit your interests and influence to promote a proposition of that kind, which I do not presume on, only so far as you may deem it to consist with the public good; in connexion with which, I flatter myself you will not be wanting, independent of any other claim I may have, or pretension to public patronage.
>
> However I need not omit informing you, that, on the first establishment of the mint, I relinquished a profession, at least equally productive and beneficial as that of the engraver's place in the Mint, which I have filled, and I believe without reproach, ever since; by the loss of which, I shall be left without resource, being so long out of the practice of my former profession, that I feel an incapacity to prosecute it with anymore effect. I, therefore, submit the following proposition to your consideration, to the consideration of Congress, or to the Department where it may properly belong:
>
> That I may be vested with the exclusive privilege, according to law, of coining cents of the United States, as well from abroad as within the realm, under such restrictions and provisions, either with respect to time or quantity, as Congress, in their wisdom, may deem proper; that the cents shall be of the present weight and quality, and that they shall be coined free of all expense to Government, excepting that of receiving them when coined, and paying the nominal amount.

Should the above propositions meet with your approbation or otherwise, I should still be happy to know your determination to forward them or not; if the former, I would beg to know the most proper mode of introducing it to Congress, whether by petition, and how conceived, or otherwise.

On March 26, 1802, Director Boudinot wrote to Hon. Mr. Giles, chairman of the Committee on the Subject of the Mint:

I am honored with your letter of the 10th instant, and hasten to give you the best answer that I can, with regard to the real and personal estate of the Mint establishment, &c. This consists of—

Two lots on Seventh street, between Market and Arch streets, 20 feet each on Seventh street, and extending back about 100 feet, with a dwelling house on the north lot, and a shell of a house on the south lot, which last lot widens on the rear to about 60 feet, on which the stable stands. These lots pay a ground rent of $27 50 per annum.

A lot on Sugar alley, at the rear of the above, 20 feet front on the alley, and about 100 feet deep. A frame building, improved for a large furnace, in the commons at the north end of Sixth Street, of little value, the ground being merely loaned to us.

As to personal estate, this consists wholly of the copper planchettes on hand amounting to about 22 tons. Three horses, good for little but for the use of the Mint. The machinery of the Mint, of no value but for the use of the Mint. Five striking presses with machinery. Three cutting presses. One milling machine. Five pair of rollers, great and small. One drawing machine. Three pair of smith's bellows. A set of blacksmith's tools. A large number of hubs and dies, on hand, of different denominations. Carpenter's tools. Seven stoves. One turning lathe. Six scale beams, scales, and weights. Two sets assay scales, and sundry adjusting scales. Furniture in the clerk's rooms. Various implements used in the several departments. About 2,000 bushels of charcoals. Engraver's tools, pots, bottles, &c.; an old horse, cart, and gears. About 2,000 fire brick; a considerable quantity of old iron.

It is impossible to ascertain the value of these articles, as most of them are of but little consequence, except for the use of the Mint, or to persons who may intend to put them to the like uses; and if sold at public sale, probably will not bring half their real value. The machinery of the Mint may last a year longer, with small repair, but, after that, will cost about three hundred dollars, to put it in good repair. The horses may, also, last another year, but must then, at farthest, be replaced by others. If it should be thought best to continue the Mint, the establishment should be rendered permanent, and the machinery should be moved by steam instead of horses, which would, in some measure, reduce the annual expenses of labor, as almost the whole of it could be carried on with the same original force.

Our lots are much too small, by which we are greatly cramped as to room. They are now very valuable, being in the heart of the city; their price would purchase a very advantageous lot in a less public place, and buildings might be now planned, so as to reduce the expenses of a mint. But I am perfectly satisfied, that no modification of the Mint could be contrived to lessen them below seventeen or eighteen thousand dollars per annum; though if a larger quantity of bullion could, by any means, be provided, a greater quantity of coin could be annually made with the same expense, although I am, individually, of opinion, that its present issue, of about five hundred thousand dollars annually, in addition to the current coin of the Union, is sufficient for the present welfare of the United States.

It is the absolute necessity of strict and regular checks, throughout the whole establishment, that makes the expense of the Mint so great, and this cannot be dispensed with, under any modification that can be proposed. I verily believe, that, under no given circumstance, can the necessary coin of the United States be produced with safety to the government, at a much less expense than it is at present; and I believe, that, in the consideration of the subject, it would not be safe to estimate the expense, at any rate, much under twenty thousand dollars.

In the above estimate of expenses, it should be remembered that the copper cents may produce a profit of five thousand dollars per annum that ought to be credited against the expenditures of the Mint in future, which reduces the amount considerably.

Continuing Challenges

This commentary in the *Annual Report* for 1802 gives a reason other than export for the scarcity of federal gold coins in circulation:

> The banks are fond of keeping the [gold] coins in their vaults, as part of their capitals, on account of the ease with which they are counted, without the trouble of weighing. The Bank of the United States, indeed, having a considerable part of their specie in this coin, have been enabled, for some time past, to cancel their five dollar notes, and to substitute the payment of half eagles, by which our coins begin to be more generally dispersed among the people.

In 1803 yellow fever again caused death and illness, and the Mint was closed for six weeks beginning on September 16. In the same year President Thomas Jefferson arranged the purchase of the Louisiana Territory from France for $15,000,000, vastly increasing the area of the United States. The Corps of Exploration, familiarly the Lewis and Clark Expedition, set off from St. Louis in 1804 to explore the upper reaches of the new acquisition. They took with them a supply of peace medals, some made in England, others at the Philadelphia Mint, to distribute as gifts to Native chiefs, a long-time tradition.

By that year the exporting of federal silver dollars had resulted in very few being found in circulation. A common arrangement was to ship them to the West Indies where they could be exchanged at par for Spanish-American dollars of slightly heavier weight. The Spanish coins were then deposited at the Mint, and a slightly larger quantity of American dollars was received. To end this, President Jefferson stopped the coinage of silver dollars, a policy that would remain in effect for the rest of the time the first Philadelphia Mint was in operation.[14] The production of $10 gold eagles was similarly halted due to the massive exportation of such coins.

Silver and Gold Coins Disappear

The melting and exporting of silver and gold coins, erratic deposits of precious metal at the Mint, and other factors combined to create a continuing scarcity of such pieces in circulation. Copper cents continued to be used in commerce without difficulty, but other denominations were scarcely seen. After 1804 the largest silver coin was the half dollar. Many if not most of these were sorted in bank vaults as reserves. The $5 gold half eagle was the largest gold denomination. Massive exports of these continued, as at their destination they were worth slightly more than face value if melted. When gold coins were sent to England, for example, they were reduced to bullion, and then coined into British denominations. Otherwise the Bank of England and others would have had on hand large quantities of gold coins of different denominations from different countries that varied in their face values and metallic purity.

In the United States the Spanish silver and gold denominations continued to dominate commerce (as delineated in chapter 1). An overview from Neil Carothers, *Fractional Money*, 1930:

The coinage of quarters, dimes, and half dimes, as contrasted with the half dollar, was negligible from 1792 through 1834. In 19 years of this period there was no coinage of quarter dollars, and in 13 years no coinage of dimes, and in 26 years no half dimes. The total number of quarters, dimes, and half dimes coined before 1830 was less than one piece for each person in the country that year. The half dollar coinage was relatively large, increasing steadily from an annual average of about $1 million face value in the early years to an average of about $2 million face value after 1825. These figures were significant for they measure of the failure of the national coinage system in the first 40 years. The ratio was favorable to silver, and yet the only considerable coinage was in half dollars that did not circulate.

The Mint conditions were primarily responsible. The officials, anxious to increase the coinage and reduce expenses, discouraged depositors who asked for small silver coins. Because there was no bullion for them the depositor had to wait while the Mint coined his bullion, and the coinage of half dollars could be accomplished more quickly and more cheaply [than that of an equivalent face value amount of half dimes through quarters]. Year after year the Mint directors apologized for the scarcity of the small silver coins with the explanation that only by coining half dollars could the Mint take care of this silver bullion presented. The reports indicate that the directors refused to coin small pieces except at times when a falling off in deposits made it convenient.

Adverse conditions were that United States coins did not circulate, and in the depression era from 1812 through 1820—part of which was during the War of 1812, there was a scarcity, and this continued through 1834. Many complaints were made to the Mint. Most small coins traded in the country were a group of silver and copper pieces including small Spanish-American silver coins, English, French, and Dutch silver pieces, United States fractional silver, and United States copper coins. Spanish pieces were much of the largest element. In 1830 the Senate Finance Committee estimated that the total of Spanish silver in circulation was $5 million. If this figure is added to the Mint reports on coinage in the Treasury reports on recoinage it is possible to estimate that the total circulation of coins below the half dollar was in 1830 less than 25 cents per person, an inadequate amount. What people needed was a quantity of clean, new, and uniform small coins. The Mint was turning out half dollars that did not circulate and was doing very little else. The legal tender status of foreign coins was renewed in 1806, 1816, 1819, 1823, 1827, and 1834, and according to each of the laws the legal tender of all foreign coins except Spanish pieces was to expire after three years. However, the language was so confusing that it is difficult to analyze today.

In practice, regardless of the legal-tender status of given silver and gold coin in this era, those of most European and Western Hemisphere nations traded readily in commerce.

COINAGE AND ECONOMIC CONDITIONS

In 1804 the first significant deposit of native gold was received at the Mint, this per the *Report* for that year:

> It is worthy of attention that about eleven thousand dollars of the gold coin is the product of virgin gold, found in the county of Cabarrus, in the State of North Carolina, where, it is said, a considerable quantity has been found since, which will, in all probability, be forwarded to the Mint.

The early 1800s were characterized by many clashes in the Atlantic Ocean and Mediterranean Sea, including demands by Barbary pirates to allow ships to pass. By 1806 the war between England and France forced the closing of many ports and a general reduction of maritime trade. The Embargo Act of 1807, which took effect in 1808, restricted American vessels from going to foreign ports, as did the Anti-Intercourse Act of 1809. This legislation froze the economy of many Atlantic Seaboard cities, causing great distress. The charter of the First Bank of the United States expired in 1811 and was not renewed. Whether this had a negative effect on commerce is a point to be argued, as many state-chartered banks issued currency. However, some of them were of uncertain stability. In the meantime, Robert Patterson was appointed director of the Mint on January 17, 1806, and would serve until 1824, not to be confused with his son, Robert M. Patterson, who would go on to be director from 1835 to 1851. On July 14 of that year Samuel Moore was appointed to the post and would serve until 1835.

The War of 1812 created more chaos. The U.S. Navy was not strong, and to help fight the enemy the government gave letters of marque, as they were called, to the captains of and owners of ships, who then engaged in privateering under the American flag, capturing whatever enemy prizes they could find and taking the ships to ports where they and their contents were sold at auction—with the money divided among the ship crews and owners. This was a rich undertaking for many, and more than a few family fortunes in Portsmouth, Baltimore, and other cities were augmented in this manner.

From the onset of the war, a huge quantity of gold coins was exported, this despite the blockade of American ports at certain times by British ships. This situation lasted for the next several years.[15] Various eastern banks were accused of aiding the British by making specie (the universal Treasury and banking term for silver and gold coins, but not applicable to copper) available to them under subterfuges as well as open violations, including shipments to Canada, the latter prohibited trade with "the contiguous British provinces."

The conflict caused hoarding of silver and gold coins and the issuance of a flood of mostly worthless paper money. The American monetary system began a long period of stress, strain, and duress. After 1813, gold coins disappeared from circulation, and except for some intermittent times before 1821, they were not seen again in the channels of commerce until autumn 1834. A flood of paper bills during the war prompted the hoarding of copper and silver coins as well. Bullion on hand at the Mint varied widely.

It was expected that after the war ended (officially by the Treaty of Ghent on December 14, 1814, but the Battle of New Orleans was fought the next month as news had not reached the United States), the economy would recover.

THE MINT IN 1815 AND 1816

In 1815, due to a scarcity of copper planchets, imports from Boulton having been halted by the war, no cents were coined. The only gold coins bearing that date were just 635 half eagles. The *Mint Report* stated that 69,232 silver coins with a face value of

$17,308 were also struck. These would have been quarter dollars. Half dollars dated 1815 were not made until early 1816.

> The high price of gold and silver bullion for some time past in the current paper money of the country has prevented, and, as long as this shall continue to be the case, must necessarily prevent, deposits of these metals being made for coinage to any considerable amount.
>
> But a fresh supply of copper having lately been received at the Mint, we have again resumed the coinage of cents; and, it is believed that we shall, in the course of the year, should no failure in the expected supply of copper take place, be fully able to coin fifty tons weight, amounting to nearly 47,000 dollars; and that, with a regular supply of copper, which can readily be procured, on terms highly advantageous to Government, we can continue to coin fifty tons per annum as long as it may be judged expedient.
>
> The circulation of these copper coins, and of those heretofore issued from the Mint, amounting to 251,646 dollars, and which must be still nearly all in the country, would, it is presumed, soon supply, in a great measure, the place of the small silver coins, which have now almost totally disappeared.

On January 11, 1816, there was a fire in the facilities in the back of the main Mint buildings, destroying the wooden mill house that had stood behind the double-front building. This necessitated building a new middle building of brick, repairs and replacement of some refining and planchet-strip equipment, and the introduction of steam power for the rolling mills. In that year only one denomination bearing that date—the cent—was produced, a unique time in American coinage history. A committee in Congress reviewed the copper coinage situation and raised this question: "Would it be advisable to coin two-cent pieces, and even four-cent pieces, and discontinue the coinage of half-cents?"[16] No changes were made.

Copper continued to be the subject of much interest in Congress, in part because this coinage was the main source of profit for the Mint. No half cents had been struck since 1811, and the focus was strictly on cents. On February 1, 1816, Mint Director Patterson wrote to Secretary of the Treasury A.J. Dallas:

> William Harrold, of the house of Harrold and Beller, Birmingham, has made a proposal to me to furnish the Mint with copper planchets fit for coinage, on terms considerably lower than any on which we have heretofore been supplied. At present, he wishes to limit his contract to *five tons*, to be shipped at Liverpool sometime in the ensuing summer, or early in the fall. His terms are fifteen pence sterling per pound avoirdupois, to be paid in Philadelphia thirty days after arrival, at the *then* current rate of exchange. Insurance, freight, and other incidental charges, to be also paid by the Mint.
>
> This contract will yield a profit on coinage of about ten percent more than the average of former contracts, and may be safely estimated at from thirty to thirty-five percent the weight remaining as at present.
>
> Availing myself of your permission, I sent an order to Mr. Boulton, sometime ago, for twenty-five tons, in addition to the twenty tons expected early in the spring; and these, with the five tons from Mr. Harrold, would sufficiently supply the Mint till the spring of the year 1817.

In the same year the Second Bank of the United States was chartered by Congress for a 20-year period and was headquartered in Philadelphia. Of the $35,000,000 capital, $7,000,000 was subscribed for by the government, and the rest went to private

investors. In time, branches were opened in many cities. As was the case with its predecessor, drafts on the Second Bank of the United States were accepted by some commercial interests in foreign countries.

There were no gold coins struck in 1817. The reason was given in the *Mint Report*:

> There have been struck and emitted: In silver coins, 67,153 pieces, amounting to
> 28,575 dollars and 75 cents; and in copper coins, 2,820,982 pieces amounting to
> 28,209 dollars and 28 cents. The amount of the latter would have been consider-
> ably greater, had it not been for a disappointment in the supply of copper. Mea-
> sures, however, are now taken, to prevent such disappointments in future.
>
> The stagnation which has for some time existed in the circulation of specie cur-
> rency, has almost totally prevented the deposits of gold and silver bullion for coinage.
> But there is now a prospect that this will not long continue to be the case—the Mint
> having, at this time, in its vaults deposits, of these metals to a very considerable amount.

Gold coinage in the form of half eagles was resumed in 1818. Still, the vast prepon-
derance of silver and gold coins in circulation consisted of foreign coins, much to the
dissatisfaction of Congress, which continually investigated the matter. Assays of foreign coins were made and the Mint regularly reported. Gold coins of Great Britain
and Portugal were found to be of the same fineness as U.S. coins. The fineness of other
coins varied. Those of Spain and France, for example, were slightly less.

CONTINUING COINAGE

Domestic commerce was mostly conducted in foreign coins, legal tender and otherwise. Notices in *Niles' Weekly Register* reflect this:

May 2, 1818:

Specie: We have the pleasure to notice the arrival of several handsome lots of specie—hoping that some of the banks, which do not pay their debts, will purchase it
and do justice to their creditors. Several sums have been received in gold within the
past week,[17] as well as 450,000 crowns and 55,000 dollars, in silver, from France;
400,000 dollars from England, and 50,000 from Antigua. The three last arrived in
New York in one day.

May 23, 1818:

Specie: A vessel has arrived at New York from London, reported to have $470,000
for the Bank of the United States. Its timely importation will serve to fit out 2 or 3
of the 100 sail of vessels that we have engaged in the East India and China trade.

August 15, 1818:

Specie: A vessel has arrived at New York from France, with the value of $600,000 in
specie. It may do a little good—being in demand. But—it is in five franc pieces.

GOLD COINS DISAPPEAR FROM CIRCULATION

From 1813 and continuing to 1820 gold coins were rarely seen in circulation. In 1819
an inquiry was made in Congress about prohibiting the export of U.S. gold coins. This
idea was quickly squelched when it was realized that American trade would suffer if
importers could not pay in half eagles. In 1819 new foundry and furnaces were installed
at the Mint, speeding the processing of gold and silver. Copper continued to be purchased from outside sources. A report to the House of Representatives, February 8,
1821, included this (italics added for emphasis):

American gold, compared with silver, ought to be somewhat higher than by law at present established. On inquiry, they find that gold coins, both foreign and of the United States, have, in a great measure, disappeared; and, from the best calculation that can be made, there is reason to apprehend they will be wholly banished from circulation; and it ought not to be a matter of surprise, under our present regulations, that this should be the case.

There remains no longer any doubt that the gold coins of the United States are, by our laws, rated at a value lower than in almost any other country in comparison with that of silver. This occasions the gold to be constantly selected, when it can be obtained, in preference to silver, whenever required for remittance from this to foreign countries; and, at the same time, prevents those who have occasion to remit to the United States from doing it in gold. Hence, there is a continual and steady drain of that metal from this country, without any correspondent return, which must continue while there remains any of it among us. The importations of it will be confined to small quantities, and from countries from which nothing better can be obtained.

There have been coined at the Mint of the United States nearly $6,000,000 in gold. It is doubtful whether any considerable portion of it can at this time be found within the United States. It is ascertained in one of our principal commercial cities, quite in the vicinity of the Mint, that the gold coin, in an office of discount and deposit of the Bank of the United States, there located, in November, 1819, amounted to $165,000, and the silver coin to $118,000; that since that time the silver coin has increased to $700,000, while the gold coin has diminished to the sum of $1,200, *$100 only of which is American.*

From 1821 through the end of the existence of the first Philadelphia Mint in 1832, no federal gold coins at all were in domestic circulation, but were exported. This from the 1824 *Annual Report:*

Deposits of gold have been for the last three years inconsiderable. While gold bullion is in demand at a premium on its standard value, for the purpose of exportation as a remittance, instead of bills at the current exchange, no adequate inducement exists to bring it to the Mint, if its value can be otherwise satisfactorily ascertained. It is obvious that, if coined and issued under such circumstances, it cannot be retained in circulation.

Of the gold coined within the year, about five percent was received from North Carolina, thirty-five from Africa, and fifty from South America and the West Indies. The remainder, about ten per cent., is of uncertain origin.

And from the 1825 *Annual Report:*

The gold coinage of the last year has exceeded that of 1824 by the sum of $63,185. Of this excess it may be interesting to observe a very sensible proportion consists of an increase of gold bullion derived from North Carolina; the value of the deposits received from that quarter within the year having been nearly $17,000—more than three-fold the amount from the same source in any previous year since 1804. Within that year, the first in which deposits of gold from North Carolina are noticed on the records of the Mint, the amount received was about $11,000. The whole amount received to the present time is about $68,000. This gold has very generally been found superior in fineness to the standard of our gold coins.

Of the amount of silver brought to the Mint within the year, nearly one-half has been received from the United States Bank in foreign coins, and the remainder, chiefly from Mexico and South America, in various forms of unwrought bullion.

The comment about the Bank of the United States is exceedingly important in the history of the first Mint. Time and again when congressmen requested that the facility be moved to Washington, the most powerful reason for keeping it in Philadelphia was that the city had the home office of the bank, and the bank was *by far* the largest depositor of gold and silver for coinage. Coinage of half cents resumed in 1825, after a hiatus since 1811.

Gold coins did reappear in commerce until autumn 1834, when coinage of reduced weight under the Act of June 28 of that year, effective August 1, reached circulation in significant numbers.

COINING AT THE MINT

The coining process and attendant activities furnished the subject of reminiscences by George Escol Sellers:

> The building used for the Mint had very much the appearance of an ordinary three-story brick dwelling house of that period, the back building and yard extending on the alley. In a rear room, facing on the alley, with a large, low-down window opening into it, a fly press stood; that is a screw-coining press mostly used for striking the old copper cents. [This was one of the four shops where coinage was mainly struck.][18]
>
> Through this window the passersby in going up and down the alley could readily see the bare-armed vigorous men swinging the heavy end-weighted balanced lever that drove the screw with sufficient force so that by the momentum of the weighted ends this quick-threaded screw had the power to impress the blank and thus coin each piece. They could see the rebound or recoil of these end weights as they struck a heavy wooden spring beam, driving the lever back to the man that worked it; they could hear the clanking of the chain that checked it at the right point to prevent its striking the man, all framing a picture very likely to leave a lasting impression, and there are no doubt still living many in Philadelphia who can recollect from this brief notice the first mint . . .
>
> The little yard in the rear of the old Mint was a very attractive place to us youngsters; its great piles of cordwood, which by the barrow load was wheeled into the furnace room and thrust full size in the boiler furnace, which to my young eyes appeared to be the hottest place on earth. There almost daily was to be seen great lattice-sided wagons of charcoal being unloaded, and the fuel stacked under a shed to be used in the melting and the annealing furnaces . . .[19]

Another view of Mint operations is provided by this excerpt from narrative in December 1818, in which Mint Director Robert Patterson replies to Honorable John Eppes, Chairman of the Committee of Finance:

> SIR—Having consulted the other officers of the Mint, and deliberately considered the subject of the queries which I have had the honor of receiving in your letter of the 16th, I shall now attempt their answer, without, however, vouching for any very great degree of accuracy.
>
> Query 1st. "What number of eagles and half eagles can the mint, in its present situation, coin per day?"
>
> Answer. With the aid of a new foundry and refining furnace, which are now nearly completed, With the aid of a new foundry and refining furnace, which are now nearly completed, the Mint, in its present situation, and coining gold without silver, would be able to prepare and strike about 7,000 pieces per day, or 35,000 per

week, reckoning only five working days in the week, to allow for unavoidable accidents and interruptions.[20]

Query 2nd. "What number of dollars?" 3d. "What number of half dollars?" 4th. "What number of twenty, ten, and five-cent pieces?"

Answer. In making silver, without gold, the weekly coinage in dollars may be rated at about 60,000; in half dollars 85,000; and in smaller coins, 100,000 pieces respectively. And beyond this, which would be fully double of the last year's coinage, the operation of the mint could not be much extended, without erecting a new building, with some additional apparatus; for in our present situation, we are very much limited for want of room, having to rent two small lots for our necessary accommodation . . .

What few American gold coins remained in private and bank hands were mostly melted.

On September 27, 1819, the Bank of the United States issued a statement of the silver and gold coins held by it ($2,197,941) as of that date and of the various branches reported earlier in the month. Specie included foreign and American coins:

Baltimore $278,498, Boston 79,936, Charleston $261,253, Chillicothe (Ohio) $28,870, Cincinnati $81,485, Fayetteville (North Carolina) $87,760, Lexington $70,035, Louisville $106,306, Middletown (Connecticut) $76,642. New Orleans $320,389, New York City $313,611. Norfolk $79,479, Pittsburgh, $10,242, Portsmouth $9,723, Providence $30,085. Richmond $110,320, Savannah $84,628 Washington $22,270.

At any given time a branch of that bank was apt to have more specie than other banks in its immediate district.

In late spring 1821 the melting of $1,000 face value in gold coins would yield $49 in profit in terms of bullion value.[21] In 1824 William Kneass, a well-known Philadelphia engraver of printing plates, including those for bank notes, was named engraver at the Mint to replace the recently deceased Robert Scot. By this time significant amounts of gold were being received from North Carolina, but most metal for coinage was still in the form of foreign coins. In this year Mint Director Robert Patterson died. Samuel Moore succeeded him in the post and would remain until 1835.

SELLERS DESCRIBES THE MINT

George Escol Sellers, born in 1808, who became well known in mechanics, lived near the Mint and told this in his reminiscences years later:

The old U.S. Mint in Philadelphia was on the east side of Seventh Street, in one of those areas called in Philadelphia a city block, these blocks being bounded on their four sides by the principal streets, and perhaps subdivided into smaller blocks by alleys or courts. The particular block in question was between Sixth and Seventh streets on the east and west and Market and Arch on the south and north, and in point of fact the building was about midway of the block on the corner of a small street named Sugar Alley, which ran from Sixth to Seventh streets, bisecting the block. The building used for the Mint had very much the appearance of an ordinary three-story brick dwelling house of that period, the back building and yard extending on the alley . . .

The impression made upon me as a boy was the more enduring as it was one of almost daily occurrence. The block on which the old Mint stood, besides being divided by Sugar Alley, had on Sixth Street near Market the entrance to what was known as Mulberry Court. This court extended nearly halfway to Seventh Street, and at the head of the court was a dwelling house facing the entrance to the court. This house separated Mulberry Court from another alley or court that entered from Seventh Street, known as St. James Street. The difference between the terms alley and court in this case was that the name alley was given to a narrow street of uniform width, either entirely passing through the block or entering it for a short distance, while the term court was applied more particularly to a narrow entrance from the main street widening into a broader area, around which area the more pretentious houses were frequently erected.

On the north side of Mulberry Court were three dwelling houses, in one of which I first saw the light. The lot where stood the house in which I was born cornered with one on Sixth Street occupied by Mr. Frederick Graff, who followed Latrobe as engineer of the Philadelphia water-works, and who designed and constructed the Fairmount water-works. A gateway connected our yards. Mr. Graff was one of my father's most intimate acquaintances, who with Dr. Robert Patterson, then in charge of the Mint, and Adam Eckfeldt, chief coiner, were together frequent visitors at our house on the court; it was a clannish neighborhood, gates connecting all our yards, even to the yard of the fire-engine shops carried on by Jacob Perkins and my father at the end of St. James. From this yard was an opening into Sugar Alley, which to us as youngsters had other attractions than the coining press, for there stood the little shop of the best molasses candy maker in Philadelphia. The house at the end of the court was eventually removed, the street being then called St. James Street, now Commerce Street, of which street it is a continuation.[22]

COINS SAVED FOR POSTERITY

Beginning around the year 1824, Adam Eckfeldt and perhaps one or two others working at the Philadelphia Mint began saving current coinage.[23] It is said that copper cents and other items were kept on hand for sale or exchange with interested collectors. Interesting rare coins were picked out of incoming deposits and saved. Many old dies were still at the Mint, and on occasion restrikes would be made to supply needed pieces. No records were kept of such activities, and today we can only speculate as to what occurred.

Marquis de Lafayette, French hero of the American Revolution, visited America in 1824 and 1825 and was designated by Congress as "the Nation's Guest." He toured every state in the Union, visited with prominent officials as well as everyday citizens, and was featured in parades and ceremonies. Cents and half dollars were counterstamped with his likeness on one side and Washington's on the other (from dies by Joseph Lewis[24]), and many medals, ribbons, and badges were made. Lafayette's portrait was subsequently used on many bank notes.

A half dollar counterstamped in 1824 for Marquis de Lafayette's return visit to America that year.

MINT RULES AND REGULATIONS

On January 1, 1825, these rules and regulations were put into effect (excerpted):

The operations of the Mint throughout the year, are to commence at 5 o'clock in the morning, under the superintendence of an officer, and continue until 4 o'clock in the afternoon, except on Saturdays, when the business of the day will close at 2 o'clock, unless on special occasions it may *be* otherwise directed by an officer . . .

The allowance under the name of *drink money* is hereafter to be discontinued, and in place of it *three dollars extra wages* per month will be allowed for the three summer months to those workmen who continue in the Mint through that season. No workman can be permitted to bring spirituous liquors into the Mint. Any workman who shall be found intoxicated within the Mint must be reported to the Director, in order that he may be discharged. No profane or indecent language can be tolerated in the Mint. Smoking within the Mint is inadmissible. The practice is of dangerous tendency; experience proves that this indulgence in public institutions, ends at last in disaster.

Visitors may be admitted by permission of an officer, to see the various operations of the Mint on all working days except Saturdays and rainy days; they are to be attended by an officer, or some person designated by him. The new coins must not be given in exchange for others to accommodate visitors, without the consent of the Chief Coiner. Christmas day and the Fourth of July, and no other days, are established holidays at the Mint.

The pressmen will carefully lock the several coining presses when the work for the day is finished, and leave the keys in such places as the Chief Coiner shall designate . . . The watchman of the Mint must attend from 6 o'clock in the evening to 5 o'clock in the morning, and until relieved by the permission of an officer, or until the arrival of the door-keeper . . . If an attempt be made on the Mint he will act conformably to his secret, instructions on that subject. In case of fire occurring in or near the Mint, he will ring the Alarm Bell if one has been provided, or sound the alarm with his rattle, and thus as soon as possible bring someone to him who can be dispatched to call an officer, and in other particulars will follow his secret instructions. . . .

TWILIGHT OF THE FIRST MINT

On May 28, 1826, Karl Bernard, the Duke of Saxe-Weimar-Eisenach, left Lancaster traveling at the rate of about five miles per hour and reached Philadelphia, 64 miles distant. He was traveling with Robert Vaux, a "philanthropic Quaker," and Mr. Niederstetter, *Chargé d'Affaires* of the King of Prussia.

I saw the Mint of the United States, which is established here. In the year 1793, when Philadelphia was still the seat of government of the United States, this mint was located in a newly-built private house, and it is as yet the only one in the United States. The processes in this Mint are very simple, and but few improvements are yet adapted, which so greatly distinguish the mints of London and Milan.

They were doing but little when we came; we saw nothing but the stretching of the bars of silver between cylinders, like those in the rolling mills at Pittsburgh, and the stamping of the pieces, which was done by means of a contrivance similar to that by which rivet-holes are made in the iron plates for steam-engine boilers. We saw, moreover, the cutting of half dollar pieces, which is done by means of a stamp, worked by two men. A third stands by to place the uncoined pieces in a box, which are then brought under the stamp by a particular contrivance. After they are coined, they fall by means of this contrivance into a box which stands below . . .

The mint itself is very small, and its boundaries are still more limited by a twelve horse-power steam-engine. No application, however, is made to Congress for a larger and better building, as it is feared that congress might then propose to remove the whole establishment of the Mint to Washington.[25]

As other accounts related, the Mint was antiquated and somewhat rustic, especially in comparison to the mints of Europe, most of which were in impressive buildings and had much better equipment. The private Soho Mint operated by Boulton & Watt in Birmingham, England, had steam-powered presses that operated automatically at high speed, so efficiently that according to one account, a boy could tend to four at a time. In Philadelphia the use of steam to strike coinage was still a decade away.

North Carolina continued to be a source for native gold. In 1828 $46,000 worth was received from the state, as compared to $60,000 received in coins from Mexico, South America, and the West Indies combined, $13,000 from Africa, and $21,000 from unattributed sources.[26]

Director Samuel Moore prepared a report on the Mint to accompany a bill in the House of Representatives, dated January 25, 1829. While some elements were incorrect and are deleted here, it gives a view of the various buildings at the time (adapted):

> At first a house and lot on 7th Street and one other lot fronting on a fourteen foot alley, were purchased. Under a subsequent Act of January 2, 1795, a third lot, marked C, was purchased, fronting on the alley and extending northward to a private court and separated from the lots before mentioned by a four foot alley, and all the property bounding it. Other areas were rented but no others were purchased.
>
> Buildings 1 and 2 on the chart are built of brick three stories high, combined as a single unit. Under it are vaults. On the first floor is a receiving and weighing room and one press room. A second floor is all of the officers except the assayers, three of whom have one apartment. The third floor furnishes the office of the assayer, a location exceedingly inappropriate for the character of the chemical agents employed, but no other arrangement has been practical.
>
> This building is after the most ordinary manner as customary 50 years ago without any pretentions as design or arrangement. The stairway by which all must reach the office is a dark winding passage in which a lamp is kept burning through the day.
>
> Number 3 is a two story brick building erected in 1816 when steam power was first substituted for that of horses for the purpose of rolling the ingots etc. It contains the steam engine and the machinery pertaining to it thereto skillfully arranged. This department of the works has hitherto been found competent to its design and is even capable of more than had been required. The engine is of the power of about ten horses.
>
> Number 4 is a small frame building rudely constructed, formerly a stable and of little value. It is, however, a place for deposit for nearly all of our coined and uncoined copper.
>
> Number 5 is a two story brick building which contains melting and refining furnaces. So efficient in its dimensions although the refining furnace occupies a cellar that all dependent operations are often retarded by the inefficiency of its accommodation. How silly it was that this building had to be erected on a lot leased which expired some time since. The building and lot are now rented year to year . . .

Number 6 is a small two story frame building that's lightly constructed. It contains on the first floor the press by which most of our silver is coined and is in daily use for this purpose.

Number 7 is a small two story brick house, old and dilapidated in which the cent coinage is executed. Above these two is a shop for making and repairing machinery.

Number 8 is a smith shop and number 9 is a coal house, both roughly boarded and covered in the same manner.

The whole range presents an aspect singularly unsuited to the important uses which some parts of it have been for so many year appropriated. Lot E has no building on it but is employed as a wood yard. . . .

You recommended that enlarged operations be done with machinery devised and executed by Mr. Boulton and adopted in some of the mints in Europe should be had in view. This topic has been the subject of some preliminary inquiries addressed to Mr. Boulton in 1827. From his reply it appears that so much of his system of machinery as might be desirable to introduce would cost about seven thousand pounds at Liverpool. It is understood that machinery of this description, constructed by Mr. Boulton for a Mint in South America, is in New York for sale. If this should be found adapted to our purposes it would be judicious to embrace the opportunity of securing it, the expense it is presumed would be low as the price charged by Mr. Boulton at Liverpool.[27]

On March 2, 1829, Congress provided for the construction and furnishing of a new Mint building as production could no longer be easily accommodated in the present facilities. On July 4 the cornerstone was laid, with Director Samuel Moore officiating. For the occasion, half dimes of the Capped Bust type were struck for the first time, there having been no coinage of this denomination since 1805. At the time the officers and their annual salaries were:

Samuel Moore, director—$3,500

Adam Eckfeldt, coiner—$2,000

Joseph Cloud, melter and refiner—$2,000

Joseph Richardson, assayer—$2,000

William Kneass, engraver—$2,000

From this time onward the days of the first Mint were numbered. In 1829 it was resolved that in the House of Representatives "a select committee be appointed to inquire into the expediency of establishing a branch of the United States Mint in the gold regions of North Carolina."

Anne Newport Royall visited the Mint in 1829 and entered this in her travelogue:

I called, for the first time, on the United States Mint. This is called the United States Mint for the sake of pride, as the U.S., it appears, has little or nothing done in it. The specie coined here is principally owned by individuals. The building is in 7th above Market Street and is quite a small and very indifferent building. No one would take it from anything but an old wash house . . . Its appearance, however, is worse in that the principal part of the building, such as work rooms and forge houses extending some distance back. The keeper or head man was by no means extraordinary, but he did happen to know a bar of gold from one of lead. But he was unqualified to give an accurate and satisfactory account of the process . . .

I saw not more than a half a dozen hands. Some were cutting the bars into square pieces; others were rounding them off, and some were engraving the face of the coins. It appears that in order to ascertain whether the proper weight is in the coins, a number of pieces are deposited together—one or perhaps more is taken from the sum and weighed, and if it lacks the smallest particle the manager levies a heavy fine and perhaps prosecution . . .

The chief of the money coined belongs to the merchants of Philadelphia, who purchase bullion for the purpose. It has no guard in day time, nor am I clear that it has one at all, and is only enclosed in the rear with a slight wall, being quite open to the street in front. It is an old yellow building of brick.[28]

The *Annual Report of the Director of the Mint*, 1830, included this:

Of the amount of gold coined within the last year, about $125,000 were derived from Mexico, South America, and the West Indies; $19,000 from Africa, $466,000 from the gold region of the United States, and about $33,000 from sources not ascertained. Of the gold of the United States above mentioned, $24,000 may be sated to have been received from Virginia; $204,000 from North Carolina, $26,000 from South Carolina, and [for the first time ever] $212,000 from Georgia.

Mint Director Samuel Moore reported that the construction of the new Mint building was progressing satisfactorily. In 1832, the last year of operation of the first Mint, denominations produced included the half cent, cent, half dime, dime, quarter dollar, half dollar, and gold quarter eagle and half eagle.

At eight o'clock in the evening of Thursday, July 19, 1832, the site and buildings of the Mint and related properties were offered at auction by C.J. Wolbert at the Merchants Coffee House. An announcement included this equipment:

The machinery comprises a steam engine of 10 horse power, together with a complete system of rolling apparatus for hot or cold rolling, together with a drawing table connected with the steam power, all now in daily use, with a variety of other implements, furnaces, &c. [29]

The first Mint was not sold. Finally, on October 8, 1835, most of the property was sold in two parcels. The first Mint buildings were purchased by Michael Kates, who rented them to workers in metal. They were later inherited by members of the Kates family.

George Escol Sellers in his reminiscences, 1893, said this:

When the present [second] U.S. Mint in Philadelphia was built it was furnished throughout with entirely new machinery that the old mint might be kept in full operation during its construction. The new machinery did not differ in any essential points from the well-tried of long service in the old, only differing as to amount to meet the requirements in increased coinage. With this new machinery we had nothing to do. I felt great interest in the building as it progressed, inasmuch as it was under the charge of my friend, J.C. Trautwine, then a pupil of William Strickland, its architect.

After the new mint went into operation the machinery of the old mint was sold under the auctioneer's hammer, mostly by weight as old metal. We became the purchasers of the rolling mill department with its shafting and connected machinery. The housings, rolls, etc., did not go into the melting furnace, but were refitted as a train of rod rolls and went into service in a Pennsylvania rolling mill.

Through the years the structures passed through several other hands. *The Numismatist* for August 1899 printed an article titled "The Old Philadelphia Mint," which noted that the old building had been nearly destroyed by fire recently:

> Fortunately the firemen were near at hand, and by their prompt response to the alarm and untiring efforts, the old historic building was saved from complete destruction, and might yet be restored in place in such condition that it would be a splendid object lesson to the present generation, illustrating the wonderful growth of the wealth of the republic from a very small and modest beginning. It seems almost incredible that within the recollection of living men this plain old structure represented the entire personal holdings in real estate security of our national government.

The first Mint buildings were eventually sold by the Kates family to Frank H. Stewart, who demolished them in stages from 1907 to 1911 to make room for a new facility for the Stewart Electric Company.

In 1914 Stewart commissioned Edwin Lamasure to create a painting showing the Mint as it appeared in the 1790s. With information supplied by Stewart, some of it historically incorrect, his rendition is problematical in many ways. First, it shows only three of the ten buildings on the first Mint "campus." Second, it places them in a bucolic setting, not in downtown Philadelphia, with other buildings cheek-by-jowl. Third, the three buildings it depicts were there together only from 1818 until the premises were abandoned by the minters in January 1833. Fourth, during those 15 years, all three were connected by covered walkways, not depicted in the painting. It's an imaginative reconstruction, well-painted, but it's not a good guide to what was actually there.

Photograph of the front of the first Mint, 1854, and an adjacent building to the right. (*American Journal of Numismatics*, November 1868)

"Ye Olde Mint" in 1907, as designated by Frank H. Stewart, whose Stewart Electric Company was to its right.

The front building, two old structures joined, was used for administration, assaying, and bullion storage. The large brick building in the middle held the Mint's steam engine (after 1816), and furnaces for melting and refining. The rear building was used for punching and processing planchets. Most coinage was struck in two small shops, not visible in this picture, behind the rear building.[30]

In 1924, seemingly very interested in Mint history, but without a touch of remorse, Stewart wrote *The History of the First United States Mint.* The volume remains a key reference for details of Mint operations. In 2011 Whitman Publishing issued *The Secret History of the First U.S. Mint*, by Joel J. Orosz and Leonard D. Augsburger. This comprehensive study included most of the information in the Stewart study, plus a wealth of additional illustrations, biographies, and narrative, making it invaluable to collectors.

COINAGE OF THE FIRST MINT, 1792–1832

The operations of the first Mint have been well chronicled in multiple texts (see the bibliography) as well as by studies in journals published by Early American Coppers, Inc. and the John Reich Collectors Society. The National Archives (NARA) is a rich repository for early documents, and several scholars have been exploring these in recent years.

The Philadelphia Mint in 1792 was at best a crude arrangement of existing buildings and new construction, not at all comparable to the national mints of Europe. Processes were antiquated, especially in comparison to the leading-edge technology used in Birmingham, England, by Boulton & Watt, including a steam-driven press with automatic feed and edge lettering, four of which could be attended by a boy. Striving to improve technology, the staff at the first Mint turned out a marvelous panorama of coins from half cents to eagles, while at the same time coping with periodic closures due to yellow fever in the early days and dealing with continuing criticism from members of Congress who wanted to move the facility to Washington (the federal capital beginning in 1800) or close it down completely.

The coins of the first Mint, struck from hand-finished dies that all differed from each other, are a unique specialty in early American numismatics.

Edwin Lamasure's conception of the first Mint, painted in 1914.

Half Cents

Liberty Cap half cents were issued in July 1793, with Miss Liberty facing right. The motif was changed to Liberty facing right in 1794, continuing through 1797. In 1795, a large quantity of Talbot, Allum & Lee copper tokens was purchased and half-cent planchets were cut from them, yielding a supply that lasted into 1797. Scrap was melted and converted into strip, then planchets for cents. Misstruck cents often had half-cent planchets punched from them, eliminating the need to reprocess the metal, except for the scrap. Although weights were specified by statute for half cents and cents, it seems that the rules were observed casually at times, for some coins of this era varied in this regard. No half cents were struck with dates of 1798 or 1799. The next coinage was dated 1800, these being of the Draped Bust type by Robert Scot.

John Reich, who was hired at the Mint as a full-time assistant engraver, created what we call the "Classic Head" design today, featuring Miss Liberty facing to the left, with a band inscribed LIBERTY across her head, with hair tresses visible above. On half cents the Classic Head first appeared in 1809, and it was used for the next two years. Mintage of half cents began anew in December 1825 in response to a significant order placed in November by Jonathan Elliott & Sons of Baltimore. The Mint Treasurer informed the Elliott Company on November 23 that no half cents were on hand at the time, but perhaps would be in January. Mint records, however, indicate that none were paid out until the second quarter of 1826.[31] By that time, the Classic Head motif had been discontinued years earlier on the *cent* pieces and, beginning in 1816 in that denomination, had been replaced with what Ken Bressett designated in modern times (for use in the *Guide Book of United States Coins*) as the Matron Head. Classic Head half cents were produced in 1828, 1829, 1831, and 1832, and then at the second Mint continuing to 1836.

As half cents were produced to the Mint's own order and not for outside depositors, and as the Mint made a profit on copper coins, it was more efficient to make one cent than two half cents. For this reason, half-cent mintages were much smaller than for cents.

DESIGN TYPES AND DATES MINTED

Liberty Cap, Head Facing Left: 1793.

Liberty Cap, Head Facing Right: 1794, 1795, 1796, 1797.

Draped Bust: 1800; 1802, 2 Over 0; 1802; 1803; 1804; 1805; 1806; 1807; 1808, 8 Over 7; 1808.

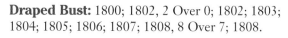

Classic Head: 1809, 1810, 1811, 1825,1826, 1828, 1829, 1831, 1832.

Large Cents

The first cents were delivered by the coiner on March 1, 1793, and were of the Chain AMERI. type, soon changed to spelling AMERICA in full. Then followed the Wreath cents with the obverse and reverse dies in exquisite high relief by an unknown artist. In the summer Joseph Wright designed the Liberty Cap cent, a motif that was continued into 1796. Wright died of yellow fever in September 1793, after which Robert Scot was appointed engraver and did most of the work, assisted at times by John Smith Gardner and others.

The copper problem eased when arrangements were made to buy casks filled with planchets made by Boulton & Watt at the Soho Manufactory in Birmingham, England. From time to time it was proposed that coins be made on contract. A memorandum from Mint Director Elias Boudinot, February 27, 1802, noted: "As to importing the cents complete from Europe, it can certainly be done for a trifling sum above the price of the planchettes, say about £20 sterling per ton, did the policy of government admit of it. Of this I would not venture to determine, the Legislature alone being competent to that purpose. I once stated it to a Committee of both Houses, but they determined that it would be a dangerous measure, and would not hearken to it."

By 1800 a conventional press for one-cent pieces could turn out about 14,000 coins per day, a capacity that on a yearly basis was never fully utilized. In 1808 the Classic Head by assistant engraver John Reich was adopted for the cent and was continued afterward.

George Escol Sellers told of a visit in 1812 when he was a pre-schooler, to the Mint:

> In a rear room, facing on the alley, with a large, low-down window opening into it, a fly press stood; that is a screw-coining press mostly used for striking the old copper cents. Through this window the passersby in going up and down the alley could readily see the bare-armed vigorous men swinging the heavy end-weighted balanced lever that drove the screw with sufficient force so that by the momentum of the weighted ends this quick-threaded screw had the power to impress the blank and thus coin each piece. They could see the rebound or recoil of these end weights as they struck a heavy wooden spring beam, driving the lever back to the man that worked it; they could hear the clanking of the chain that checked it at the right point to prevent its striking the man, all framing a picture very likely to leave a lasting impression, and there are no doubt still living many in Philadelphia who can recollect from this brief notice the first mint. One day in charge of my elder brother I stood on tip-toe with my nose resting on the iron bar placed across the open window of the coining room to keep out intruders, watching the men swing the levers of the fly press; it must have been about noon, for Mr. Eckfeldt came into the room, watch in hand, and gave a signal to the men who stopped work. Seeing me peering over the bar, he took me by the arms and lifted me over it. Setting me down by the coining press, he asked me if I did not want to make a cent, at the same

time stopping the men who had put on their jackets to leave the room. He put a blank planchet into my hand, showed me how to drop it in, and where to place my hand to catch it as it came out; the lever and weights were swung, and I caught the penny as we boys called cents, but I at once dropped it. Mr. Eckfeldt laughed and asked me why I dropped it. Because it was hot and I feared it would burn me; he picked it up and handed it to me, then certainly not hot enough to burn; he asked if it was not cold when he gave it to me to drop into the press; he told me to look and see there was no fire, and feel the press that it was cold; he then told me I must keep the cent until I learned what made it hot; then I might if I liked spend it for candy . . . I have no recollection of ever having seen the copper planchets for cents being made in the Mint, but I have a vivid recollection of small iron hooped casks filled with copper planchets for cents and half cents. I have the impression that they were imported as copper in that condition and only stamped or coined in the Mint. These casks were similar to the casks in which card wire was imported from England at that period.[32]

The War of 1812 ended the supply of planchets from Boulton & Watt, but coinage continued into 1814 when the supply ran out. Coinage resumed in January 1816, thought to have been from earlier dies. Later in that year the Matron Head cent was introduced, to be continued through 1832 when coinage at the first Mint ended. Planchets were obtained again from Boulton & Watt (until 1837) and also from domestic suppliers, beginning with the Taunton (Massachusetts) Manufacturing Company, also known as Crocker Brothers & Co., with a shipment of samples in October 1832.

Meanwhile, large quantities of cents continued to be struck. In fact, there were so many that they could not be effectively placed into circulation. The *Annual Report* for 1820 noted, "The press employed in the copper coinage did not continue in operation more than six months; as the quantity of copper coins had accumulated far beyond the public demand."

The Mint paid out cents to all comers, in exchange for silver coins, including foreign issues that were legal tender, as well as at par for notes from any bank whose currency was accepted at face value by the Bank of the United States.[33]

As the Mint earned a profit on half cents (the coinage of which was resumed in 1825) and cents, it continued to offer free shipping to buyers, as described in this statement in the *Annual Report of the Director of the Mint* for 1827: "The amount of copper coins distributed within the last year is $21,910. They are forwarded, as heretofore, at the expense and risk of the government, to all parts of the United States, accessible by regular means of transportation, on receiving the value thereof here, or a certificate of deposit, to the credit of the treasurer of the United States, for the requisite amount, in any of the banks authorized to receive deposits of public money."

And this from the 1828 report: "Copper coins have been distributed within the last year to the amount of $27,566.34, with a profit thereon of nearly $8,000. The profit thus accruing on the issue of copper coins is regularly paid into the Treasury of the United States; and is an effective reimbursement of so much of the amount annually expended for the support of this establishment."

For this reason, the production of cents continued to be steady and extensive, limited only by the ability of commerce to receive them. In contrast, silver and gold mintages varied widely, depending on deposits.

Design Types and Dates Minted

Flowing Hair, Chain Reverse: 1793
(AMERI. and AMERICA reverses).

Flowing Hair, Wreath Reverse: 1793.

Liberty Cap: 1793, 1794, 1795, 1796.

Draped Bust: 1796; 1797; 1798, 8 Over 7;
1798; 1799, 9 Over 8; 1799; 1800; 1801;
1802; 1803; 1804; 1805; 1806; 1807.

Classic Head: 1808; 1809; 1810, 10 Over 09;
1810; 1811; 1812; 1813; 1814.

Liberty Head, Matron Head: 1816;
1817; 1818; 1819, 9 Over 8; 1819;
1820, 20 Over 19; 1820; 1821; 1822;
1823, 3 Over 3; 1823; 1824, 4 Over 2;
1824; 1825; 1826, 6 Over 5; 1826; 1827;
1828; 1829; 1830; 1831; 1832.

Half Dimes

The 1792 half disme, most of which were made outside of the Mint, but with additional coinage in the Mint in October, is discussed previously in the narrative. The *disme* spelling was used for many years in government and other documents. Only gradually did *dime* become popular. Production of half dimes of the Flowing Hair and Draped Bust types was small, as most depositors of silver bullion requested dollars in exchange. These were easier to store and count. For the Mint the dollar coins were more efficient to make, a win-win situation. However, dollars were last coined in 1804.

After 1805 no half dimes were made for nearly a quarter century as depositors of silver had not requested them. The half dollar became the silver coin of choice. It was not until 1829 that Capped Bust coins of this denomination were made, the first being struck in the wee hours of the morning on July 4 in a ceremony to lay the cornerstone of the second Mint. Capped Bust half dimes continued to be made each year from that point to 1837.

DESIGN TYPES AND DATES MINTED

Half Disme: 1792.

Flowing Hair: 1794, 1795.

Draped Bust, Small Eagle Reverse: 1796, 6 Over 5; 1796; 1797.

Draped Bust, Heraldic Eagle Reverse: 1800, 1801, 1802, 1803, 1805.

Capped Bust: 1829, 1830, 1831, 1832.

Dimes

Similar to half dimes, dimes were made in relatively small quantities in the first Philadelphia Mint. The first coinage took place in 1796 and was continued intermittently until 1820, when production was yearly except for 1826. The earlier designs were by Robert Scot, engraver at the Mint, and the Capped Bust motif was by John Reich.

Dimes were not particularly popular in commerce during this era and were often traded on an equal basis with the Spanish real or bit worth 12-1/2 cents.

Design Types and Dates Minted

Draped Bust, Small Eagle Reverse: 1796, 1797.

Draped Bust, Heraldic Eagle Reverse: 1798, 8 Over 7; 1798; 1800; 1801; 1802; 1803; 1804; 1805; 1807.

Capped Bust, Variety 1, Wide Border:
1809; 1811, 11 Over 09; 1814; 1820; 1821; 1822; 1823, 3 Over 2; 1824, 4 Over 2; 1825; 1827; 1828.

Capped Bust, Variety 2, Modified Design:
1828; 1829; 1830, 30 Over 29; 1830; 1831; 1832.

Quarter Dollars

Paralleling the situation for the half dime and dime, relatively few silver quarters were requested by depositors in the early years. After an initial coinage of only 6,146 coins in 1796, no more were struck until 1804 when 6,738 were made. After 1807 no coins of this denomination were made until 1815, when John Reich's Capped Bust motif was first employed. These were made to fill a request received from the Planters Bank of New Orleans received on July 8, accompanying a shipment of silver coins to use as bullion, mostly Mexican dollars. Director Patterson resisted the idea and asked if the same amount in dimes would be satisfactory. He received a reply stating that dimes were not popular in that city. "On December 16 chief coiner Adam Eckfeldt delivered 69,232 quarter dollars. Of these 57,485 pieces were sent by sea to the New Orleans bank. The remaining 12,000 or so coins were paid out to other silver depositors, primarily Jones, Firth & Co., a local Philadelphia firm."[34] This vividly demonstrates the Mint's preference to coin silver in larger denominations, the highest at the time being the half dollar. The next coinage of twenty-five cent pieces was in 1818, after which quarters were steadily made through 1828, excepting 1826.

Although 4,000 quarters were reported to have been struck in calendar year 1827, they may have been from earlier-dated dies that were still serviceable. All but one coin known today is with Proof finish and was made for numismatic or presentation purposes. There was no coinage in 1829 or 1830. In 1831 a reduced-size version of the Capped Bust motif was introduced with different-style denticles at the borders and without the motto E PLURIBUS UNUM. Quarters of this style were made continuously to 1832 and later at the second Mint into 1838.

DESIGN TYPES AND DATES MINTED

Draped Bust, Small Eagle Reverse: 1796.

Draped Bust, Heraldic Eagle Reverse:
1804; 1805; 1806, 6 Over 5; 1806; 1807.

Capped Bust, Variety 1, Large Diameter:
1815; 1818, 8 Over 5; 1818; 1819; 1820;
1821; 1822; 1823, 3 Over 2; 1824, 4 Over 2;
1825, 5 Over 2; 1825, 5 Over 4; 1827; 1828.

Capped Bust, Variety 2, Reduced Diameter, Motto Removed: 1831, 1832.

Half Dollars

The first federal half dollars were of the Flowing Hair type and were delivered by the coiner in November 1794, the same month that the first dollars were struck. Coinage of the design was continued in 1795, following in 1796 and 1797 with a sharply reduced quantity of the new motif with Draped Bust obverse and Small Eagle reverse. As nearly all depositors requested silver dollars, no more half dollars were struck until 1801, at which time the Draped Bust obverse was combined with the Heraldic Eagle reverse (first used on the quarter eagle of 1796). Coinage of the design was continuous through 1807, except for 1804, although an 1804 die was made and later used in 1805, over-dated 1805, 5 Over 4. The design did not strike up well, and many coins, particularly those dated 1807, were weak in areas.

In 1807 John Reich was hired as an assistant engraver to work with Robert Scot. The first order of business was to develop new motifs for the coinage. In 1807 Reich's "Capped Bust" design, as we call it today, was used on the half dollar and half eagle, after other designs were used and then retired earlier in the year. The Capped Bust half dollars were followed within the decade by use on the quarter eagle (1808), dime (1809), and quarter (1815). Miss Liberty faces to the left and wears a soft cloth cap, sometimes called a *mob cap*.

In 1828 a New Hampshire citizen went on a trip to Philadelphia and entered this in his diary: "I and Mr. Fisk went to see the United States Mint. There they were, striking off half dollars in one room, and cents in another. The other part of the works were not in operation. I enquired how many half dollars they struck off in a minute, and they said 43.[35] The pieces of silver were plated out and cut to the right size previously, so that they were only given the impression. It took three men to do this; all of them had hold of the machine at a time, and it appeared to be pretty hard labor."[36]

DESIGN TYPES AND DATES MINTED

Flowing Hair: 1794, 1795.

Draped Bust, Small Eagle Reverse: 1796, 1797.

Draped Bust, Heraldic Eagle Reverse: 1801; 1802; 1803; 1805, 5 Over 4; 1805; 1806, 6 Over 5; 1806; 1807.

Capped Bust, Lettered Edge, First Style: 1807; 1808, 8 Over 7; 1808.

Capped Bust, Lettered Edge, Remodeled Portrait and Eagle: 1809; 1810; 1811, 11 Over 10; 1811; 1812, 2 Over 1; 1812; 1813; 1814, 4 Over 3; 1814; 1815, 5 Over 2; 1817, 7 Over 3; 1817, 7 Over 4; 1817; 1818; 1819, 9 Over 8; 1819; 1820, 20 Over 19; 1820; 1821; 1822, 2 Over 1; 1822; 1823; 1824, 4 Over 1; 1824 over various dates; 1824; 1825; 1826; 1827, 7 Over 6; 1827; 1828; 1829, 9 Over 7; 1829; 1830; 1831; 1832.

Silver Dollars

Silver dollars were first coined at the Mint in November 1794 on a press designed to strike half dollars as the largest diameter. The obverse and reverse dies were not aligned in parallel, with the result that the lower left of the obverse and the corresponding part of the reverse was weakly struck on all examples. After examining the output, net 1,758 pieces of the Flowing Hair design were found satisfactory for distribution. It was not until May 1795 that a dollar-size press was installed and ready for service. From then onward the production of dollars was extensive. Toward the end of the year the Draped Bust obverse design with Small Eagle reverse was introduced with two obverse dies combined with a single reverse (a long-lived Small Letters die that would continue in intermittent use until 1798).

Many silver dollars were sent to the treaty port of Canton, China, and were lost forever. Others were sent to the West Indies to be exchanged for Spanish-American dollars of slightly heavier weight, which were melted to yield a small profit. The net result was that by 1804 very few silver dollars were seen in domestic commerce, and the coinage of the denomination was suspended because of it. The half dollar became the largest silver coin. It was hoped that these would remain stateside and be useful in trade. That happened, as few were shipped abroad. It was not until 1836 that silver dollars were again made for circulation (these being the limited production of the new Gobrecht Liberty Seated design).

DESIGN TYPES AND DATES MINTED

Flowing Hair: 1794, 1795.

Draped Bust, Small Eagle Reverse: 1795, 1796, 1797, 1798.

Draped Bust, Heraldic Eagle Reverse: 1798; 1799, 9 Over 8; 1799; 1800; 1801; 1802, 2 Over 1; 1802; 1803; 1804 (first struck with this date in 1834: Class I, II, III varieties).

$2.50 Quarter Eagles

Quarter eagles, the smallest denomination in the gold series, were first struck in 1796. In that year two obverse designs were made—Draped Bust with plain obverse field and Draped Bust with stars. It is thought that stars were not used on the first obverse die as there were stars on the reverse die. In contrast, half eagles were made with stars on the obverse from the outset as the reverse had a perched eagle and no stars. After stars were added to the quarter eagle later in 1796 they were maintained for the rest of the series.

Mintages of quarter eagles were small in comparison to half eagles and eagles as most depositors of gold requested the larger denominations, for a given face value of gold was easier to count. Quarter eagles were minted through 1798, then again in 1802 and 1804 through 1807. In 1808 John Reich's Capped Bust was used, after which no quarter eagles were minted until 1821, at which time a modified Capped Bust was used. Later coinages at the first Mint included 1824, 4 Over 1; 1825 to 1827; and 1829 to 1832. Quarter eagles did not circulate in commerce after 1820 as they cost more than face value to produce. There was a provision that members of Congress could receive their salary in such coins, yielding a small profit when spent—in trade, gold coins were valued at their intrinsic worth. Senator Thomas Hart Benton, nicknamed "Old Bullion," was one of the members who took advantage of this.

DESIGN TYPES AND DATES MINTED

Capped Bust to Right, No Stars on Obverse: 1796.

Capped Bust to Right, Stars on Obverse: 1796; 1797; 1798; 1802; 1804; 1805; 1806, 6 Over 4; 1806, 6 Over 5; 1807.

Draped Bust to Left, Large Size: 1808.

Capped Head to Left, Large Diameter 1821;
1824, 4 Over 1; 1825; 1826; 1827.

Capped Head to Left, Reduced Diameter: 1829,
1830, 1831, 1832.

$5 Half Eagles

Commentary: The first half eagles were delivered at the end of August 1795, these being the first federal gold coinage. The design by engraver Robert Scot featured the Draped Bust obverse with a reverse depicting an eagle seizing a wreath while on a palm branch, said to have been inspired by an ancient Roman onyx cameo. At the Mint more dies were made in 1795 than were called for by coinage requirements, with the result that serviceable dies, including 1795-dated obverses, were used in later years. In 1798 the reverse was replaced by the Heraldic Eagle design. An unused 1795 obverse was used this year to create an anachronistic 1795-dated coin with the Heraldic Eagle reverse, an oddity among coinage of the era. The designs went through several transitions.

In 1804 the mintage of $10 eagles was stopped, as nearly all were being exported. This positioned the $5 as the largest-denomination American gold coin, a status it maintained through the end of production at the first Mint. After 1820 it cost more than face value to coin a half eagle, and they disappeared from domestic commerce. Depositors who requested half eagles, who were required to pay more than $5 in bullion, used them in the export trade, mainly to Europe, where they were valued for their gold content. Most of these exported coins were melted. As a result, half eagles of later dates are extreme numismatic rarities today, led by the 1822, of which just three are known, with two in the National Numismatic Collection at the Smithsonian Institution.

On December 28, 1829, Benjamin Leonard Covington Wailes of Natchez, Mississippi, visited the Mint, where he toured the facility, entering this in his diary regarding a coining press in the process of striking half eagles:

> This is a very powerful, ingenious, simple (though very perfect) piece of machinery. It consists (like the cutting machine) of a very powerful upright screw, to the top of which is affixed a heavy and strong lever worked with great apparent ease by one man at each end, and by which the screw is made to make about one-fourth of a revolution and returning instantly to its former position. At the lower end of the screw is affixed the die which gives the impression on the upper side, and immediately under it, the die considering the impression of the reverse of the coin. . . . Near the lower end of the screw stands a tube sufficiently large to admit a considerable number of the coins, one on top of the other, which may be termed the hopper, at the bottom of which is an apparatus sufficiently large to admit the passage of the coin, one at a time. The lower end in the file or hopper is struck out with great accuracy by a thin piece of iron made to strike edgewise at each movement of the lever and is conveyed by a channel formed for this purpose and is conveyed directly on and under the die. The screw is brought down and the nippers close with such

a force that makes the impression. The lever is instantly brought back, the nippers open, the stamped coin is struck out of its place and conveyed to a box by a spout (or conductor) as the screw rises, and another unstamped coin takes its place, and the screw comes down again.[37]

DESIGN TYPES AND DATES MINTED

Capped Bust to Right, Small Eagle Reverse: 1795; 1796, 6 Over 5; 1796; 1797; 1798.

Capped Bust to Right, Heraldic Eagle Reverse: 1795; 1797, 7 Over 5; 1797; 1798; 1799; 1800; 1802, 2 Over 1; 1803, 3 Over 2; 1804; 1805; 1806; 1807.

Draped Bust to Left: 1807; 1808, 8 Over 7; 1808; 1809, 9 Over 8; 1810; 1811; 1812.

Capped Head to Left: *Large Diameter:* 1813; 1814, 4 Over 3; 1815; 1818; 1819; 1820; 1821; 1822; 1823; 1824; 1825, 5 Over 4; 1826; 1827; 1828, 8 Over 7; 1828; 1829; 1830; 1831; 1832.

$10 Eagles

The first $10 gold eagles were delivered in 1795. The designs from that point followed those used on contemporary half eagles. Mintage of eagles was continuous through 1804, after which production ceased. These coins had been intended to be used in domestic commerce, eventually to replace foreign legal-tender gold. This was an exercise in futility: nearly all were exported, resulting in their discontinuation. It turned out that $5 gold coins were exported as well. The net result is that at any time during the operation of the first Philadelphia Mint very few gold coins were seen in everyday transactions in the United States. After 1820, when all gold coins became worth more than face value due to the rise in price of gold on the international market, none were seen at all.

DESIGN TYPES AND DATES MINTED

Capped Bust to Right, Small Eagle Reverse: 1795, 1796, 1797.

Capped Bust to Right, Heraldic Eagle Reverse: 1797; 1798, 8 Over 7; 1799; 1800; 1801; 1803; 1804.

SECOND PHILADELPHIA MINT (1833–1901)

LAYING THE CORNERSTONE

On March 2, 1829, Congress appropriated funds for a new mint. Land on Chestnut Street above 13th Street, near the corner of Juniper Street, was purchased. The frontage was 150 feet and the depth 204 feet. The cornerstone was laid on July 4 of the same year.

Designed by well-known architect William Strickland, a personal friend of chief engraver William Kneass (who named one of his children after Strickland), the building was in the Greek Revival style so popular at the time.[38] It was inspired by what remained of an ancient Greek temple on the River Ilisos near Athens. Six large columns graced the front and upheld the pediment. **William Strickland.** Many other government and bank buildings were made to look like incarnations of Greek temples, the Bank of the United States (in Philadelphia) and the Sub-Treasury (in New York City) being but two of numerous examples.

Niles' Weekly Register, July 18, 1829, carried this account:

> The Mint of the United States. From the *Philadelphia Gazette*.
>
> The foundation stone of the edifice about to be erected, under the provisions of the law for extending the Mint establishment, according to a plan thereof approved by the president, was laid, on the morning of the 4th of July, at 6 o'clock, in the presence of the officers of the Mint, and a number of distinguished citizens.
>
> Within the stone was deposited a package, securely enveloped, containing the newspapers of the day, a copy of the Declaration of Independence, of the Constitution of the United States, and of the farewell address of general Washington; also, specimens of the national coins, including one of the very few executed in the year 1792, and a half dime coined on the morning of the 4th, being the first of a new emission of that coin, of which denomination none have been issued since the year 1805.[39]

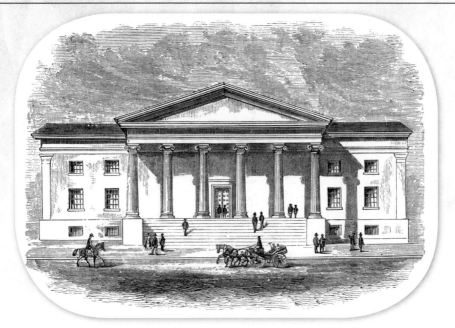

The second Philadelphia Mint opened in early 1833.

No half dimes at all had been made since 1805, and now, in the wee hours of the morning of July 4, 1829, the Mint was aglow with activity—as sparkling new coins were stamped out in time for the special cornerstone laying. Why the half dime? It was a logical choice, as the smallest authorized silver coin of the United States. These would make ideal souvenirs for those attending, it was thought. And they did! In attendance were various Mint officers and employees and local dignitaries.

THE NEW MINT

The building was ready for occupancy in January 1833. While balance scales, serviceable screw presses, and certain scientific apparatus were brought from the old Mint to the new, most equipment was newly manufactured.[40] In this way the equipment at the old Mint, much of which was in poor condition or obsolete, could be continued in use at the old site while the new Mint was being built and set up for operation. Afterward the old machinery was sold at auction.[41] A steam plant provided power for rolling ingots into strips and other processes in planchet-making, but coining continued to be done by hand-operated presses.

On January 13, 1833, Mint Director Samuel Moore reported:

> Operations, I have the satisfaction to say, have been commenced in the new Mint for the proof of the machinery; and all the departments of the institution will be transferred in a few days to that edifice.[42]

The Philadelphia *Commercial Herald* told of the construction of the Mint:

Mint of the United States

In accordance with the Act of March 2, 1829, a lot was purchased, extending from Chestnut Street to Penn Square, on the West side of Juniper Street, containing 150 feet front on Chestnut Street by 204 feet on Juniper Street.[43] On the Fourth of July, 1829, the cornerstone of the Mint of the United States, was laid with appropriate ceremonies.

The building is of white marble, erected from designs furnished by William Strickland. It fronts on Chestnut Street, Penn Square, and Juniper Street, and is 123 feet front, by 189 deep, exclusive of two porticos each 27 feet, making the whole depth 193 feet. Porticos are each 60 feet front, supported by 6 columns, of the Ionic order, 3 feet in diameter, fluted, and bound at the neck of the capital with an olive wreath. The entablature of the porticos extends entirely round the front and flanks of the building, supported at the corners, and surmounted at the extremes by the flanks, by four pediments.

The building consists of a basement, principal and attic stories. The officers' rooms, vaults, &c. are on Chestnut Street, a part of the western flank, and are arched and rendered completely fire proof. The roof is of copper. In the centre of the interior, there is a court 55 by 84 feet, for the purpose of affording additional light, to the various apartments, and a more ready access to each story, by means of piazzas.

The entrance from the south portico, is into a circular vestibule, communicating with the apartments of the director and the treasurer, and by arched passages, with those of the chief coiner, melter, and refiner, and with the rooms for receiving bullion and delivering coins. These passages communicate with the attic story by means of marble stairways, where the apartments of the assayers and engravers are situated. Rooms are appropriated for the apartments of the chief coiner, melter and refiner. The important process of assaying is carried on in rooms 50 feet by 20; those of melter and refiner occupy a range extending 95 by 35 feet. The principal melting room is 37 by 32 feet; and the process of gold and silver parting is carried on in a room 53 feet by 32.

The preparatory operations of the chief coiner are carried on in two rooms 55 by 40 feet, opening to the north portico; the propelling steam power being placed in the basement story. The immediate operations of coinage occupy a range of apartments 120 feet by 32. The principal coining room is 37 feet by 32 feet, and is large enough to contain 10 coining presses.

In a distinct suite of rooms, in the attic story, the standard weights of the Mint and the balances for adjusting them are kept.

The operations of coining were commenced in this building early in the present year.

On May 23, 1833, the Mint was opened for the general public to visit. The Philadelphia *Saturday Evening Post* ran this story:

Corner view of the new Mint.

United States Mint

The new mint appears to be a favorite place of resort for the curious among our fellow citizens.[44] Visitors pass in by the Chestnut Street front, at all hours of the morning, and are at once ushered into a beautiful and capacious building, well adapted for the important purposes for which it was erected. When we look round its ample dimensions, we wonder how it was possible to accommodate so extensive a business as was done in the miserably confined apartments of the old coining house in Seventh Street, and fail not, at the same time, to admire the neat and simple beauty of the present building.

The first object that attracts attention on entering is a huge steam engine, at the opposite end of the building, the noise of which, added to the incessant jarring of the dies, gives token of the laborious purpose to which it is applied. This engine, of 30 horse power, is the most highly finished specimen of steam engine, we have ever witnessed. The shafts, upright and horizontal, are of polished metal, and most of the cog-wheels are of brass. The huge fly-wheels run with the precision of a watch-wheel, while the various and totally different purposes to which its power is applied, strike the beholder with admiration of the skill and ingenuity of the machinist. Rush and Muhlenburgh, of this city, constructed this engine; its cost was about eight thousand dollars.

From the hot rooms in which the bullion is converted into ingots, we entered the rooms where the ingots are passed through a succession of steel rollers, until they assume the flatness and thinness of a common iron hoop. Thence we ascended into a room where these thin bars are passed through a steel gauge, to give them a uniform thickness, equal to that of the half dollar. A punch, worked by the said engine, cuts out the silver of a proper size; the scraps of silver are melted over again into ingots.

From this room the prepared bits are taken down to the die room, where they are passed on their edges, through a machine which gives them the impression they bear on the edge. They are thence handed over to the coiners, by whom they are placed in a tube in a pile a foot high whence they drop, one at a time, on a slide which conveys them directly to the dies. Here they receive the proper impression on each side, from dies forced together by means of an iron bar, ten or twelve feet long, worked horizontally by three men. The instant the coin receives the proper impression, it is forced off the die into a box ready to receive it, and gives place to another, which immediately occupies the same position, and undergoes the same operation.

After having gone through the whole establishment, the impression left upon my mind is that of astonishment and wonder, that an end of such immense importance as the supply of coin for a whole nation, can be attained by means so apparently simple, and of such ready comprehension. The spectator, going through alone, needs no one to explain this or that operation. Everything explains itself on the instant; for everything is free from mystery or concealment, while the excellent condition of establishment, and the extreme politeness to strangers, manifested by every person about it, materially enhances the pleasure of a visit to the Mint of the United States.

On July 9, 1833, *Niles' Weekly Register* ran "City of Philadelphia," giving "as a proof of the prosperity of Philadelphia," the cost of various buildings, including $175,000 for the Mint. Among others listed were the Bank of the United States, $413,000; Bank of Pennsylvania, $235,000; Girard Bank, $250,000; and the Philadelphia Bank, $50,000.

In addition to the Mint, other important buildings were on Chestnut Street including the Arcade (within which was Peale's Museum), the Academy of Arts, the Bank of the United States, and four theaters.

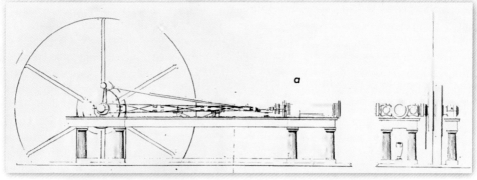

Peale's sketch of the steam engine he designed for the Philadelphia Mint.

In 1833 the coinage of the Mint, mostly if not entirely effected in the new building, amounted to:

Half cents: 154,000

Cents: 2,739,000

Half dimes: 1,370,000

Dimes: 485,000

Quarter dollars: 156,000

Half dollars: 5,206,000

Quarter eagles: 4,160

Half eagles: 193,630

Half eagles were nearly all made for the export trade, sent abroad and melted. Gold did not circulate in domestic commerce.

In 1833 gold for coinage came from these sources:

North Carolina: $475,000

Georgia: $216,000

Virginia: $104,000

South Carolina: $66,000

Tennessee: $7,000

Mexico, South America, and the West Indies: $85,000

Africa: $12,000

From sources not ascertained: $13,000.

These figures total $978,000, of which $868,000 came from domestic mines.

IMPROVEMENTS IN TECHNOLOGY

In 1833 Director Samuel Moore arranged for Franklin Peale, a son of local artist and museum proprietor Charles Wilson Peale, to go to Europe and visit the various mints there. He left in May and was abroad for nearly two years. While there he examined coining equipment and assaying processes in England, France, and elsewhere. Upon his return the Mint ordered certain equipment, including a Contamin portrait lathe, or *Tour à Portrait de Contamin*. This device aided in the making of metal hubs by transfer or pantograph from larger images in brass or iron. The Contamin unit remained in use past the era of copper half cents and large cents.

The Coinage Act of June 28, 1834, produced a dramatic change. Engineered by Senator Thomas Hart Benton of Missouri, nicknamed "Old Bullion," it changed the official ratio of gold to silver from *1 to 15* to *1 to 16*. At the time in Europe the ratio was about 1 to 15-1/2. The weight of gold coins was reduced slightly as a result of the new ratio. The practical result was that after August 1, 1834, when the legislation took effect, $2.50 and $5 gold coins were coined in quantity and went into domestic circulation. It was no longer profitable to melt them because their face value was worth slightly more than their bullion value. For the first time since 1820 gold coins were seen in commerce. In 1838 the $10 gold eagle, the coinage of which had been suspended in 1804, was resumed. Silver half dollars, already plentiful, were joined by increased mintages of other denominations. All the while cents were common. Now, for the first time in American monetary history there was a sufficient and readily available supply of all except the silver dollar (which had not been coined since 1804). Taking cues from extant laws in some places, several states passed laws forbidding banks to issue paper money of a face value below $5—an effort to encourage the circulation of coins, including the new gold issues. In this way merchants could not give change in $1, $2, or $3 notes, the most popular denominations of the time, but would have to use coins instead.

In the summer of 1835 Samuel Moore tendered his resignation as Mint director and was replaced on August 1 by Robert Maskell Peterson, the son of an earlier director. In August 1835 chief engraver William Kneass suffered a stroke, after which he became partially incapacitated. An accomplished artist and engraver in the private sector, Christian Gobrecht, was hired to work under the title of "second engraver." After Kneass's death in 1840 Gobrecht assumed the title of chief engraver. He created many notable designs, including the Braided Hair portrait used on half cents, cents, quarter eagles, half eagles, and eagles, and the Liberty Seated motif for silver coins from the half dime to the dollar.

Mint Director Robert Maskell Patterson.

On October 17, 1835, *Niles' Weekly Register* reported:

> Spanish dollars are at a four percent premium at New York—probably for the Chinese market—Canton. In the "palmy days" of the Bank of the United States, payments at Canton were made by its drafts, and without a risk to anyone.[45]

By September 1835 a model steam-powered coining press had been made in Philadelphia under Peale's supervision. In the same city the firm of Merrick & Agnew, makers of heavy equipment including fire engines, was commissioned to construct such a machine for the Mint. This employed the knuckle-action mechanism invented by Uhlhorn in Europe and went into operation in March 1836. Power was from a vertical "steeple" type steam engine located remotely within the building, transmitting power to the press via a system of shafts and pulleys.[46] Steam was also used to drive an improved model of a milling machine that imparted raised rims to blank planchets.

In time, other coining presses were made, in three capacities, the medium size being suitable for striking eagles, quarter dollars, and cents, as demonstrated on March 23. The units had automatic feed by a vertical tube that stored planchets.[47] Dollars, the broadest or widest denomination, were first struck by steam in 1837. By that time the production of silver dollars, the first since 1804, had resumed with a modest coinage of just 1,000 pieces on a screw press in December 1836.[48] In 1836 some pattern gold dollars were made as well.

The Coinage Act of January 18, 1837, standardized the silver and gold content in coins at 90% and set up a $1,000,000 bullion fund so that depositors of precious metals could receive coins immediately, rather than waiting for their metal to be coined and calling for it later.

Mint Director Robert Maskell Patterson wrote to Secretary of the Treasury Levi Woodbury on June 30, 1837, commenting: "One of the steam-presses has coined 400,000 cents within the last few weeks, without interruption or accident, and our largest press has been coining half dollars with perfect success."[49]

Steam-powered coining press of 1836.

THE PANIC OF 1837

In 1836, by many outward appearances the American economy continued to be prosperous. However, business reports increasingly told of problems with banks, manufacturers, and credit. Many large loans were past due. The land boom in the West had slowed to nearly a standstill that summer after President Andrew Jackson issued the Specie Circular, which mandated that land purchases be paid for only in gold or silver coins, and that paper money was no longer acceptable for those transactions. The second Bank of the United States wound down its operations, sold its buildings, and closed. The slack was taken up by state-chartered banks, without any problems.[50]

As the year 1837 commenced, the price of everyday goods kept rising. On February 15 a crowd of several thousand people gathered in New York City in a park near City Hall to protest the high costs of bread, meat, and fuel. Unemployment was increasing, and many families were having difficulty making ends meet. Matters worsened, there was a tightening in credit, and beginning May 10, continuing for the next several days, banks stopped paying out specie (silver and gold coins) at par in exchange for bank notes. Coins were hoarded by the public, and most commerce was conducted with bank notes. Millions of privately issued "Hard Times tokens" filled the gap for cents and circulated widely. The economy recovered in a strong way beginning in the spring of 1843.

THE MINT CABINET

In 1838 the coinage of $10 eagles, suspended since 1804, was resumed. A new Braided Hair design by Christian Gobrecht was featured on the obverse. In 1839 this motif was extended to half eagles and in 1840 to quarter eagles.

In June 1838 the Mint Cabinet, a collection of coins, medals, and minerals, was formed at Philadelphia with a stipend of $1,000 in the Treasury Department's budget for that year. Mint assayers Jacob Reese Eckfeldt and William E. DuBois took charge of gathering specimens, including some said to have been furnished by old-time employee Adam Eckfeldt (Jacob's father). However, most were culled from incoming deposits of silver and gold coins. In addition to numismatic specimens the Cabinet included ores and some items wrought from precious metals.

In the same year the first branch mints were opened. Located in Charlotte, North Carolina; Dahlonega, Georgia; and New Orleans, Louisiana, the first two produced only gold coins, and the last struck both silver and gold (see subsequent chapters).

CHANGES AT THE MINT

Adam Eckfeldt, who had been appointed as chief coiner by President James Madison on February 15, 1814, and who had been on the staff since 1795 and by contract earlier, resigned early in 1839, and Franklin Peale was named in his place on March 29. In 1840 the production of silver dollars in quantity commenced. The Liberty Seated design was standard and was used until the denomination was discontinued by the Coinage Act of February 12, 1873. At the outset these dollars were mostly exported and were available only by paying a small premium to banks or bullion exchanges. After 1849 nearly all were exported. As a result, the half dollar was the largest coin regularly seen in circulation, as had been the case for earlier years as well.

Christian Gobrecht died on July 23, 1844. On September 6 he was replaced as engraver by James Barton Longacre, who was well known as an engraver of printing plates, including for bank notes, and for his production of the volumes comprising *The National Portrait Gallery of Distinguished Americans*.

CALIFORNIA GOLD

The discovery of gold in quantity at Sutter's Mill on the American River in California on January 24, 1848, precipitated the Gold Rush. In 1849 tens of thousands of fortune seekers arrived in California by land and sea. In time, the vast quantity of new gold on the international market (with prices usually calculated in London) disturbed the traditional ratio of gold to silver, set as 1 to 15 by the Mint, but closer to 1 to 15.5 on world markets. Silver became more valuable. New York City was the destination of much gold shipped from California to the East. Proposals in Congress to locate a branch mint there were defeated.

The Coinage Act of March 3, 1849, authorized two new denominations: the gold dollar and the $20 double eagle. The double eagle in particular, first introduced in 1850, was a convenient way to convert large amounts of bullion to legal-tender coin form.

The *Mint Report* for 1849 included this:

> By an act passed at the last session of Congress, two new coins—the gold dollar and the double-eagle—were added to those previously authorized. Of the former, 936,789 have been coined, namely: 688,567 at the Mint in Philadelphia, 215,000 at New Orleans, 21,588 at Dahlonega, and 11,634 at Charlotte. I regret to say, that in consequence of difficulties in the execution of the dies, double-eagles have not yet been coined.

During the past year the deposits of California gold amounted at the Philadelphia Mint to $5,481,439, and at the New Orleans Branch Mint to $666,080. Near the close of the year 1848, the first California gold was brought to the mint; and it amounted to $44,177.

More from the 1850 *Mint Report:*

Prior to the discovery of the California mines, the deposits at this and the New Orleans Mint were almost entirely in foreign coins, which, being already refined, were fit to pass into ingots for coinage, without other preparation than a proper admixture to bring them to our own standard. The apparatus and coining arrangements were quite ample for converting into our own money any probable amount of deposits of this character. California gold, however, contains a large proportion of silver, above one-ninth of the mass being in that metal. This is much more than we are allowed by law to leave in the gold coin as alloy. A separation of the superfluous silver, therefore, became necessary.

This department of labor, although it forms, generally, in other countries, no part of the functions of their mints, has been made obligatory upon the mints of our own country. Separating (or refining) departments had accordingly been organized upon a scale quite moderate, indeed; but much more than sufficient for all the business of that kind prior to the influx of the California gold. The refinery of the Philadelphia mint at the close of 1848, when the first deposits from California were received, was capable of separating about $100,000 per month of argentiferous [silver-bearing] gold bullion. Since then, the amount of that bullion received has gone on increasing, month by month, until it reached the sum of $4,600,000 in December last.

To meet the demands thus made upon us very extensive enlargements became necessary and have been effected, in the refining department of the Mint. By arrangements now on the point of completion, the capacity of this establishment for refining will have reached to form six to seven millions of dollars per month, by the separating process now in use. This large increase of power has not been attained without many necessary delays and interruptions in the progress of our work, while the stream of our deposits was constantly on the rise; so that, although the changes indicated were prosecuted with the greatest energy, we have not been able to avert the accumulation of the large uncoined balance to which I have referred.

I see no reason to doubt, however, that, with the means now at our disposal, and such further enlargements as are in our power, the Mint will be enabled in a few months both to free itself from the debt now accumulated, and to secure prompt payment of all future deposits. . . .

In the coining department of this mint no changes of a marked character were required. In consequence of the increasing demand for the smaller gold coins, we have, however, found it necessary to add largely to the adjusting force, whose duty it is to test the weights of the separate pieces before they are struck. At the suggestion of the chief coiner I authorized him to employ females for this purpose, the labor being entirely suited to their capacity. Nearly forty have already been introduced into that service; and the number will be still further increased. We are consequently enabled, in addition to the necessary supply of the larger pieces, to extend very greatly our coinage of the smaller gold pieces, with a view to meet the deficiency created by the withdrawal of silver from circulation.

Large amounts of gold continued to arrive at the Mint, usually via steamship from San Francisco to Panama City, across land by pack animals and small boats to Chagres (soon renamed Aspinwall) on the Atlantic side of Panama, then by steamship to New York City. Most of this gold was in registered commercial shipments, but returning miners and other adventures also carried gold.

By 1850 it cost more than face value to strike the current half dime, dime, quarter dollar, half dollar, and silver dollar denominations. Silver coins disappeared from circulation and were available only at a premium from banks and exchange brokers. The needs of commerce were filled by Spanish and Spanish-American silver coins, which had been dominant for generations, but which now comprised nearly all of the circulating coinage. The 2 reales coin valued at 25¢ was by far the most popular denomination.

In 1850, in response to political pressure and a changing administration in Washington, Dr. Robert Maskell Patterson resigned as director of the Mint. There ensued a long string of political rather than merit-based appointments to the post, which continues to this day. Unlike Patterson, most later directors entered the office with very little knowledge of coinage, manufacturing, or monetary history.

"The morning after the arrival of the last Chagres steamer some six or seven returned Californians were seen on the steps of the United States Mint, Philadelphia, at a very early hour, waiting for the Mint to open. Each one had his bag of gold dust, and all seemed most impatiently awaiting the time when the doors should be thrown open. One or two seem to be asleep, others are lying on the steps." (*Ballou's Pictorial Drawing Room Companion*, July 19, 1851)

Receipt for the deposit of California gold at the Mint, July 23, 1852.

ASSAY OFFICES

In 1851 the first of many assay offices was opened. In San Francisco, California, Augustus Humbert was named as the United States Assayer of Gold and was in charge of a coinage and assaying office within the premises of Moffat & Company (see chapter 12). This was continued through 1853, after which it was replaced by the San Francisco Mint.

In 1854 the New York City Assay Office was opened. At the time the city was the principal arrival port for raw gold, ingots, and privately-made coins from California. It also was a storage facility for gold belonging to the government. It remained in operation until 1982. It was extensively updated and renovated after 1914, but not before the façade was saved for use as the front of the Metropolitan Museum of Art, where it is still in place today.

An assay office known as the Denver Mint was opened in Denver in 1863 after the facilities of Clark, Gruber & Company were purchased. This assay office continued in use until 1906. Despite the name, no coins were ever struck there. The Charlotte Assay Office operated in the old Charlotte Mint premises from 1867 to 1913. The Boise Assay Office in Idaho was in operation from 1871 to 1933. For a brief period from 1876 until coinage operations resumed in 1879 the New Orleans Assay Office was active. The Helena Assay Office was in Montana from 1877 to 1933. The St. Louis Assay Office operated from 1881 to 1911. Over the years there were several calls in Congress to establish a coining mint in St. Louis, but that never happened. The Deadwood Assay Office was in South Dakota from 1898 to 1927. The Seattle Assay Office operation extended from 1898 to 1955. The Salt Lake City Assay office ran from 1909 to 1933.[51]

THE PHILADELPHIA MINT IN 1852

In the 1850s several illustrated weekly newspapers made their debut, including *Gleason's Pictorial Drawing Room Companion*. In its issue of July 17, 1852, this article was featured:

United States Mint, Philadelphia

On this and the following page, we give some very fine scenes descriptive of the United States Mint at Philadelphia.[52] They are given with great accuracy and beauty by our artist, Mr. Devereaux. The United States Mint was founded in 1792, and the business of coining commenced in 1793, in the building occupied at present by the Apprentices' Library. It was removed in 1833 to the fine building it now occupies in Chestnut Street above Olive Street.

The edifice is of white marble, and the north front opposite to Penn Square is 123 feet long, with a portico 60 feet long, of six Ionic columns, and the south front on Chestnut Street has a similar portico. Since the enormous influx of gold from California, the United States Mint has become an object of more than common attention and interest, and the place is usually filled with visitors watching the various processes which the metal goes through before it comes out a finished coin.

The machinery and apparatus by which these are accomplished are of the most complete and perfect character. The rooms in which the smelting, refining and alloying are done, are spacious apartments, in which a large number of workmen are employed. Heaps of the rich ores are to be seen laying around, as they were extracted from the mines, or gathered in dust from the sands of the mountain streams of California. Bars of the pure metal, representing many thousand dollars

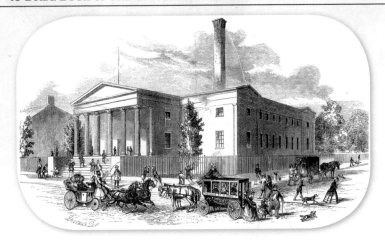

in value, arc passing through hands which, like those of Midas, seem to turn what they touch into gold. The heat of this place is insufferable; fires glow with the intensity of those in a foundry; the men are as dust-begrimed as those in a smithery; there is a suffocating sensation of hot air, steam and perspiration penetrating the atmosphere, which is anything but pleasant to experience, when the thermometer is palpitating under a summer temperature. Crucibles arc handled with iron tongs and cotton mittens, the metal is shaped into bars and then reduced to the requisite fineness. All this takes place in one apartment.

In another, there is a most beautiful steam-engine which drives the rolling and stamping machinery. This engine is of one hundred horse power, and works the rolls, draw-benches, and cutting presses. It is called a steeple-engine, and has two cylinders. It is worked by boilers forty feet in length, and forty inches in diameter, which also works a ten horse and five horse engine in the separating and cleaning apartments. This main engine is of the most elegant workmanship, polished like a piece of cutlery, and works without the least perceptible jar.

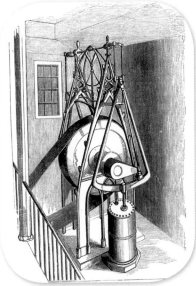

From this room the visitor walks into that where the rolling machines are at work, turning out the metal to the proper degree of thickness which each particular denomination of coin requires. The metal is cast into ingots 14 inches in length, and about 5/8 in thickness; they are then rolled to very near the proper thickness, when they are passed through the draw-benches to equalize them; the strips are then cut at the presses; these presses cut out from two hundred to two hundred and sixty a minute. There are fourteen men employed in this room—two for each pair of rolls.

The pieces cut then pass to the Adjusting Room. Here each piece is weighed separately and adjusted with a file. Light and imperfect pieces are re-melted. There are fifty-four

The steam engine at the Mint. females employed in this room.

Rolling mills that reduce ingots to long flat strips are shown in the right foreground and against the wall. To the right are machines that punch planchets from the strips.

Adjusting Room where blank planchets are weighed and if too heavy are filed. If too light they are set aside to be melted.

The pieces are then taken to the Milling and Coining Room. There are from two hundred to four hundred milled in a minute, according to their size. In another apartment the coins are cut with a punch the desired size and then stamped. The coins are placed by a person seated at the machine, in a perpendicular tube, down which they descend, one at a time, being seized as they drop, by a part of the machinery, which pushes the coin under the stamp, whence it falls under the machine into a glass-covered box. This part of the process used in former years to be performed by a press, which still remains in the building, worked by a lever and screw, requiring eight men to laboriously work at it; now the process requires scarcely any manual labor but handling the pieces of coin.

The rapidity with which the pieces are executed is surprising—being at the rate of from seventy- five to two hundred per minute. Cents, dollars, eagles, double eagles are turned out with equal facility, the process being the same in all. Some idea of the extensiveness of these operations, and of the metalliferous fecundity of the gold possessions of the United States, may he had when it is stated that in one month, lately, nearly three millions of pieces, gold, silver and copper, were coined, and that nearly four millions in value are coined every month.

In addition to the other attractions of the Mint, there is a most extensive cabinet of coins, ancient and modern (Roman, Greek, Chinese, European), which is one of the greatest of curiosities, probably to be met with nowhere else in the country.

The officers of the Mint are polite and attentive to visitors and endeavor to make their visit one of instruction as well as amusement. It is under the very effective management of Mr. Dale, the director.

An October 1853 article in *Banker's Magazine* included this:

In these days, when the simple dogma, that to the victors belong the spoils, governs appointments to office, and partisanship outweighs all higher claim, it is a public misfortune that the leaven of tried worth cannot remain in some departments of the public service.

In view of the vast business of the Mint, the scientific and technical nature of its operations, the government should have made it consistent with Mr. Patterson's interest to remain. He takes out of it, however, a reputation for high talents and perfect probity; and, personally, he must be the gainer by the voluntary withdrawal from a place, which is no longer secure against the periodical squabbles of needy politicians.

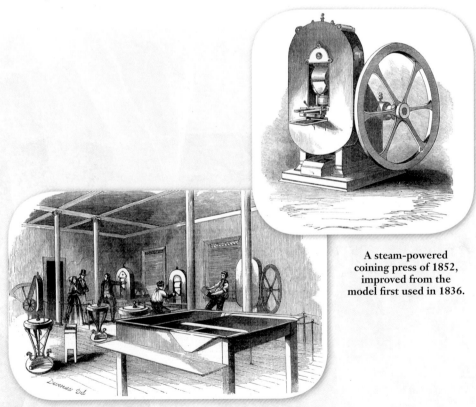

A steam-powered coining press of 1852, improved from the model first used in 1836.

The Milling and Coining Room where planchets are run through a milling machine to give them upset rims, after which they are coined in the presses shown against the wall.

In the meantime, federal silver coins continued to be hoarded. The Coinage Act of February 21, 1853, changed that by lowering the silver content of all coins except the dollar. After that time small silver coins readily circulated and became common in the channels of trade. An 1855 account told this:

> A letter from Washington says that the Treasury is now burthened with the custody of over five millions of dollars in small silver change, from half-dollars to three cent pieces. Mr. Hunter's coinage bill was passed, slightly reducing the actual value of our silver coin, and providing for its own rapid manufacture. The expected results have followed. The wants of circulation have been fully supplied; but another less desirable consequence has ensured, to wit, this small change has become a drug. People will not take it and the law makes it a legal tender in sums of not over five dollars. Though the inconvenience of an inadequate supply of small change was a serious one, prudent financiers expressed doubts of the soundness of the remedy adopted, at the time it was proposed.—Orders have been issued to suspend the coinage of quarters and halves, and the operations of the Mint are much reduced.[53]

Silver dollars remained at their old weight and were used in foreign trade as bullion coins. They cost over a dollar to make, and at their foreign destinations, mainly Canton and Liverpool, they were valued for their silver content, not the denomination lettered on them. These dollars of the Liberty Seated design continued to be minted into early 1873. For the first time in the history of the Mint, depositors could no longer deposit silver bullion and receive an equivalent amount in coins. From this time onward the United States was *de facto* on the gold standard, but that would not be made official until 1900. In June 1853 James Ross Snowden became director of the Mint and would remain in the post until 1861.

Mint Director James Ross Snowden.

PROBLEMS AT THE MINT

In the spring of 1854, the Philadelphia *North American* reported that the Mint was becoming secretive:

> We are unable to present to-day our customary exhibit of the operations of the Mint for the month of April, in consequence of an order, from the Treasury Department, forbidding the officers of the institution furnishing the information for publication. In future, we understand, the information is only to be communicated to the Secretary of the Treasury, who will exercise his own discretion to the time and form of publication.
>
> This, we cannot but think, is a reform backward, and apparently without any object, unless it be the object of the honorable secretary to withhold facts from the business and financial, portions of the country at large, and in which they have a deep interest.[54]

Part of the above may have had to do with a major investigation within the Philadelphia Mint, in which many irregularities and illegal acts were discovered. It was learned that the weigh clerk had complete control of incoming shipments, including gold bullion and coins minted and without audit, and others were similarly unsupervised. A devastating report many pages in length included this:

> In the melting-room there are always several operatives at work, and they, to a certain extent, would seem as a check against depredations, but not a certain preventive, as each deposit is under the charge of a single workman, who, if so disposed, might readily slip out of the pan a California slug or other valuable piece of gold. Thus, from the time the deposit is first made until placed in the crucible to be melted, it is exposed to the depredations of the several parties who have access to the weigh-room, to the vault, and to the melting room.
>
> An abstraction of one-fourth of one percent only from an annual deposit of fifty millions of bullion, would produce $125,000; and yet, from the variable character of the deposits, differing, as they do, in their results, from the one-half of one percent up to six and seven per cent., it will readily be perceived that an abstraction of one, two, and even three per cent., might be made from the deposits without detection, and almost without suspicion; for as the loss will fall on the depositor, they might readily be made to believe, or to suppose, that the bullion contained a larger portion of the baser material than was supposed.
>
> From this it will be seen what an enormous sum might be abstracted from the deposits; and as the character of the Mint is at stake, every possible effort should be made to preserve its good name. As the vault is open to several persons connected with the weigh-room, it would be an easy matter for any of these persons to take from the boxes containing the deposits, pieces of California slugs, or lumps of gold; and to equalize the loss among the several depositors, they might filch a small portion of the dust in the different boxes and deposit the same in the boxes from which the slugs or lumps of gold were taken.[55]

During the investigation it was found that from a shipment that arrived in April 1854, Mint weigh clerk J. Engle Negus had "filched a portion from each of the sixteen boxes," and 72.80 ounces of gold was found in "the bag in his private vault." His assistant, a boy named Henry Cochran, had become suspicious and reported his thoughts. Upon being confronted Engle said he had done this to see if it would be found missing, at which time he planned to return it and suggest heightened security at the Mint. This explanation was not accepted. Upon close questioning, Engle confessed to stealing gold. To settle matters he paid back the sum he said he extracted. On May 21, 1854, while the Treasury Department was still debating what to do with Negus, he blithely sailed off to Southampton aboard the United States Mail Steamer SS *Washington*. His family sailed to join him later.

Secret observations and over 100 pages of testimony revealed for *several years*, dating back to at least December 1851, he had been stealing in quantity and had become a wealthy man. Along the way he had sent much money to England. There were other thieves as well, but none had stolen to the extent that Negus did. Without raising suspicions, Negus had lived in a $19,000 house, the price of a mansion, and a few months before his treachery was discovered "had given an entertainment last winter, at which most of the officers of the Mint were present." The testimony also revealed that there was little security regarding precious metals at the Mint. In a sad end to the story, young Cochran was promoted and worked at the Mint for several decades, until he himself was caught stealing gold ingots. He confessed to stealing about $100,000 worth and was sentenced to prison for six years and seven months.[56]

For a long time chief coiner Franklin Peale had been using Mint employees and equipment during regular working hours to conduct his private business of making dies for and striking medals, and he had misused Mint assets and money in other ways. He also feuded with engraver Longacre, agitating the artist in ways that, while not actionable, were uncomfortable for the recipient. In 1853 charges were preferred against Peale. He survived this, but in late 1854 there was an incident involving medal dies, and he was fired. Peale then petitioned Congress for $30,000 for all the work and improvements he had done at the Mint. Many if not most of the Mint employees felt his dismissal was incorrect. Peale was widely liked, was prominent in Philadelphia society and scientific circles, and for years afterwards he was mentioned by his former fellow workers as being responsible for many of the better innovations at the Mint.[57]

NEW POLICIES

In August 1854 new policies were put in place for the Philadelphia and branch mints. These were precipitated in great part from the fidelity problems experienced in Philadelphia. From Monday through Friday the working hours were from seven o'clock in the morning until five o'clock in the afternoon, with an hour off for dinner (the term at the time for what we call lunch today). On Saturday the hours were 7:00 a.m. to 2:00 p.m. Holidays were the Fourth of July, Election Day, and Christmas. The last was not uniformly observed through state and local governments. Some felt Christmas was a popish holiday for Catholics, and many offices and businesses remained open as usual. The change prompting more people to view Christmas as a national holiday came with the growing popularity of Clement Moore's "The Night Before Christmas" and the popular of Santa Claus with the younger set.

Each month the superintendents of the branch mints were required to report to the director the habits, fidelity, and skills of each workman. Watchmen and messengers were to be middle-aged, able-bodied men of good character and habits. No newspapers or magazines were to be brought to any mint by workmen, lest they become distracted and misapply their time.

Visitors to the mints were discouraged as they distracted the workmen and officers and exposed coins and precious metal to the risk of theft. The superintendents of the branches were encouraged to have no more than one weekday set aside for visitors, plus Saturday. Exceptions could be made by permission of the superintendents. In practice, visits to mints continued more or less as earlier, but with fewer members of the public involved.

On July 19, 1855, all coinage operations were closed down for much of the remainder of the year, while attention was paid to building repairs, installation of fireproofing material, and machinery maintenance. The receipt of bullion, assaying, and refining continued without interruption. By this time it had been Mint policy to encourage, and successfully, the use of gold ingots for export, rather than gold coins. This point was partly moot, as the lighter-weight gold coins produced under the Act of June 28, 1834, were not in particular demand for that trade. A machine to adjust the weight of gold half eagles had been imported and showed good promise. Suggestions that the double eagle be abolished in order to encourage the circulation of half eagles in commerce were dismissed. The $20 gold piece remained the most efficient way to convert a large amount of gold bullion into coins.[58]

The discovery of vast deposits of silver in the Comstock Lode in the late 1850s and the subsequent development of mining vastly increased the domestic supply of that metal. Earlier, such quantities were scarce, one source being a byproduct from copper refining.

THE RISE OF NUMISMATICS

Under the Act of February 21, 1857, the copper half cent and cent were discontinued and a small-diameter, 72-grain copper-nickel cent of the Flying Eagle design took the place of the larger denomination. These were released in quantity starting in May. This catalyzed a strong wave of sentiment across the nation when citizens realized that the large copper "pennies" of childhood would soon disappear. Thousands of people set about saving the old cents, many of them seeking one of each date back to 1793. Numismatics, sometimes called numismatology at the time, became popular. In December 1857 the Numismatic and Antiquarian Society of Philadelphia was formed. In March 1858 teenaged Augustus B. Sage and his friends launched the American Numismatic Society in the Sage family apartment on the second floor of 121 Essex Street, New York City.

WASHINGTON TOKENS AND MEDALS

In 1859 Sage cataloged three auctions. By that time the collecting of tokens and medals depicting George Washington led the market, and prices rose sharply. Colonial coins, copper cents, patterns, and Proof coins (first widely sold to the public in 1858) were sought. Mint Director James Ross Snowden was besieged with requests for patterns, restrikes, and more. Often he traded rarities for Washington items, per this circular letter he sent out on May 24, 1859:

> I desire to obtain for the Cabinet of the Mint, if practicable, one or more copies of every medal, medallet, coin, or token on which the head or name of Washington appears.
>
> The possessor of any such memorial of Washington will confer an obligation by sending me a description of it, and state whether he is willing to present it to the Mint, or dispose of it, either for cash, and if so, at what price, or exchange it for other coins or medals. In the course of a recent investigation I ascertained the existence of sixty different memorials of the above character, and there are doubtless others which have escaped my notice.
>
> A few of these medals, are now at the Mint, but I desire if possible, to supply the Cabinet with a full set of these interesting memorials of the Father of his Country. To accomplish this I invite the cooperation of my fellow citizens.

About the same time he turned this into a secret business. Without publicity, Mint officers and others struck tens of thousands of patterns, restrikes of early rarities, and more, and sold them for private profit through favored local dealers such as William K. Idler and John W. Haseltine. This largesse was kept up steadily until the summer of 1885 under several directors and, after 1873, superintendents, when an end was put to the practice by a new superintendent, Daniel M. Fox.[59]

CONTINUING COINAGE

Returning to the narrative of regular coinage, in the summer of 1857 there were bankruptcies and financial difficulties that evolved in what became known as the Panic of 1857. In 1858 the economy regained strength. By 1860 most commercial transactions were with paper money issued by state-chartered banks. The Panic of 1857 and its after effects were history, and banks in operation were mostly solid. Few new ones were formed however. The economy remained more or less on hold, due to the uncertain political situation.

By 1860 most commercial transactions were with paper money issued by state-chartered banks. The Panic of 1857 and its after effects were history, and banks in operation were mostly solid. Few new ones were formed however. The economy remained more or less on hold, due to the uncertain political situation. Foreign coins were completely gone from circulation, their legal-tender status having expired in September 1859. Copper-nickel cents, silver three-cent pieces, half dimes, dimes, quarters, and half dollars, and gold $1, $2.50, $3, $5, $10, and $20 coins circulated freely in commerce. Not often had the coinage situation been so ideal. Silver dollars were the exception. They still cost more than face value to mint and were treated as bullion coins in overseas trade, mainly with Canton and Liverpool.

MINTING PROCESSES IN 1861

In its issue of December 1861 *Harper's New Monthly Magazine* published this article from which we give edited excerpts.[60] The operation of coining press is especially detailed. Much of the information is also applicable to the operations of the other mint of the time—San Francisco. The New Orleans, Charlotte, and Dahlonega mints had closed earlier in the year after their states had seceded from the Union. Illustrations are from the article.

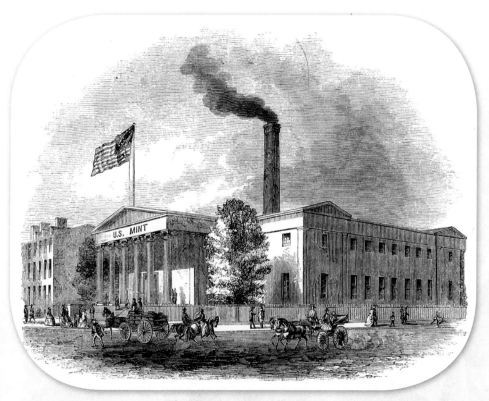

The Mint as it appeared in 1861.

Making Money

Visiting the Mint

Leaving our hotel we walk up Chestnut Street, and between Thirteenth and Fourteenth streets come to a fine, substantial, two-story marble building, entirely fireproof, and enclosing within its quadrangular walls a spacious courtyard. Ascending the massive marble steps, we enter an airy hall, freshened by a gentle breeze which sweeps through into the courtyard beyond. Between the hours of nine and twelve visitors are admitted, who are escorted about the building by gentlemanly conductors, of whom there are seven. Passing through the hall, on one side are the weighrooms for bullion and the office of the chief clerk of the treasurer, and on the other the offices of the cashier and treasurer.

We cross the paved courtyard, spacious and orderly, with boxes piled neatly around, and stacks of copper and nickel ingots ready for rolling. The well-proportioned chimney, 130 feet high—somewhat bullet-marked by pistol practice of the night watchmen—towers above the surrounding roofs, which look low by contrast.

Processing Metal

Thus we are conducted into the melting, refining, and assaying rooms. These ingots are bars sharpened at one end like a chisel blade, and are about a foot long, three-fourths to two and half inches broad, and half an inch thick, according to the coin to be cut from them.

Continuing our walk through a short entry, we come to the Rolling Room. Those massive machines are the rolling mills—four of them in a row, with their black heavy stanchions and polished steel rollers. The pressure applied is so intense

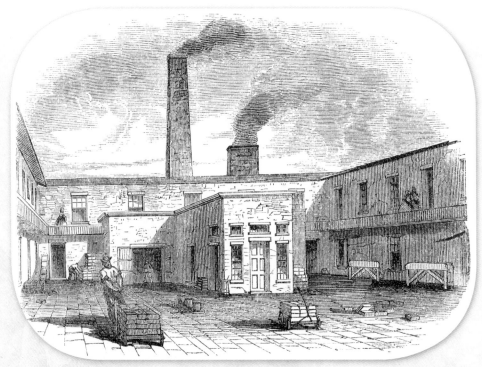

The courtyard within the Mint.

that half a day's rolling heats not only the strips and rollers, but even the huge iron stanchions, weighing several tons, so hot that you can hardly hold your hand on them. Every mill can be altered to roll to any degree of thinness, but usually the ingot passes through several mills, each reducing it slightly. When the rolling is completed the strip is about six feet long, or six times as long as the ingot. The strips are then placed into annealing boxes, sealed to prevent air from discoloring the metal, and heated.

Thus softened, the silver strips are greased and the gold strips waxed, after which they are pulled through the drawing bench to reduce them to the exact thickness desired to make planchets.

**Rolling mill that reduces
ingots of metal into strips.**

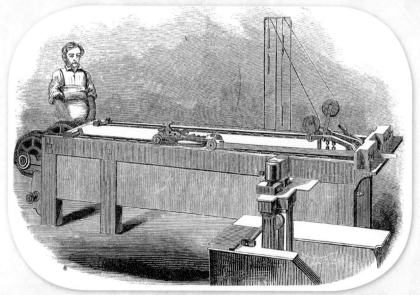

The drawing bench reduces planchet strips to precisely the desired thickness.

Planchets are then cut from the strip.

We now go to the Adjusting Room to see the sorting of the planchets. They are thrown upon a table with two holes in it, and a woman picks out all the imperfect pieces or chips, which are slipped into one hole, and the perfect ones into the other, where they fall into different boxes. It is not much to see; so come into the entry, up the marble stairs to the second story, past the director's room, out upon a gallery looking down upon the court-yard below. At the further end of the gallery we pass through a small entry, and enter a room. What a peculiar noise, like a young ladies' school at recess, only a strange filing sound withal! Nearly sixty females, some young and pretty, some—middle-aged and fine looking. Each operator has on the table before her a pair of assay scales. Seated close to the table, a leather apron, one end tacked to the table, is fastened under her arms to catch any gold that may fall. In short sleeves, to avoid sweeping away the dust, and armed with a fine flat file, she is at work, chatting and laughing merrily. She catches a double eagle planchet from a pile by her side and puts it into the scale. It is too heavy. She files it around the edge, and weighs it. Still too heavy. Files it again, and weighs it. Almost right. Just touches it with the file. Right; the index is in the centre. She tosses it into the box, and picks up another to undergo the same operation.

The proper weight of the double eagle is 510 grains, and the smaller gold coins are in the same proportion. Absolute perfection is impossible in the weight of coin, as in other matters, and the law therefore allows a variation of one half of a grain in the double eagles; therefore, between a heavy and a light piece, there may be a difference of one grain. This is so slight, however, not two cents in value, as to be deemed sufficiently correct. The weight of the silver half dollar is 192 grains, and

Machines with punches make circular planchets from the metal strips. An attached device weighs silver planchets, but not precisely.

The Adjusting Room where women adjust the weights of gold planchets as to meet standards.

smaller pieces in proportion, with the exception of the cent, which, being composed of 88 per cent, copper and 12 per cent, nickel, the weight is 72 grains.

To adjust a coin so accurately requires great delicacy and skill, as a too free use of the file would quickly make it too light. Yet by long practice, so accustomed do the operators become, that they work with apparent recklessness, scarcely glancing at planchet or scales, but seemingly guided by their touch. The whole process, however, is behind the times. Hand-work cannot compete with machinery. Sixty adjusters cannot keep the coining presses supplied, and genius must find a quicker way of performing the work. It is here that the delay occurs, keeping depositors waiting from twenty to thirty days for the coin they should receive in a week. It is astonishing that our Mint has not made the advancement here that it has in every other department.

Only the gold pieces are adjusted in this manner. The silver has merely the adjustment of the two planchets weighed at the cutting press. A greater allowance is made in the weight of silver coin, as it is less valuable, and it would be almost impossible to have such a vast number of small pieces separately examined.

The females in the adjusting room are paid $1.10 a day for ten hours' work. They look happy and contented. Behind the screens, at each end of the room, are dining-halls, where they eat the dinners they bring with them. On the whole, it is the pleasantest work-shop for women we have yet seen, and the pay, in comparison with that ordinarily given to women, is good.

If you examine a double eagle, or, lacking one, a quarter of a dollar, a slight rim will be noticed around the edge, raised a little higher than the device. It is done to prevent the device being worn by rubbing on counters, etc., and also that the coins may be piled one on another steadily. This edge is raised by a very beautiful piece of mechanism called a Milling Machine, the invention of Mr. Peale, and vastly superior to any other in use.

Some 20 or 30 planchets are placed in one of the brass vertical tubes, of which there are three, for different sized coins. At the bottom of the tube the lowest planchet is struck by a revolving feeder, which drives it horizontally between the revolving steel wheel (marked A in the engraving) on one side, and the fixed segment (marked B) on the other. The segment is on the same curve as the wheel, though somewhat nearer to it at the further end. The planchet is caught in a narrow groove cut in the wheel and segment, and the space being somewhat less than the diameter of the planchet the edge is crowded up about the thirty-second part of an inch. The planchet makes four revolutions when it reaches the end of the segment, and being released from the grooves falls into a box below. The edge is perfectly smooth, the fluting or "reeding," as it is termed, being put

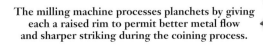

The milling machine processes planchets by giving each a raised rim to permit better metal flow and sharper striking during the coining process.

on in the process of coining. The work is so nimbly performed that about one hundred and twenty double eagles, or five hundred and sixty half dimes, can be milled in one minute. This is a vast improvement on the English milling machine, worked by hand, and operating on but two planchets at a time.

The planchets being milled are called blanks. They are very dirty and discolored by the processes they have undergone, requiring to be polished before coining. This is done in the Whitening Room, and an exceedingly hot place it is. Sometimes in summer the thermometer will indicate 120°, though the tall man by the furnace declares that it is often at 175°. The room is too small and poorly ventilated for the use to which it is put. There are two furnaces for annealing the blanks, they being placed in a copper box, with a cover sealed on air-tight with clay. Boxes and blanks are heated red-hot, and the blanks tipped into a vat containing a weak solution of sulphuric acid and water, to cleanse them. The stream of water in the other vat is hot, in which the blanks are washed free from the acid, leaving them a beautiful white color, almost like silver.

The curious copper machine, looking like a large revolving squirrel-cage, is the drying-drum. About half of it is a tight copper drum, into which the blanks from the hot water are placed with a quantity of basswood saw-dust. Steam is introduced through the axis to heat the interior, and the drum made to revolve, causing the blanks to roll among the heated saw-dust and dry themselves. Basswood dust is used because of its freedom from sap, pitch, or gum of any kind. It is extremely pure. In the language of one of the men, "It ain't got nothing about it but just wood."

When the blanks are dry a door in the end of the drum is opened, allowing them to fall into the sieve, where they tumble about, the dust gradually sifting out, leaving the coin clean. To brighten them they are kept revolving for an hour, and the friction of one upon another gives them a beautiful lustre. It is in this way that pins, brass buttons, and the like are polished.

They are taken from the drying-drum, and heated in a large warming-pan, with steam-pipes running under it, until Jessie would hardly hold a handful for the gift of them, though they are all double eagles. They are now ready for coinage, and that prettily painted truck is taking a couple of hundred thousand dollars to the other room for the purpose.

Dies

Before examining the coining we must visit the Die Room, to learn how the dies are made. A coin has an impression on both sides, requiring, of course, a die for each. These are to be made with extreme care, to be of the finest workmanship, and all exactly alike. Their manufacture is one of the most important operations in the Mint.

Look at the bas-relief of Liberty on one side of a coin. It would be exceedingly difficult to design this in hard steel and of so small a size; so they first make the design in wax, probably six times as large as the coin, by which means the beautiful proportions can be obtained. From this a brass cast is taken, and reduced on steel to the size of the coin by a transfer or reducing lathe.

The brass cast is fastened to the large wheel at the right-hand side of the lathe. On the small wheel to the left of the cast is fastened a piece of soft steel, on which the design is to be engraved. Both of these wheels revolve in the same way and at the same speed. There is a long iron bar or lever fastened by a joint to an iron support at the extreme left, which runs in front of the two wheels. A spring at the upper end draws it in toward the wheels. Fastened to the lever is a pointed steel stub, which touches the cast. A very sharp "graver" is fastened to the lever below, which touches the steel. The wheels revolve, and the stub, when it is pushed back by the heavy relief of the cast, forces back the lever, which draws back the graver, and prevents it cutting the steel. So where there is a raised place in the cast the graver is prevented from cutting into the steel, but where there is a depression in the cast the graver cuts the same in the steel.

As the lever is jointed at the left, the nearer the graver is placed to that end the less motion it will have. So that the distance of the steel from the joint regulates the proportion of the reduction from the cast.

After the graver has cut one small shaving around the steel, a screw is turned, which lowers the right end of the lever slightly, just enough to allow the graver to cut another shaving, and the stub to touch the cast a very little further from the centre. Thus the graver cuts a very little at a time; but the work is cut over several times, until the design is sufficiently blocked out. This machine will not finish off the die perfect enough to use; but it reduces the design in perfect proportion, and performs most of the rough work. The original dies for coins being now made, the lathe is used mostly for medals, of which a great many are struck, by order of Congress, for various purposes. A very fine one was presented to the Japanese while they were in this country. There is now in the machine a cast of Washington's bust, merely to show how the cast is placed.

The Japanese Embassy visiting the Philadelphia Mint on June 13, 1860. They were presented with medals and a cased set of Proof coins from the cent to the double eagle.

Transfer lathe for making hubs, from which master dies are made.

After the die comes from the lathe it is carefully finished off by hand, and when all polished is a beautiful piece of work. It is still very soft, requiring to be hardened before it can be used, which is done by heating it very hot, and holding it under a stream of water until cold. The relief is exactly like the coin—that is, the device is raised as in the coin. It will not do to use this in stamping, as it would reverse the appearance on the coin. Therefore this "hub," or "male die," as it is named, is used only to make other dies.

Coining dies.

Round pieces of very soft steel, a little larger than the die, are smoothed off on the top, the centre being brought to a point a little higher than the sides. It is placed on a solid bed, under a very powerful screw-press, and the hub placed on top of it—the centre of the hub on the point of the steel, like a seal on the sealing-wax. The screw is turned with great force by several men, and presses the hub a little into the steel. It is necessary to have the steel higher in the centre, as if the centre impression is not taken first, it cannot be brought out sharp and distinct. The steel is softened again by being heated and allowed to cool slowly, and the operation is repeated. This is done several times, until the whole impression is full and distinct. If there is any little defect it is rectified with the engraver's tool. The surplus steel around the edge is cut off, and the date put in by hand, when it is hardened and ready for use. The date is not cut on the hub or on the first die— which is called a "female"—as perhaps the hub will last for two years, and the date cannot be altered. This die is never used to stamp with, but preserved, so that if the hub breaks it can be used to make another.

The dies for use are prepared in the same way. About thirteen hundred a year are made for the various branch mints, and those for the New Orleans Mint were sent on just before the State seceded, which the authorities have not yet had time to return. Sometimes a die will wear for a couple of days, and again they will break in stamping the first coin. Steel is treacherous, and no dependence can be placed in its strength. As nearly as can be ascertained their cost is $16 a pair.

The Coining Room

We will now enter the Coining Room, a light, airy hall, filled with brightly polished machinery, kept as clean as the milk-pans in a New England dairy. A passageway in the middle of the room is separated from the machines on both sides by a neat iron fence. The quantity of gold and silver lying about would make it unwise, especially in these times, to allow strangers to mingle among it. Visitors can see everything from this passageway, but the pleasure of handling is denied.

There are two styles of coining presses, both working on the same principle, but some more compact and handsome than the others. They are the invention of Mr. Franklin Peale, the plan being taken from the French press of Thonnelier's. It seems to be as nearly perfect as anything can be. There are eight presses, all turned by a beautiful steam engine at the further end of the room.

The power of the press is known as the "toggle," or, vulgarly, "knee" joint, moved by a lever worked by a crank. The arch is a solid piece of cast iron, weighing several tons, and unites with its beauty great strength. The table is also of iron, brightly polished and very heavy. In the interior of the arch is a nearly round plate of brass, called a triangle. It is fastened to a lever above by two steel bands, termed stirrups, one of which can be seen to the right of the arch.

The stout arm above is also connected with the triangle by a ball-and-socket joint, and it is this arm which forces the end of the lever above by a joint somewhat like that of the knee. When the crank lifts the further end of the lever it draws in the knee and forces down the arm until it is perfectly straight. By that time the crank has revolved and is lowering the lever, which forces out the knee again and raises the arm. As the triangle is fastened to the arm it has to follow all its movements.

Under the triangle, buried in the lower part of the arch, is a steel cap, or technically, a "die stake." Into this is fastened the reverse die. The die stake is arranged to rise about the eighth of an inch, but when down it rests firmly on the solid foundation of the arch. Over the die stake is a steel collar or plate, in which is a hole just large enough to allow a blank to drop upon the die. In the triangle above the obverse die is fastened, which moves with the triangle; and when the knee is straightened the die fits into the collar and presses down upon the reverse die.

The Coining Room. Visitors are allowed to watch behind iron railings.

Improved coining press.

Just in front of the triangle will be seen an upright tube made of brass, and of the size to hold the blanks to be coined. The blanks are examined by the girl in attendance, and the perfect ones are placed in this tube. As they reach the bottom they are seized singly by a pair of steel feeders, in motion as similar to that of the finger and thumb as is possible in machinery, and carried over the collar and dropped upon the die. The knee is straightened, forcing the obverse die to enter the collar and press both sides of the blank at once. The sides of the collar are fluted, and the intense pressure expands the blank about the sixteenth of an inch, filling the collar and producing on the coin the fluted or reeded edge. It is put on to prevent any of the gold being filed away.

After the blank has been dropped upon the die, the feeders slide back on the little platform extending in front of the machine, in readiness to receive another. The knee is bent, which raises the die about half an inch above the collar. The die stake is raised at the same time, so as to lift the newly-born coin from the collar, and the feeders coming along with another blank, push the coin over into a sloping channel, whence it slides into a box underneath.

The pressure on the double eagle is about 75 tons; yet so rapid are all these complex motions that 80 double eagles are coined in a minute. The smaller pieces, such as dimes and half dimes, are coined at the rate of 150 a minute. While usually only 75 tons pressure are applied, the large presses will stand a strain of 150 tons. Sometimes government and other large medals are struck, which require this heavy power.

It is a beautiful sight, as the bright glistening coins drop in a golden stream, with the peculiar metallic clink so pleasant to hear. It is as pretty a cascade as one often sees.

The chief coiner's room where coins are checked and put into cloth bags prior to being delivered to the Mint treasurer.

After being stamped the coins are taken to the chief coiner's room, and placed on a long table, the double eagles in piles of 10 each. The gold coins—as small as quarter eagles being counted, and weighed to verify the count—are put up in bags of $5,000 each. The $3 pieces are put up in bags of $3,000, and $1 pieces in $1,000 bags. The silver pieces, and sometimes small gold, are counted on a very ingenious contrivance called a "counting-board," somewhat resembling a common washboard. They are all subsequently weighed, however, to verify the correctness of the counting.

For the various duties of the Mint there are about 200 persons employed as clerks, workmen, etc.—say 140 [men] and 60 women—the number depending, of course, upon the amount of work to be done.

At the start of the year, the director of the Mint was James Ross Snowden, who served from June 1853 to April 1861. Since May 1861 James Pollock has been director.

THE CIVIL WAR

By the summer of 1860 slavery had been a burning social issue for a long time. When Harriet Beecher Stowe's novel, *Uncle Tom's Cabin*, was published in 1852 it became a runaway best seller second only to the Bible in terms of copies in print. Describing the life of Uncle Tom and other slaves suffering on a plantation and their effort to flee to the safety of the North, it tugged on the heartstrings of many people. In 1860, four candidates jousted for the presidency—Abraham Lincoln, John C. Breckinridge, James Bell, and Stephen A. Douglas. The winner with 40% of the popular vote and 180 electoral votes was Lincoln on the Republican ticket, a dedicated and vocal abolitionist. In 1860 the decennial census put the United States population at 31,443,321, or twice that of 1820.

The prospect of Lincoln in the White House was unacceptable to many Southerners. South Carolina seceded from the Union soon afterward. In early 1861 Mississippi, Florida, Alabama, Georgia, Louisiana, Texas, Virginia, Tennessee, Arkansas, and North Carolina withdrew as well. Delegates from seven of these states met in Montgomery, Alabama, on February 4 to form the government of the Confederate States of America. Jefferson Davis was named president on a provisional basis; in October a general election confirmed the choice. Arkansas, Tennessee, North Carolina, and Virginia soon joined as well. It was the desire of many in the South that the Confederacy would exist and prosper as a new country, maintaining slavery, but trading with the North. Federal facilities including the New Orleans, Charlotte, and Dahlonega mints came under Confederate control.

Co-existence did not happen. On April 11, 1861, Confederate batteries at Fort Moultrie began shelling Fort Sumter in the harbor of Charleston, South Carolina, reducing it to ruins. The War of the Rebellion, later known as the Civil War, began a few days later. Northerners envisioned an easy win. President Lincoln called for men to enlist for 90 days, by which time it was believed that the war would surely be over, and the seceding states would return to the Union. This did not happen either. The 90 days passed, and it was not until July 21 that the first major battle took place—at Bull Run in Manassas, Virginia, not far from the nation's capital.

COINS AND COMMERCE DURING THE WAR

By late December the outcome of the war was uncertain. Both sides claimed victories, and there was the possibility that Great Britain, the largest market for Southern cotton, would back the Confederacy. Concerned citizens withdrew gold coins from banks. By early 1862 all had disappeared from circulation. In March the Union, short of funds, authorized Legal Tender Notes—currency that was only redeemable with other bills and not at par with coins. Silver coins began to be hoarded, and by late spring all were gone. Although they had little intrinsic value, copper-nickel cents disappeared by the second week of July. For the first time in history, there were no coins in circulation! On July 17 ordinary postage stamps were made legal tender, followed by Postage Currency, and then in 1863 by Fractional Currency. Private small-denomination bills, encased postage stamps, and countless millions of cent-size tokens facilitated commerce.

The Philadelphia Mint continued to strike silver and gold coins, but in reduced quantities. These were available only at a premium in terms of Legal Tender notes. For a time it took over $200 in currency to buy $100 in coins from the Mint or a bullion broker.

In the meantime, on the West Coast silver and gold coins circulated at par, and any Legal Tender Notes brought into the area were accepted only at a deep discount in inverse ratio to the premium paid for coin in the East. The State Constitution of California, passed in 1850, forbid the issuance of currency by banks there.

The value of Legal Tender Notes went from bad to worse, and on July 11, 1864, a $10 paper note was worth just $3.90 in silver coins. In the South a $10 Confederate States of America note was worth just 46¢ of federal coins.

The war ended in April 1865. Treasury officials thought that normal monetary conditions would be resumed, but the public remained wary of the government's financial stability. Monetary reserves were low, and circulation was flooded with Legal Tender Notes that could not be exchanged at par with coins.

In 1867 Dr. Richard Henry Linderman became the latest director of the Mint. Following his illustrious predecessors in office, he continued the secret minting of restrikes, rare die combinations, patterns, and the like for secret sale. No records were kept. Linderman found the Mint to be in need of significant repairs. Coinage operations ended on August 1 and did not resume until renovations were complete on October 1. The Engraving Department began using a new portrait lathe to reduce models to hubs and dies. The invention by J.C. Hill of London was superior to the Contamin portrait lathe in use since 1836.

SOURCES OF GOLD

In 1867 the *Annual Report of the Director of the Mint* included information on the grand total of all native gold deposits at the mints and assay offices since 1804. Gold from melted coins domestic and foreign and foreign deposits were not included. Although the figures appeared to be precise, in fact most combined facts with estimates. The total of $682,941,318.65 was from these figures to which gold parted from silver during the refining process at the Mint and about $12 million in ingots was added:

Alabama: $202,172.

California: $597,899,964. This state accounted for the vast majority of gold deposits and was still a strong gold producer in 1867.

Colorado: $13,382,232. This territory was still being explored, and rich deposits continued to be discovered.

Dakotas: $9,759.

Georgia: $7,000,440. From the 1820s onward this state was an important producer of gold.

Idaho: $13,164,433. In 1867 this state was a leading active producer of gold.

Montana: $13,867,876. In 1867 this state was a leading active producer of gold.

Nebraska: $3,645.

Nevada: $171,926. By this time the state had registered $5,171,476 in *silver* deposits, far more than any other state. California was next with several million dollars (figure not given) in silver obtained as a byproduct of gold refining.

North Carolina: $9,344,933. From the 1820s onward this state was an important producer of gold.

Oregon: $9,203,014.12. In 1867 this state was a leading active producer.

Tennessee: $81,407.

Utah: $82,886. Although the Mormons produced gold coins dated 1849, 1850, and 1860, nearly all of the metal came from California (see chapter 12).

Vermont: $614.

Virginia: $1,580,389.

Washington: $61,260.

In Paris a monetary convention was held in 1867 and was attended by representatives from Austria, Baden, Bavaria, Belgium, Denmark, Spain, France, Great Britain, Greece, Italy, Holland, Portugal, Prussia, Russia, Sweden and Norway, Switzerland, Turkey, Württemburg, and the United States. It was recommended that a universal coin be adopted, about the size of the American half eagle, to be acceptable in many nations. The idea of an international coin would result in various pattern gold coins being struck at the Mint in 1868, 1874, 1879, and secretly for numismatic sale in 1880. These attempts disregarded the seemingly obvious fact that exchange rates between countries varied continually.

On January 1, 1869, engraver James B. Longacre died. He had been at the Mint since 1844. William Barber succeeded him in the post and would serve until 1879. In 1869 speculation in gold bullion disturbed the market for that metal. Silver coins, absent from Eastern and Midwestern commerce since the spring of 1862, might circulate again if lighter-weight coins were made. Standard Silver patterns, as they were called, were struck in many varieties beginning in 1869, but no such coins were ever made for commerce. Countless thousands were struck in various metals and die combinations for sale to numismatists, mostly privately.

In his *Annual Report* for 1870 director James Pollock commented:

> Mint work is necessarily hindered and restricted by the continued suspension of specie payments. We are doing less than was done many years ago, when there was a much smaller population and far less wealth. Certainly there is no need of creating any more coining establishments.[61]
>
> Emerging from a tremendous Civil War which shook every social interest to the very foundation, it is no wonder that our currency continues in an abnormal condition. Most of our people rarely get the sight of a gold or silver coin. . . .[62] The people at large will never give up the idea that the real money is made of gold and silver; made of definite weight and fineness, and certified by government stamp.

In 1871 a new denomination, the commercial dollar weighing 420 grains, was proposed. This evolved into the trade dollar (see listings on page 152) struck from 1873 to 1885. The Coinage Act of February 12, 1873, abolished the two-cent piece, silver three-cent piece, half dime, and silver dollar, and created the trade dollar. The bill also created a bureau within the Treasury Department in Washington, headed by the director of the Mint, to supervise all of the assay offices and mints in America. Previously, the director's office was within the Philadelphia Mint. The act also put an end to the ability of citizens to have silver bullion coined at the Mint—this in advance of the expected drop in value of silver following the discovery of the Comstock Lode.

PARITY FOR COINS AND PAPER MONEY

In 1876 the Centennial Exhibition was held in Philadelphia. Commemorative medals were struck at the Mint. It was not until April 20 of this year that silver coins achieved parity with paper money, by that time in exchange for Fractional Currency. The Act of July 22, 1876, provided that Legal Tender Notes could be exchanged for specie, at which time the Treasury turned loose a flood of silver. By autumn commercial channels were awash in such coins. In the same year, Mint Director Henry R. Linderman felt that the current designs produced by Mint engraver William Barber were substandard from an artistic viewpoint. He recruited a British engraver, George T. Morgan, who came to America and soon produced an illustrious suite of pattern half dollars.

One of several pattern half dollar designs made by George T. Morgan in 1877. (Judd-1512)

The Treasury anticipated that there would be a tremendous demand for new silver coins, and production was ramped up to high levels. The opposite happened. Long-hoarded silver came out of hiding, and there were so many silver coins that mintages of the denominations from the dime to the half dollar dropped precipitately and would remain low through the 1880s.

In July 1877 *Banker's Magazine* reported on the Treasury's effort to distribute silver coins in quantity:

> The Treasurer of the United States gives notice through a circular, that he will cause to be forwarded from some Mint of the United States, to any point in the United States reached through established express lines by continuous railway or steamboat communication, subsidiary silver coin in return for national bank notes forwarded to him for redemption in sums of $1,000, or any multiple thereof not exceeding $10,000. The expense of transportation will be paid by the Mint.
>
> The total disbursements of silver to date have been as follows: For currency obligations, $12,250,000; for fractional currency, $19,100,000; of Fractional Currency there remains outstanding $22,500,000. It is stated that Secretary John Sherman claims the right to issue silver to the amount of $54,000,000, the extra $4,000,000 being the amount on hand in the Treasury previous to the passage of the act fixing the limit of silver issue at $50,000,000.

In 1877 automatic weighing machines were introduced at the Philadelphia Mint (which used machines made by Seyss & Co., of Atzgersdorf, near Vienna, with a capacity of 120 blanks per minute) and the San Francisco Mint (which used machines by Napier

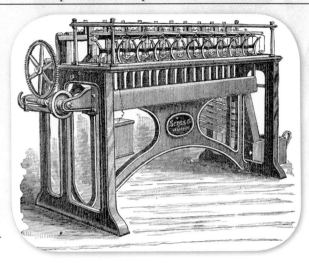

Seyss automatic planchet-
weighing machine.

& Son of London, with a capacity of 40 blanks per minute). These were mainly used on half dollar and trade dollar planchets. The economy experienced a setback this year, extending into 1878, and tens of thousands of businesses failed.

On December 17, 1878, gold coins achieved parity with paper money. The Treasury anticipated that gold coins would become common in commerce. This did not happen. Citizens, knowing they could exchange paper for gold or silver kept using paper bills. By year's end Morgan silver dollars, first minted in March, were piling up in vaults by the millions.

THE SILVER QUESTION

The year 1876 saw the rise of what was called the "Silver Question." In view of the continuing fall of the price of that metal due to expanded output in America and vast quantities of silver available overseas, the Western mining and political interests were in a quandary. What to do to raise the price of the metal they were newly rich in?

Many articles were written, a conference was held in Saratoga Springs, and other actions were taken. The Coinage Act of 1873 was now painted as the root of the market evil, despite the fact that the act was written as a *response* to the predictable decline in the value of silver due to the mining of the immense Comstock Lode. Many argued that it had been passed hastily and had been poorly constructed, especially the clause eliminating the regular silver dollar. The fact that more silver trade dollars were made after that point was ignored. Now called the "Crime of 1873," the bill was portrayed as a wolf in sheep's clothing. Senator John Sherman reminded people that the bill had been more carefully studied than just about any other piece of legislation he had seen in his many years of congressional service. The public ignored him. History was rewritten so effectively that even today historians who do not study the matter in depth believe this act caused the silver market to fall.

The movement continued. In 1878 the Bland-Allison Act (see silver dollars on page 150) was passed under the influence of Western mining and other interests. The trade dollar was discontinued, and the 412.5-grain Morgan silver dollar was instituted, to be coined continuously through and after the end of the second Mint. The prospect that American debtors might pay their obligations in silver dollars that were not worth their melt-down value caused a great rush to import double eagles.

The Mint in 1880.

MINT PROCESSES IN 1880

The issue of June 19, 1880, *Harper's Illustrated Weekly* included "Coining Silver Dollars at the Philadelphia Mint," selected images from which are shown below.

Melting precious metals preparatory to casting ingots.

Vault for the storage of silver ingots in a year in which millions of silver dollars were coined.

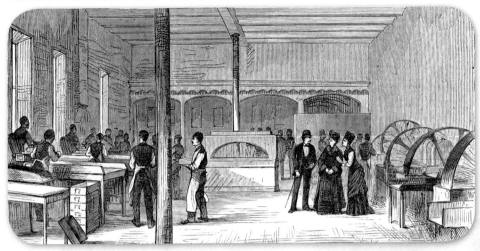

Room where ingots are rolled into thin strips, from which planchets are cut.

Punching planchets from thin metal strips.

Putting raised rims on blank planchets with a milling machine.

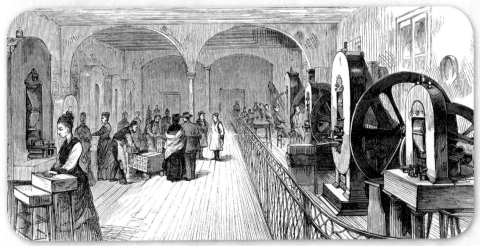

Coining Room at the Mint.

Counting coins by stacking them in piles.

GOLD AND SILVER COINS

On October 1, 1880, Mint Director Horatio C. Burchard gave this estimate:

> The total amounts estimated as in the country are $369,881,000; gold and $149,799,335 silver. Of these amounts are $67,204,293 gold in the Treasury, and $302,676,707 in circulation, or held by banks; and $72,454,600 silver in the Treasury and $77,344,735 in circulation. In addition the Treasury holds $68,040,540 gold bullion and $5,557,759 silver bullion, which the Mints are turning into coin as rapidly as the facilities will admit. This is a larger amount of specie than has ever before been in the country. The increase of gold bullion is from imports of foreign gold during the past year.

Continuing a recent trend the commercial and banking interests of Europe in the 1880s were becoming increasingly concerned that the American government would settle its accounts in silver rather than gold. In 1891 the average cost of minting silver and gold coins at the Philadelphia Mint was $0.0175+ or nearly two cents. This was the lowest figure for the four mints in operation at the time.

Visitors could watch the striking of coins from behind a wrought iron railing. Six tour guides were on hand to assist the public.

An 1881 Report on the Second Mint

The condition and status of the mint were subjects of congressional discussions in 1880 and 1881. From a report:

The lot on which the building is situated has a front of 148 feet on Chestnut Street and extends 204 feet on Juniper and 148 feet on Market Street, and has an area of 30,192 square feet.[63] The building itself occupies 21,360 square feet, and was originally built in the form of a hollow square, with an interior court. This court originally afforded both light and ventilation to rooms facing the courtyard, but the necessities for more room have compelled the superintendent to erect within this court additional buildings, until nearly every foot of this square is now occupied by workshops and depositories of the precious metals, cutting off both light and ventilation and rendering many of the inner rooms dark and unhealthy for the workmen.

The basement was occupied chiefly by the six boilers, also by steam engines, several vaults and rooms for the painter, carpenter, and blacksmith shops, plus a washing room for sweepings.

The first story contains, the deposit or weighing room, the deposit melting room, rooms of the melter and refiner, the rolling room, with annealing furnaces, the cleaning room, and coining room, and some small offices.

The second story is occupied with offices of superintendent, coiner, assayer, &c.; a cabinet of specimen coins of different nations, the adjusting room, the separating rooms, engraving room, &c.

The Adams Express Company occupies a small temporary wooden structure lately built in the inner court. The superintendent has also been compelled recently to put up an additional frame room near the stack, away from the jar of the machinery, for the adjustment of the scales.

Nearly all the rooms, which were designed when the work to be performed was far less than at present—only a little over $3,000,000 being coined in 1833, while during the fiscal year ending June 30, 1879, there were coined at this mint $23,552,032.50, and during the month of January last there were coined $5,000,000 in gold and $1,350,000 in standard silver dollars and 5,856,000 pieces of minor coins—

The operator took blank planchets from a bin to her left and dropped them into the top of a brass feeding tube. From there planchets were automatically fed into the press one-by-one.

Visitors at a coining press.

The Mint Cabinet was a prime attraction for visitors.

are small, and now so crowded with machinery that much of the work is done under great disadvantage. The deposit room, for instance, is but 12 x 16 feet, and contains two large scales and often piles of bullion, besides being encumbered with trucks and other appliances for handling and working the metals. The deposit melting room, in which there are four furnaces, is constricted and poorly ventilated. The rooms occupied for melting and refining are still more crowded. In this apartment are eleven furnaces, which, with the bad ventilation, resulting in part from roofing over the inner court, make the temperature of the room almost unendurable in hot weather.

The rolling room is so crowded with machinery that the workmen are more or less in one another's way, and work in consequence is done at great disadvantage. In this room is a large engine which drives the machinery, and on one side are the annealing furnaces, which, for want of better ventilation, raise the temperature of the room to a degree uncomfortable and unhealthy for workmen. In the cleaning room adjoining the temperature in summer runs up to 136°F.

The coining room on the east side of the building is the most comfortably and the most conveniently arranged for work of any in the building, but even here there is no room for more presses, if they should be needed, and not even room to set up and work a valuable and costly machine for separating light, standard, and overweight coins.

Of course, the capacity of the mint depends largely upon the proper proportioning of the various parts, as furnaces, rolls, presses, &c. Your committee was informed by the superintendent that the capacity for melting, refining, rolling, and annealing was not equal to the present capacity of other parts of the mint. But there is no room to add other furnaces or rolls. . . .

Despite obvious inadequacies, no major changes were made at the time.

INTO THE 1890s

On September 20, 1890, this policy of Mint procedures became U.S. Code, Title 31. Section 270:

The engraver shall prepare from the original dies already authorized all the working dies required for use in the coinage of the several mints, and, when new coins, emblems, devices, legends, or designs are authorized, shall, if required by the Director of the Mint, prepare the devices, models, hubs, or original dies for the same.

The Director of the Mint shall have power, with the approval of the Secretary of the Treasury, to cause new designs or models of authorized emblems or devices to be prepared and adopted in the same manner as when new coins or devices are authorized. But no change in the design or die of any coin shall be made oftener than once in twenty-five years from and including the year of the first adoption of the design, model, die, or hub for the same coin: Provided, that no change be made in the diameter of any coin.

But the Director of the Mint may nevertheless, with the approval of the Secretary of the Treasury, engage temporarily for this purpose the services of one or more artists, distinguished in their respective departments of art, who shall be paid for such service from the contingent appropriation for the mint at Philadelphia.

In 1891 $2,824,146 in gold coins was imported while exports rose to a frightening $67,704,900. The *Mint Report* told this:

In the summer of 1890 a movement of gold from this country occurred [and] aggregated in a period of less than two months the sum of $15,672,982. In February of this present year, 1891, another movement of gold to Europe commenced which did not cease until near the close of July, exceeding in amount the exports of the fiscal year 1889, and causing by far the most serious loss which this country has sustained for many years. The total amount exported from the port of New York was $70,223,494.31.

In the meantime, gold coins by the millions, mostly in the form of double eagles, continued to be exported. In 1891 the Liberty Seated design—inaugurated on the silver dollar of 1836 and later used on the half dime, dime, twenty-cent piece, quarter, half dollar, and silver dollar—was discontinued. The Liberty Head design by Chief Engraver Charles E. Barber made its appearance in 1892.

In 1892 a powerful hydraulic press was installed in the Medal Department. Rated at up to 2,000 tons of pressure, it was used to strike medals.

In 1893 the World's Columbian Exposition opened to the public in Chicago, a year behind schedule. The Mint struck commemorative half dollars for the event in 1892 and 1893 and a quarter in 1893, the first in what would become a long series of commemoratives. By August 1894 the gold reserve was only $52,200,000, a crisis averted by issuing $50,000,000 in interest bearing gold bonds. This resulted in an influx of coins to the point at which the reserves were about $61,000,000 in October of that year. In November another $50,000,000 in such bonds was issued, increasing the reserve to $105,424,000. This did nothing to stanch the drain of gold, and by year's end the reserve was down to $86,000,000.

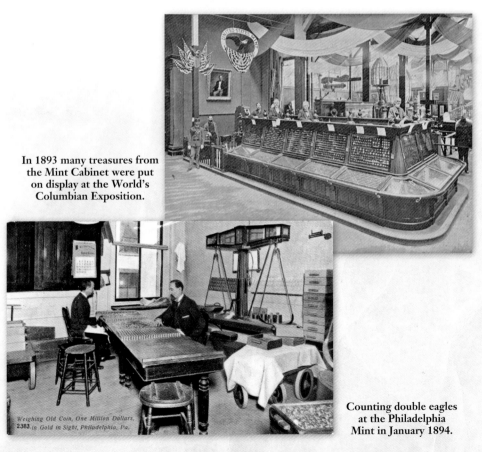

In 1893 many treasures from the Mint Cabinet were put on display at the World's Columbian Exposition.

Weighing Old Coin, One Million Dollars, 2383 in Gold in Sight, Philadelphia, Pa.

Counting double eagles at the Philadelphia Mint in January 1894.

The crisis deepened, with no end in sight. On February 1 President Cleveland and Secretary of the Treasury John J. Carlisle met with financier J.P. Morgan to attempt to solve the problem. By February 9, 1895, only $41,393,212 was held by the Treasury for the redemption of its obligations payable in gold. Arrangements were quickly made, and J.P. Morgan & Co. and August Belmont & Co. of New York, in cooperation with N.M. Rothschild & Sons of London, made an offering aimed at foreign investors who held gold. 30-year bonds, at least half of which were required to be sold in Europe, yielded 4% interest. They sold out in 22 minutes!

$200,000 in double eagles in 1895, a year of crisis.

The March 1895 issue of *The Numismatist* told how Chief Engraver Charles E. Barber prepared coinage designs:

> Though various materials may be employed, Mr. Barber uses a wax composition for designing coins. He takes a quantity of an extremely pure quality of beeswax, with which he mixes a small proportion of resinous gum and a little of the finest vermilion. The mixture, after being melted and stirred so as to make it homogenous, is poured into cold water. When cool, it is thoroughly kneaded like so much dough. Having thus made ready for business, the designer draws on a slate a circle a foot in diameter. This is the outline of the coin model, which is made on a large scale for convenience. The circular space is then covered with a layer of the wax mixture, perhaps three quarters of an inch in thickness and on this surface the design in relief is made with delicate modeling tools of boxwood. The method is much the same as that employed for modeling in clay, but the material is finer and smoother. If hair is to be represented a brush of fine bristles is something employed, while the over-long thumbnail of the operator serves occasionally as a supplementary instrument, being a sensitive organ of touch.
>
> From the finished wax model is made an electrotype, and from the latter is produced the die for the coin by a simple process. A machine is used which operates on the principle of the 'pantograph' employed in drawing. While a blunt point is guided by hand over every curve and intricacy of the wax medallion, another point, tipped with a diamond, revolves rapidly and cuts away the surface of the steel disk, which is just the size of the future coin. Thus an exact reproduction is obtained of the original model, but in it every course detail of the wax medallion becomes fine and delicate. This is the cameo—that is, a bas-relief. The die, of course, must be in intaglio. To get the intaglio from the cameo is easy enough. The cameo, which is of hard steel, is squeezed against a disk of soft steel under enormous pressure. The soft steel disk, having received a perfect impression, has only to be tempered, and it is then ready to serve as a die.

In the 1896 presidential campaign Democratic candidate William Jennings Bryan, representing silver interests and promoting the unlimited coinage of that price-depressed metal, faced Republican William McKinley. Bryan lost by a large margin, the Silver Question was largely moot after that time. During the 1890s there were rich gold strikes in Cripple Creek, Colorado, and, starting in 1896, in the Yukon Territory in Canada.

With gold streaming in from Cripple Creek, the Klondike, and other scattered areas, and with a large flow of double eagles and other coins returning from Europe, the Treasury had large gold reserves by 1899. During the year, the face value of gold coins at the mints totaled $111,344,220, setting a record and crossing the $100,000,000 mark for the first time.

There were gold coins totaling $257,306,000 snug in Treasury vaults, while National Banks held an estimated $203,701,000, and the public and others owned $293,388,000. The grand total of gold in America, including bullion, was pegged at $897,473,000. In the meantime, over 400 million unwanted silver dollars were stored in various mint and Treasury vaults.

In 1899 a coining press, made in the shops of the Virginia & Truckee Railroad in Nevada and used to strike silver dollars at the Carson City Mint, was shipped to the Philadelphia Mint. It had remained idle after the Nevada facility has ceased coinage in 1893. Later the press was moved to the third Philadelphia Mint, then in 1963 to the San Francisco Mint, then in 1958 back to Carson City, where the old Mint building was home to the Nevada State Museum (opened in 1942).

On March 14, 1900, Congress passed the Gold Standard Act. Although gold coins had maintained very close to their full weight and intrinsic value since the first federal half eagles were struck in the summer of 1795, gold had not been official as a fixed standard. The new legislation made it so. The act was a non-event because most citizens thought America had been on such a standard for a long time.

In 1902 the old building was sold for $2 million dollars.

COINAGE OF THE SECOND MINT, 1833–1901

The second Philadelphia Mint in its long history saw many changes, including conversion to striking coins by steam power, innovations in making dies, and more. Beginning in 1838 it furnished dies for the branch mints, including New Orleans (first in 1838), Charlotte (first in 1838), Dahlonega (first in 1838), San Francisco (first in 1854), and Carson City (first in 1870).

The second Philadelphia Mint
at the turn of the 20th century.

Packing cloth bags filled with coins
into wooden barrels for shipment.

Half Cents

Half cents had piled up in storage at the Mint after coinage of the denomination resumed in the 1820s. Between 1829 and a report of June 19, 1833, 12 kegs with $1,170 face value, or 234,000 half cents, had been melted for use as alloy in the silver and copper coinage. Nine kegs (141,000 coins) were still at the Mint as of June 19.[64] As to why the Mint struck more half cents in 1835, when several-hundred thousand were still on hand from earlier years, is a mystery. After this time no half cents were struck for circulation until 1849. In the meantime Proofs were made for cabinet purposes.

In 1849 circulation strikes were again made, these having large date numerals, in contrast to Proofs with small numbers. Production for circulation use continued through early 1857, except for 1852 when only Proofs were made. Although 35,180 half cents were struck in 1857 (delivered on January 14), Mint records seem to indicate that only 10,000 were shipped. It is not known if all were dated 1857. Requests for coins, including from collectors, were honored by the director until February 28, 1857, after which the remaining coins were melted.[65]

DESIGN TYPES AND DATES MINTED

Classic Head: 1833, 1834, 1835, 1836.

Braided Hair: 1840, 1841, 1842, 1843, 1844, 1845, 1846, 1847, 1848, 1849, 1850, 1851, 1852, 1853, 1854, 1855, 1856, 1857.

Large Cents

Coinage of the Matron Head design continued at the new Mint, this year with a production of 2,739,000 pieces. As before, copper cents produced a profit, unlike gold and silver. As a result, coinage was continuous and generous through January 1857 when the last of these were coined. Late in 1834 the government of Venezuela ordered 100,000 cents, and these were shipped. In 1835 the government ordered a million more cents. It is not known how many, *if any*, were shipped of the 1835 coins.[66] If any were, likely they consisted of, say, up to a half dozen different die varieties, perhaps accounting for the rarity of certain of these varieties today.

In March 1836, following the ceremonial striking of medalets on March 23, a steam press was first employed for coinage, with the cent denomination. In 1837 a new portrait was introduced, attributed to Christian Gobrecht. Further modifications were made to the design from that time through 1839, when the Braided Hair type was introduced. In 1851 coinage of large cents hit a record high of 9,889,707, a figure not closely approached by any other to that point. Large cents were last made in January

1857 to the extent of 333,456 pieces. Most probably, many were held back at the mint and melted. It is unlikely that even half the mintage was released. Craig Sholley suggests 140,000 coins may have left the Mint.[67]

In 1868—by which time the Mint had been creating rare patterns, restrikes, and other issues and secretly filtering them into the numismatic market—a large copper cent was created of the exact Braided Hair type last regularly issued in 1857. It is estimated that 12 to 15 were struck in copper.[68]

DESIGN TYPES AND DATES MINTED

Liberty Head, Matron Head: 1833, 1834, 1835.

Liberty Head, Matron Head Modified: 1836; 1837; 1838; 1839, 1839 Over 1836; 1839.

Braided Hair: 1839, 1840, 1841, 1842, 1843, 1844, 1845, 1846, 1847, 1848, 1849, 1850, 1851, 1852, 1853, 1854, 1855, 1856, 1857, 1868 (created as a rarity).

Small Cents

The Act of February 21, 1857, abolished the half cent and cent and provided for new cents of small diameter, made of copper-nickel and weighing 72 grains. Flying Eagle cent patterns of this format were made in 1856, followed by circulation quantities in 1857 and 1858. In 1859 the Indian Head cent was introduced. With some modifications, including changing to bronze in the spring of 1864, coinage was continuous through 1901.

DESIGN TYPES AND DATES MINTED

Flying Eagle: 1857; 1858, 8 Over 7; 1858.

Indian Head, Variety 1, Laurel Wreath Reverse: 1859.

Indian Head, Oak Wreath With Shield Reverse:
Variety 2, Copper-Nickel: 1860, 1861, 1862, 1863, 1864.
Variety 3, Bronze: 1864; 1865; 1866; 1867; 1868; 1869;
1870; 1871; 1872; 1873; 1874; 1875; 1876; 1877; 1878;
1879; 1880; 1881; 1882; 1883; 1884; 1885; 1886; 1887;
1888, Last 8 Over 7; 1888; 1889; 1890; 1891; 1892; 1893; 1894; 1895; 1896; 1897;
1898; 1899; 1900; 1901.

Two-Cent Pieces

Bronze two-cent pieces were made at the second Philadelphia Mint from the inception of the series until the denomination was discontinued by the Act of February 12, 1873. In the last year only Proofs were struck. These were the first circulating coins to bear the motto IN GOD WE TRUST.

DESIGN TYPE AND DATES MINTED
Shield: 1864, 1865, 1866, 1867, 1868, 1869,
1870, 1871, 1872, 1873.

Three-Cent Pieces

In 1851 the silver three-cent piece or trime was authorized by the Act of March 3, 1851, implemented on June 30 of that year. With just 75% silver content, it required less silver than face value to produce and helped supply small change when regular 90% silver coins were absent from circulation. These coins also facilitated the purchase of three-cent postage stamps. Only five million trimes were coined in 1851, but this figure moved up strongly in 1852 to nearly 19 million pieces. Another 11.4 million pieces followed in the first quarter of 1853. Coupled with a heavy coinage of gold dollars (first struck in 1849) and copper cents, the nation was able to make do with the supply of coins in the marketplace, though not well. It was clear that drastic change was still required to save the monetary and economic system from eventual collapse. After the Act of February 21, 1853, lowered the weight of all silver coins except the dollar, half dimes and dimes came into wide use, and the mintage of trimes diminished. In 1865 when nickel three-cent pieces were introduced the trime became redundant. Mintages were very low during the Civil War. The silver three-cent piece was abolished by the Act of February 12, 1873.

The Annual Report of the Director of the Mint, 1882, stated:

> The United States Mint in Philadelphia has lately received over two tons of three-cent silver pieces, which by act of Congress have been abolished from the national

coinage. This is said to be, however, only about one-fifth of the quantity of these small coins which are to be brought there from the various sub-treasuries in the country, there having already been redeemed more than ten tons. The three-cent pieces now in the Mint would fill three large wagons. They are to be recoined into dimes, as the three-cent nickel pieces are to take the place of the old silver three cent coins.

Nickel three-cent pieces were made each year at the second Mint from the inception of the denomination in 1865 through to the discontinuation of the denomination in 1889. Those of 1877 ad 1878 were only issued in Proof format, with no related circulation strikes. The 1887, 7 Over 6, is one of only a few Proofs in American coinage to have an overdate.

DESIGN TYPES AND DATES MINTED

Silver, Variety 1: 1851, 1852, 1853.

Silver, Variety 2: 1854, 1855, 1856, 1857, 1858.

Silver, Variety 3: 1859; 1860; 1861; 1862, 2 Over 1; 1862; 1863; 1864; 1865; 1866; 1867; 1868; 1869; 1870; 1871; 1872; 1873.

Nickel: 1865; 1866; 1867; 1868; 1869; 1870; 1871; 1872; 1873; 1874; 1875; 1876; 1877; 1878; 1879; 1880; 1881; 1882; 1883; 1884; 1885; 1886; 1887, 7 Over 6; 1887; 1888; 1889.

Nickel Five-Cent Pieces

Nickels were made each year at the second Mint from the inception of the denomination in 1866, through 1901. Those of 1877 and 1878 were only issued in Proof format, with no related circulation strikes.

DESIGN TYPES AND DATES MINTED

Shield, Variety 1, Rays Between Stars: 1866, 1867.

Shield, Variety 2, Without Rays: 1867; 1868; 1869; 1870; 1871; 1872; 1873; 1874; 1875; 1876; 1877; 1878; 1879, 9 Over 8; 1879; 1880; 1881; 1882; 1883, 3 Over 2; 1883.

Liberty Head, Variety 1, Without CENTS: 1883.

Liberty Head, Variety 2, With CENTS: 1883, 1884, 1885, 1886, 1887, 1888, 1889, 1890, 1891, 1892, 1893, 1894, 1895, 1896, 1897, 1898, 1899, 1900, 1901.

Half Dimes

Half dimes were produced each year at the second Philadelphia Mint from 1833 until the denomination was discontinued by the Act of February 12, 1873.

DESIGN TYPES AND DATES MINTED

Capped Bust: 1833, 1834, 1835, 1836, 1837.

Liberty Seated, Variety 1, No Stars on Obverse: 1837.

Liberty Seated, Variety 2, Stars on Obverse: 1838; 1839; 1840; 1841; 1842; 1843; 1844; 1845; 1846; 1847; 1848; 1849, 9 Over 6; 1849; 1850; 1851; 1852; 1853. *Variety 3, Arrows at Date, Reduced Weight:* 1853, 1854, 1855. *Arrows Removed, Weight Remains:* 1856, 1857, 1858, 1859, 1859 (transitional pattern), 1860 (transitional pattern).

Liberty Seated, Variety 4, Legend on Obverse: 1860, 1861, 1862, 1863, 1864, 1865, 1866, 1867, 1868, 1869, 1870, 1871, 1872, 1873.

Dimes

The designs of dimes made at the second Mint closely follow those of half dimes. The Liberty Seated motif made its debut in 1837, this year without stars on the obverse. Later issues had stars, then the stars were removed in favor of lettering. After half dimes were discontinued in 1873, dime types included 1873 and 1874, With Arrows at Date. Mintages of the denomination were continuous from 1833 to 1891. After 1892 the Liberty Seated dime reverse was continued on the Barber type. The latter motif was designed by Charles E. Barber, featured the head of Miss Liberty on the obverse, and was produced at the second Mint through 1901.

DESIGN TYPES AND DATES MINTED

Capped Bust, Variety 2, Modified Design: 1833, 1834, 1835, 1836, 1837.

Liberty Seated, Variety 1, No Stars on Obverse: 1837.

Liberty Seated, Variety 2, Stars on Obverse: 1838, 1839, 1840, 1841, 1842, 1843, 1844, 1845, 1846, 1847, 1848, 1849, 1850, 1851, 1852, 1853. *Variety 3, Arrows at Date, Reduced Weight:* 1853, 1854, 1855. *Arrows Removed, Weight Remains:* 1856, 1857, 1858, 1859.

Liberty Seated, Variety 4, Legend on Obverse: 1859 (transitional pattern), 1860, 1861, 1862, 1863, 1864, 1865, 1866, 1867, 1868, 1869, 1870, 1871, 1872, 1873. *Variety 5, Arrows at Date, Increased Weight:* 1873, 1874. *Arrows Removed, Weight Remains:* 1875, 1876, 1877, 1878, 1879, 1880, 1881, 1882, 1883, 1884, 1885, 1886, 1887, 1888, 1889, 1890, 1891.

Barber or Liberty Head: 1892, 1893, 1894, 1895, 1896, 1897, 1898, 1899, 1900, 1901.

Twenty-Cent Pieces

The twenty-cent piece, occasionally called the double dime, was primarily made for the West Coast. Introduced in 1875, most were struck in San Francisco, a lesser production occurred in Carson City, and only 38,500 circulation strikes were made in Philadelphia. At the time silver coins were not in general circulation in the East and Midwest. The public confused the new coins with the somewhat similarly sized quarters. Production dropped sharply in 1876, with just 14,400 made in Philadelphia, many of which are presumed to have been melted. Proofs for collectors were made of all years from 1875 to 1878.

DESIGN TYPE AND DATES MINTED

Liberty Seated: 1875, 1876, 1877, 1878.

Quarter Dollars

Capped Bust quarters were made each year from 1833 through 1838, the last striking being somewhat anachronistic, as by that time the other silver denominations had changed to the Liberty Seated motif. Liberty Seated quarters were produced each year from 1838 through 1891, with slight variations in design details and weight. Barber quarters were produced at the second Mint from 1892 to 1901.

DESIGN TYPES AND DATES MINTED

Capped Bust, Variety 2, Reduced Diameter, Motto Removed: 1833, 1834, 1835, 1836, 1837, 1838.

Liberty Seated, Variety 1, No Motto Above Eagle: 1838, 1839, 1840, 1841, 1842, 1843, 1844, 1845, 1846, 1847, 1848, 1849, 1850, 1851, 1852, 1853.

Liberty Seated, Arrows at Date: *Variety 2, Rays on Reverse:* 1853; 1853, 3 Over 4. *Variety 3, No Rays:* 1854, 1855. *Arrows Removed, Weight Remains:* 1856, 1857, 1858, 1859, 1860, 1861, 1862, 1863, 1864, 1865, 1866.

Liberty Seated, Variety 4, Motto Above Eagle: 1866, 1867, 1868, 1869, 1870, 1871, 1872, 1873. *Variety 5, Arrows at Date, Increased Weight:* 1873, 1874. *Arrows Removed, Weight Remains:* 1875, 1876, 1877, 1878, 1879, 1880, 1881, 1882, 1883, 1884, 1885, 1886, 1887, 1888, 1889, 1890, 1891.

Barber or Liberty Head: 1892, 1893, 1894, 1895, 1896, 1897, 1898, 1899, 1900, 1901.

Half Dollars

Capped Bust, Lettered Edge, half dollars were made in quantity from 1833 to 1836, these being the largest-denomination silver coins in general circulation at the time. In November 1836 a smaller-diameter Capped Bust half dollar with a reeded edge, and with the denomination given as 50 CENTS, was introduced, and it was made through 1837. In 1838 and part of 1839 a new reverse with FIFTY CENTS was used.

The Liberty Seated motif made its debut in 1839, the first coins lacking drapery at Miss Liberty's elbow. With variations, Liberty Seated half dollars were made continuously through 1891. From 1891 to 1901 Barber half dollars were struck.

DESIGN TYPES AND DATES MINTED

Capped Bust, Lettered Edge, Remodeled Portrait and Eagle: 1833, 1834, 1835, 1836.

Capped Bust, Reeded Edge, Reverse 50 CENTS: 1836, 1837.

Capped Bust, Reeded Edge, Reverse HALF DOL.: 1838, 1839.

Liberty Seated, Variety 1, No Motto Above Eagle: *No Drapery From Elbow:* 1839. *Drapery From Elbow:* 1839; 1840; 1841; 1842; 1843; 1844; 1845; 1846; 1847, 7 Over 6; 1847; 1848; 1849; 1850; 1851; 1852.

Liberty Seated, Arrows at Date:
Variety 2, Rays Around Eagle: 1853.
Variety 3, No Rays: 1854, 1855. *Arrows*
Removed, Weight Remains: 1856, 1857,
1858, 1859, 1860, 1861, 1862, 1863, 1864,
1865, 1866.

Liberty Seated, Variety 4, Motto
Above Eagle: 1866, 1867, 1868, 1869,
1870, 1871, 1872, 1873. *Variety 5, Arrows*
at Date, Increased Weight: 1873, 1874.
Arrows Removed, Weight Remains: 1875;
1876; 1877, 7 Over 6, 1877; 1878; 1879;
1880; 1881; 1882; 1883; 1884; 1885; 1886;
1887; 1888; 1889; 1890; 1891.

Barber or Liberty Head: 1892, 1893,
1894, 1895, 1896, 1897, 1898, 1899,
1900, 1901.

Silver Dollars

After a lapse following 1804, silver dollars were again coined starting in 1836. This year 1,000 of the Liberty Seated design with Flying Eagle reverse were struck on a screw press normally used to make medals. In January 1837 an additional 600 pieces were made from the same dies. In 1839 300 silver dollars were made for circulation, and in 1840 production resumed in earnest. In 1866 the motto IN GOD WE TRUST was added to the reverse. The Liberty Seated obverse, slightly modified, continued to be produced through early 1873. The Act of February 12 of that year discontinued the silver dollar in favor of the new trade dollar (see page 152).

The Bland-Allison Act of February 28, 1878, discontinued the trade dollar and mandated that the government buy from two- to four-million ounces of silver each month, a nod to the political pressure applied by western mining interests and others. A design made by George T. Morgan for a pattern 1877 half dollar was used. "Morgan dollars," as they are called, were first struck in the afternoon of March 11, 1878, at the Philadelphia Mint. The first delivery of coins in quantity for circulation occurred on March 13 to the extent of 40,000 pieces. By March 18 Mint Director Henry R. Linderman had expressed dissatisfaction with various aspects of the designs, including a request to change the number of feathers in the eagle's tail from eight to seven. A few other changes were subsequently made, so that the standard by June 1878 became that of seven tail feathers, with the top arrow feather slanting and the breast rounded. Earlier issues had the top arrow feather horizontal. The production of Morgan dollars was continuous from that time until the closing of the second Philadelphia Mint in 1901.

DESIGN TYPES AND DATES MINTED

Gobrecht, Liberty Seated, No Stars on Obverse: 1836.

Gobrecht, Liberty Seated, Stars on Obverse: 1838, 1839.

Liberty Seated, No Motto: 1840, 1841, 1842, 1843, 1844, 1845, 1846, 1847, 1848, 1849, 1850, 1851, 1852, 1853, 1854, 1855, 1856, 1857, 1858, 1859, 1860, 1861, 1862, 1863, 1864, 1865, 1866.

Liberty Seated, With Motto IN GOD WE TRUST: 1866, 1867, 1868, 1869, 1870, 1871, 1872, 1873.

Morgan: 1878; 1879; 1880, 80 Over 79; 1880; 1881; 1882; 1883; 1884; 1885; 1886; 1887, 7 Over 6; 1887; 1888; 1889; 1890; 1891; 1892; 1893; 1894; 1895; 1896; 1897; 1898; 1899; 1900; 1901.

Trade Dollars

The Coinage Act of February 12, 1873, discontinued the standard Liberty Seated silver dollar and provided for a new denomination—the trade dollar. In contrast to the 412.5-grains weight of the silver dollar, the new coin weight became 420 grains of 90-percent silver. The heavier coin was intended to compete in Asia with the Mexican 8-reales coin. William Barber designed the coin, which depicted Miss Liberty seated on bales of merchandise looking westward across the ocean, presumably toward China. Coinage was effected at the Philadelphia, Carson City, and San Francisco mints. From then until the denomination was discontinued by the Bland-Allison Act of February 28, 1878, San Francisco was by far the largest producer, it being the closest port to China. Most silver came from the Comstock Lode in Nevada. The first exports of trade dollars began in June 1874 and continued through early 1878.[69]

Depositors at the mints could bring the appropriate amount of silver bullion and receive in exchange trade dollars. While most were exported, they were also legal tender in the United States. Because of overproduction and reduced demand for the metal, the price of silver declined during this period. By July 22, 1876, depositors could bring less than a dollar's worth of silver to the mints and receive a legal-tender trade dollar in exchange. To stop this, Congress demonetized the trade dollar, the only time this has been done in American coinage history. After August 22 trade dollars were valued only at melt-down value in the United States. The Coinage Act of 1965 affirmed that all previous federal coins were legal tender, inadvertently restoring that status to the trade dollar.

The *Annual Report of the Director of the Mint*, 1878, included this: "The purpose for which the trade dollar was instituted and the mode in which these coins were supplied by the mints are already well-known, and it is unnecessary to advert in this report to that subject. It is sufficient to state that wherever they have been introduced in China they have met with a favorable reception and continue to grow in the estimation of the Chinese."

By the time the last 1878 trade dollar fell from the dies, more than $36 million of the denomination had been produced, a staggering sum and an amount more than four-times greater than all of the silver dollars that had been coined from 1794 until the denomination was suspended in 1873. In a way it is curious that the trade-dollar denomination was suspended at the very height of its success, but such was (and is) politics. The Bland-Allison Act offered greater opportunities for the silver market than did the trade-dollar legislation; now, with the Bland-Allison Act and the Morgan dollar, the government was forced to buy silver and make silver dollars. It was never forced to make trade dollars.

In later years many trade dollars were redeemed at bullion value and much of the metal used to coin Morgan-design silver dollars.

DESIGN TYPE AND DATES MINTED

Trade Dollar: 1873, 1874, 1875, 1876, 1877, 1878, 1879, 1880, 1881, 1882, 1883, 1884, 1885.

Gold Dollars

Authorized by the Act of March 3, 1849, dies for the gold dollar were prepared by engraver James B. Longacre. The first examples reached circulation in mid-May of that year. Gold dollars were made in quantity in the early years but by 1860 were found to be redundant to coinage needs. Silver half dollars (dollars were not in circulation at the time) and gold quarter eagles seemed to take their place in commerce.

The Philadelphia *North American* printed this on March 4, 1873: "The United States Mint has commenced melting the first installment ($1,000,000) of twenty million $1 gold pieces, which, during the ensuing month are to be re-coined into larger denominations. These pieces were of inconvenient size, and the government has experienced trouble in issuing them in large quantities. This induced the government to take them from the Sub-Treasury in New York, where they have been idle the past few years, and place the metal in a more desirable shape. From 1849, when the first one-dollar gold pieces were coined at the Mint in this city, to 1867, when the coinage was stopped, there has been $17,709,442 made in the Philadelphia Mint alone."

Gold dollars continued to be produced through 1889, with the mintage for 1875 being just 400 circulation strikes and 20 Proofs. Most made in the 1880s went to collectors and to the jewelry trade.

DESIGN TYPES AND DATES MINTED

Liberty Head (Type I): 1849, 1850, 1851, 1852, 1853, 1854.

Indian Princess Head, Small Head (Type II): 1854, 1855.

Indian Princess Head, Large Head (Type III): 1856, 1857 1857, 1858, 1859, 1860, 1861, 1862, 1863, 1864, 1865, 1866, 1867, 1868, 1869, 1870, 1871, 1872, 1873, 1874, 1875, 1876, 1877, 1878, 1879, 1880, 1881, 1882, 1883, 1884, 1885, 1886, 1887, 1888, 1889.

$2.50 Quarter Eagles

Quarter eagles of the reduced Capped Bust motif were made in 1833 and early 1834. However, they did not circulate, as they cost more than face value to produce. After the Coinage Act of June 28, 1834, lowered the authorized weight, the Classic Head was introduced. These were made in quantity and were common sights in circulation beginning in autumn 1834. In 1840 the Liberty Head design by Christian Gobrecht was introduced, after which coins of this design were made each year through 1901. No production records have been found for the year 1841, but about a dozen quarter eagles of this date are known. In 1863 the mintage was limited to just 30 Proofs.

DESIGN TYPES AND DATES MINTED

Capped Head to Left, Reduced Diameter: 1833, 1834.

Classic Head: 1834, 1835, 1836, 1837, 1838, 1839.

Liberty Head: 1840, 1841, 1842, 1843, 1844, 1845, 1846, 1847, 1848, 1849, 1850, 1851, 1852, 1853, 1854, 1855, 1856, 1857, 1858, 1859, 1860, 1861, 1862, 1863, 1864, 1865, 1866, 1867, 1868, 1869, 1870, 1871, 1872, 1873, 1874, 1875, 1876, 1877, 1878, 1879, 1880, 1881, 1882, 1883, 1884, 1885, 1886, 1887, 1888, 1889, 1890, 1891, 1892, 1893, 1894, 1895, 1896, 1897, 1898, 1899, 1900, 1901.

$3 Gold Coins

From the 1853 *Mint Report*: "The three-dollar gold coin, authorized by the last Congress, will be issued as soon as the dies, now in progress, are completed. From the close approximation in weight and value which the coin will bear to the quarter eagle, it has been deemed expedient to make the devices upon it different from any coin heretofore issued. The device adopted for the *obverse* is an ideal head, emblematic of America, enclosed within the national legend. The *reverse* will present a wreath, indicating the most prominent productions of our soil, and enclosing the denomination and date of the coin."

These were coined continuously from the inception of the denomination in 1854 until it was discontinued in 1889. In 1875 and 1876 only Proofs were struck, with no related circulation strikes.

DESIGN TYPE AND DATES MINTED

Indian Princess Head: 1854, 1855, 1856, 1857, 1858, 1859, 1860, 1861, 1862, 1863, 1864, 1865, 1866, 1867, 1868, 1869, 1870, 1871, 1872, 1873, 1874, 1875, 1876, 1877, 1878, 1879, 1880, 1881, 1882, 1883, 1884, 1885, 1886, 1887, 1888, 1889.

$5 Half Eagles

Paralleling the scenario for quarter eagles, half eagles of the reduced Capped Bust motif were made in 1833 and early 1834 but did not circulate as they cost more than face value to produce. After the Coinage Act of June 28, 1834, lowered the authorized weight, the Classic Head was introduced. These were made in quantity and were common sights in circulation beginning in autumn 1834. In 1839 the Liberty Head design by Christian Gobrecht was introduced, after which coins of this design were made each year through 1901. In 1866 the motto IN GOD WE TRUST was added to the reverse.

DESIGN TYPES AND DATES MINTED

Capped Head to Left, Reduced Diameter: 1833, 1834.

Classic Head: 1834, 1835, 1836, 1837, 1838.

Liberty Head, Variety 1, No Motto Above Eagle: 1839, 1840, 1841, 1842, 1843, 1844, 1845, 1846, 1847, 1848, 1849, 1850, 1851, 1852, 1853, 1854, 1855, 1856, 1857, 1858, 1859, 1860, 1861, 1862, 1863, 1864, 1865.

Liberty Head, Variety 2, Motto Above Eagle: 1866; 1867; 1868; 1869; 1870; 1871; 1872; 1873; 1874; 1875; 1876; 1877; 1878; 1879; 1880; 1881, Final 1 Over 0; 1881; 1882; 1883; 1884; 1885; 1886; 1887; 1888; 1889; 1890; 1891; 1892; 1893; 1894; 1895; 1896; 1897; 1898; 1899; 1900; 1901.

$10 Eagles

The mintage of the $10 denomination, suspended since 1804, was resumed in 1838. The Braided Hair motif by Christian Gobrecht was inaugurated this year, to be later used on cents and half eagles beginning in 1839 and quarter eagles in 1840. Eagles of this design were minted through 1901. In 1866 the motto IN GOD WE TRUST was added to the reverse.

DESIGN TYPES AND DATES MINTED

Liberty Head, No Motto Above Eagle: 1838; 1839; 1840; 1841; 1842; 1843; 1844; 1845; 1846; 1847; 1848; 1849; 1850; 1851; 1852; 1853, 3 Over 2; 1853; 1854; 1855; 1856; 1857; 1858; 1859; 1860; 1861; 1862; 1863; 1864; 1865.

Liberty Head, Motto Above Eagle: 1866, 1867, 1868, 1869, 1870, 1871, 1872, 1873, 1874, 1875, 1876, 1877, 1878, 1879, 1880, 1881, 1882, 1883, 1884, 1885, 1886, 1887, 1888, 1889, 1890, 1891, 1892, 1893, 1894, 1895, 1896, 1897, 1898, 1899, 1900, 1901.

$20 Double Eagles

Authorized by the Act of March 3, 1849 (that also created the gold dollar), the $20 double eagle was specified as America's largest denomination. By this time the idea was hardly new. On December 21, 1836, the *Philadelphia Public Ledger* presciently endorsed the idea of such a coin and even called it a *double eagle*: "If we are to have large coins, let them be in gold. In addition to the eagle, which has the size of the half dollar, we would recommend the double eagle, which of the size of our silver dollar, would contain the value of twenty." Engraver James B. Longacre used the same obverse portrait of Liberty that he had employed on the gold dollar of the year before. The reverse had an eagle and shield design. From 1850 through 1901 the reverse had three lettering variations: Type I (without motto, 1850–1866); Type II (with motto IN GOD WE TRUST, denomination as TWENTY D., 1866–1876); and Type III (with motto, denomination TWENTY DOLLARS).

By 1851 the double eagle was a fixture in commercial and banking circles. In this year records were set with 2,087,155 pieces struck at the Philadelphia Mint and 315,000 in New Orleans, nearly all from California gold. These figures were about twice those of the year before.

Bankers' Magazine printed this in its December 1872 issue: "We understand the U.S. Mint is now employed in melting up one million one-dollar pieces for the purpose of converting the product into pieces of larger denominations, principally into double eagles. It is reported that twenty millions of these smallest gold coins are to be melted up. This is an indication most unfavorable to a return to coin circulation. The double eagles to be made are to be used as counters for the exchange of large sums between bankers and bullion dealers. They would scarcely at all enter into circulation, even if paper was at par with coin. The movement at the Mint is one directly away from a resumption of specie payments."

From 1850 through the end of the life of the second Mint in 1901, far more gold was coined into double eagles than for all of the other gold denominations combined. Beginning in a large way in the 1870s, when "the Silver Question" dominated politics, many congressmen and others wanted to leave the *de facto* gold standard and either have silver alone or a combination of silver and gold. Foreign treasuries and banks, fearful that Americans would pay overseas debts in silver dollars not worth their intrinsic value, began importing double eagles in huge quantities—so many that by early 1895 the Treasury nearly ran out of gold! Years later in 1933, when President Franklin D. Roosevelt mandated that gold coins no longer be paid out and soon called for the public to turn them in, foreign countries held on to American gold more tightly than ever. Beginning in the late 1940s and continuing for years afterward, millions of double eagles were brought back to the United States and sold to numismatists and investors.

DESIGN TYPES AND DATES MINTED

Liberty Head, Without Motto on Reverse: 1850; 1851; 1852; 1853; 1854; 1855; 1856; 1857; 1858; 1859; 1860; 1861; 1861, Paquet Reverse; 1862; 1863; 1864; 1865.

Liberty Head, Motto Above Eagle, Value TWENTY D.: 1866, 1867, 1868, 1869, 1870, 1871, 1872, 1873, 1874, 1875, 1876.

Liberty Head, Motto Above Eagle, Value TWENTY DOLLARS: 1877, 1878, 1879, 1880, 1881, 1882, 1883, 1884, 1885, 1886, 1887, 1888, 1889, 1890, 1891, 1892, 1893, 1894, 1895, 1896, 1897, 1898, 1899, 1900, 1901.

THIRD PHILADELPHIA MINT (1901–1969)

THE NEW MINT

Planning and Construction

In 1891 Treasury Secretary William Windom reported to Congress that the Mint opened in 1833 was doing business far beyond its planned capacity, despite the additional output from branches in Carson City, New Orleans, and San Francisco. On February 19 of that year Senator James Donald Cameron introduced a bill to provide for the purchase of a new Mint site in Philadelphia. After quick approval by the Senate, it went to the House, where there was strong opposition. Finally the bill was passed. After considering various sites, one of a block bounded by Sixteenth, Seventeenth, Spring Garden, and Buttonwood streets was selected, with the front to be on Spring Garden Street. In August 1894 the secretary of the Treasury approved the $305,000 cost of the land.

The building, designed by architect James Knox Taylor and his staff, was constructed in the Roman Ionic style and had a footprint of 58,000 square feet.

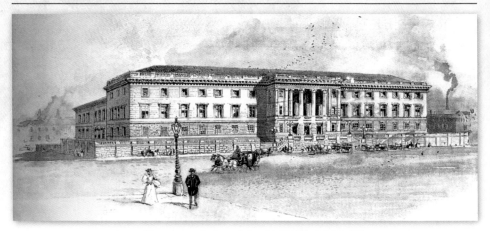

An architect's rendering of the third Philadelphia Mint.

In August 1899 in *The Numismatist*, Augustus G. Heaton told of a visit to the construction site:

> The granite walls, as seen over the present temporary fence, have reached the second floor and suggest a future imposing building. It will have the form of hollow quadrangle, with an isolated two-story building in the courtyard as the museum, approached by a gallery.
>
> Galleries will go through all the future workrooms, so that visitors can inspect coining operations from above, without interference or communication with the persons employed. The basement and its vaults are already so far finished as to permit the transfer, now going on, of millions of coins from the old Mint, but the indications are that the new Mint will not be in order for coinage for visiting for two years or so to come. . . .

By 1900 the construction, a two-story, square, white-marble building on Chestnut Street, was nearing completion. The structure was three stories high in the front and two in the back, with a mostly above-ground basement in the back which gave it the practical effect of three stories. The interior was richly decorated with colorful mosaics from the shop of Louis Comfort Tiffany that cost nearly $40,000. The official opening date was

The new Mint under construction, June 30, 1899.

June 13, 1901; however, certain operations had taken place within the building since 1899. The Mint Cabinet display had been moved there as well as millions of annoying silver collars, at least two million of which had been found stored in bags that became wet and moldy, thus discoloring all of them. It was debated whether to clean them, but the idea was discarded when it was realized that no one wanted them in circulation anyway.

The total cost of the building was $2,025,000, including $328,338.71 for the site and incidental expenses. The main contractor was Charles McCaul of Philadelphia, who passed away during the construction period, after which it was finished by the Charles McCaul Company which continued the business.

The third Philadelphia Mint.

The New Mint in Operation

The *Annual Report of the Director of the Mint*, 1901, told of the vaults:

> For the storage of bullion, coin, blanks, dies, etc., the Mint is provided with 20 steel-lined vaults, 8 of which are located in the basement and are as follows: The silver-dollar vault, which is 100 feet long, 52 feet wide, and 10-1/2 feet high, has a capacity for storing $112,000,000 in silver dollars, packed in boxes.
>
> The cashier's working vault, measuring 80 feet long, 52 feet wide, and 10-1/2 feet high, is used for storing the various denominations of coin prior to shipment. The remaining 6 vaults in the basement are 122 feet long, 21 feet wide, and 10-1/2 feet high, and are provided with four entrances, all opening into the main corridor. These are utilized for the storage of gold and silver bullion, bars, etc. These vaults are all of the most approved modern construction. The combined weight of steel used in their construction is approximately 3,250,000 pounds. . . .

The narrative went on to tell of two even larger vaults.

The Mint Collection continued to be a popular tourist attraction. For visitors, souvenir coins were available at face value. Of these, Indian Head cents were the most popular. At holiday time there was a demand for quarter eagles. There was not a great demand for current Proof coins. Only 813 silver sets were sold in 1901, accompanied by 1,985 minor sets containing the Indian Head cent and Liberty Head nickel.

The Mint Cabinet

The Mint Cabinet comprised about 11,000 coins and 2,300 medals from all eras and countries. Foreign coins were arranged by countries, with each section in chronological order. Two of the most popular coins were the "widow's mite" and "penny" mentioned in the Bible.[70] James Rankin Young in his 1903 book *The United States Mint at Philadelphia*, the source for much information given here, noted:

> The collection of American coins in the Cabinet is, naturally, the most complete in the world. Here is the double eagle of 1849, the only coin of that denomination struck during that year; and the half eagles of 1815 and 1822, both of exceedingly high numismatic value and rarity.

Display of various coins, U. S. Mint, Philadelphia, Pa.

2388

The Mint Cabinet was ornately appointed.

In the collection are many rare specimens of the early colonial coins. The designs on most of these are crude. The New England shilling, for instance, was simply a planchet of silver with a small "N.E." stamped one side, and "XII" on the other. The pine tree and oak tree shillings had rude representations of one of these trees stamped on their obverse side.

Many of the early colonial coins had such quaint inscriptions around them as "Value Me as You Please," "I cut my way through," "I am good copper," and "God Preserve Carolina and the Lords Proprietors." A number of the coins like the Bar Cent and the Rosa Americana series, were designed and stamped abroad. In the early life of the nation some of the smaller denominations of coins were made by private individuals to meet a local demand. . . . They were finally driven out of circulation by the plentiful supply of government coin. Specimens will be found here of all gold, silver, nickel and bronze pieces issued from the Mint.

Visitors were welcomed Monday through Friday, from nine in the morning until two in the afternoon. On Saturdays the hours were 9 to 11 in the morning. Visitors were allowed to buy recently minted coins for face value, as noted above. In addition, any medals were available as well, most being struck from dies made years earlier. The first floor and mezzanine were the only areas open to the general public.

Mint Facilities

Although there was much new equipment installed, the various processes at the new Mint were essentially the same as conducted in the old Mint. A new naphtha gas plant installed by the American Gas Furnace Company processed atomized oil and furnished heat for various processes, including annealing and keeping the building warm in the colder seasons. Nearly all machinery was operated by electric motors. The Mint had its own power plant with boilers that provided steam to operate generators. Time clocks (made by the International Time Recording Company of Binghamton, New York, which evolved to become International Business Machines), watchmen's clocks, 51 connected telephones, 31 fire-alarm boxes, 1 passenger elevator, 8 freight elevators, and other facilities were installed.

Although automatic planchet-weighing sufficed for silver coins, in 1901 a corps of about 60 women worked with gold at tables in long rows in a spacious room, each with a balance scale (including new ones furnished by Henry Troemner of Philadelphia). Lightweight planchets were set aside to be melted, and heavy planchets were filed by hand. There were improved milling machines in operation, one each for dimes (676 blanks per minute), quarters, half dollars, dollars (464), quarter eagles, half eagles, eagles, and double eagles. Planchets for cents and nickels were supplied by private contractors and already had raised rims.

Punching planchets from thin metal strips.

Inspecting blank planchets.

Planchet-weighing machines.

There were 25 knuckle-action presses installed on the first floor, including 10 new ones recently built by the T.C. Dill Machine Company in Philadelphia. Ladies with boxes of planchets took them into their hands and fed them into the top of a vertical brass tube, from which they were mechanically fed into the space between the dies. Cents, nickels, and dimes were struck at the rate of 100 per minute on smaller coining presses driven by 3-horsepower motors. For higher denominations the rate was 90 per minute on a large press run by a 7.5-horsepower motor. Presses operated with the following tons of pressure per denomination: cents 40, nickels 60, dimes 35, quarters 60, half dollars 98, silver dollars 160, quar-

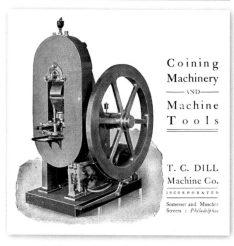

Coining Machinery — AND — Machine Tools

T. C. DILL Machine Co.
INCORPORATED
Somerset and Mascher Streets : *Philadelphia*

Adding to those already on hand, the T.C. Dill Company furnished new presses to the third Philadelphia Mint.

ter eagles 35, half eagles 60, eagles 110, and double eagles 135. These pressures could vary by plus or minus five percent. The finished coins were dumped into bins at the rear of each press and were collected at intervals.

The coins were then sent to the office of the coiner at the end of the Coining Room. There a staff counted silver dollars and all gold coins by hand, piling them in little stacks on counting tables. Other coins were put into huge counting boards with low parallel brass ridges. The board was tilted, and any coins not between the ridges slid off. The remaining coins were then lined up in neat rows and could be easily counted. It took less than 1 minute to count 1,000 cents by this method.

Gold coins were put in groups of $5,000 each, which optimally weighed 268.75 troy ounces. Silver coins were grouped in lots of $1,000, weighing 859.75 ounces. Dimes to dollars weighed 803.75 ounces per group. The coins were then placed into canvas

New coining presses.

Inspecting finished coins to remove any that were not struck properly.

Where the Metal is converted into coin, U. S. Mint, Philadelphia, Pa.
2382.

The Coining Room.

bags. From there they were delivered to the office of the cashier. They were recorded and then shipped directly to Sub-Treasury offices, banks, and businesses. The demand for coins was always stronger late in the year and in the holiday season.

Also on the first floor was the Medal Room, additionally under the supervision of the coiner. There were two hydraulic presses that struck one coin or medal at a time, each with a capacity of 300 to 400 tons. Proof coins were made there, as were medals. Proof planchets were annealed to soften them, given a bath in acid, and then polished individually by hand using fine-grain wet sand. Proof sets from the cent to the dollar were sold for $2.50, and sets of four gold coins for $38.50. Minor Proof sets with the cent and nickel were available separately for 8 cents, and Proof gold coins were available individually for 25 cents over face value.

Certain medals were given a sand-blast finish, a style popularized at the Paris Mint (the process was not used on U.S. coins until 1908, and then only on gold). This was accomplished in a wooden box with glass sides. Fine sand particles were blasted at high speed through a metal tube on to one side of a medal. The medal business occupied a lot of activity, as the Mint made them not only for the government but also for many colleges, societies, athletic groups, and other entities who supplied their own dies (current law forbids the Mint from making dies for outside interests). Over 300 medal dies were on hand at the new Mint. Some were very old, and some were made in France.

Praise for the New Mint

In his 1902 *Annual Report*, Mint Director George E. Roberts commented: "This beautiful new edifice is unquestionably the finest building ever constructed for coining purposes in the world."

The United States Mint at Philadelphia by James Rankin Young, 1903, described and illustrated the facilities and processes of the new building, including this synopsis:

> Passing through the massive bronze doors at the main entrance the visitor finds himself in the spacious lobby with its mosaic paneled ceilings and walls. Through the broad vaulted corridors he is shown the location of the offices of the superintendent and his executive staff. Mounting the main stairway to the second floor a view is obtained of the white corridor formed by a series of beautiful marble columns. Then he is given a sight of the numismatic room, where are gathered for public inspection a collection of the representative coins of the world, dating back

to a period before the Christian era—a collection most valuable and precious. From the corridors surrounding this department the visitor sees before him the workrooms, with the work of minting going on. The visitor is shown in regular order the processes through which a piece of metal must go before it becomes a coin of the Republic. First is the room where the bullion is received, weighed and deposited; next the room of the refiner, where the pure metal is separated from the base; then the room where the metal is melted, alloyed and cast into ingot bars; next back to the hands of the assayer; then to the room where the metal that is found to be up to the required standard of fineness goes through the workmanship that rolls it into long ribbons, from which the blanks or planchets are stamped; next to the room where the blanks are weighed and adjusted and where those that are found to have the legal weight are raised around the edge and cleaned; then to the room where on the finished blanks the coin is stamped, and finally into the rooms where respectively the coins are weighed, counted and packed into bags, to be transferred to the office of the cashier for shipment.

In 1903 the Philadelphia Mint began striking coins for the Philippine Islands, an archipelago the United States had captured during the brief one-sided War of 1898. Coinage for the islands continued in Philadelphia until the early 1920s, when the Manila Mint was established in the Philippines. In this era silver dollars and gold coins were rarely seen in commerce in the Midwest and the East, had scattered use in the West (except for San Francisco, where they were popular), but were common in the Rocky Mountain states. The popularity of coin-operated vending machines and amusement devices, penny arcades, nickel admission to theatres, and the like created a tremendous demand for cents and nickels, denominations that were only made at the Philadelphia Mint.

The Medal Room at the new Mint.

THE LOUISIANA PURCHASE EXPOSITION

In 1904 the Treasury Department set up a large working exhibit at the Louisiana Purchase Exposition in St. Louis. Planchet-making operations were demonstrated, as was a coining press in action. Large framed panels of currency were displayed by the Bureau of Engraving and Printing, and a separate area featured treasures from the Mint Cabinet. Also included in the Treasury area were exhibits from the Bureau of Public Health and the Marine Hospital Service.

This exhibit received relatively little notice in the press or even the official exhibition catalog. Separately, Farran Zerbe mounted his Money of the World commercial exhibit and offered items for sale, including two varieties of commemorative gold dollars, the first coined in this metal.

"Brilliant is best" was the rallying cry for nearly all collectors of and dealers in silver coins. The Mint Cabinet collection silver coins had been cleaned recently by an attendant who bought silver polish in a department store to do so. Farran Zerbe on a visit there asked an employee about this questionable practice and was told, "That is nothing. I have been here eight years and they have been cleaned three or four times in my time."[71]

In 1904 the supply of bullion authorized by Congress to coin millions of annual Morgan-design dollars ran out. A few years later most of the hubs and master dies were destroyed—with hundreds of millions of silver dollars piled up in vaults, the Treasury Department figured that no more would ever need to be made.

In 1906 the machines for punching planchets out of silver and gold strip were changed to cut two to five blanks per stroke, rather than just one. Planchets with raised rims for cents and nickels were obtained under private contract, as had been the case for many years.

Proof coins, medals, awards, and badges were produced in two main rooms. The smaller, called the Bronzing Room, was outfitted to sandblast medals and to apply bronzing, antique, or other finishes to them. The larger press room had four hydraulic presses—1 of 1,000 tons, 2 of 450 tons, and 1 of 350 tons. Coins and medals were struck there, some medals requiring multiple blows to bring up the relief. Pins, bars, and award items were also made. Ribbons and pins were affixed to badges.

Other machines included two grinders, one rotary file, one cutting press for preparing planchets, one milling machine, one lathe, and one shaper. In the fiscal year of 1906 there were 98,991 visitors to the Philadelphia Mint.

The Treasury Department display at the 1904 Louisiana Purchase Exposition in St. Louis.

The Philadelphia Mint continued to be the main coining facility and the home of the Engraving Department and the Die Department. A Janvier-style portrait lathe was installed and facilitated the translation of galvanos to metal master dies and hubs. Dies were furnished to the branches in Denver, New Orleans, and San Francisco. Related Treasury Department facilities during this era included the New York Assay Office (the most important such facility) and lesser assay offices at Boise, Carson City (in the old Mint building), Helena, Charlotte (in the old Mint building), St. Louis, Deadwood, Salt Lake City, and Seattle. Gold and silver refined at the assay offices was cast into ingots and shipped by rail to the various mints. In 1908 minor coins were struck for the first time at a facility other than Philadelphia, when the San Francisco Mint produced 1908-S Indian Head cents.

At some time in 1909 a changeover in die-making was completed. Before 1907 the dates were punched into working dies by a four-digit logotype. That changed in 1907 with the new Saint-Gaudens $10 and $20 coins and in 1908 with the $2.50 and $5 coins of the Pratt design, for which dies were made with the dates included. Sometime in 1909 this was extended to other denominations as well. In this year coinage ceased forever at the New Orleans Mint, and after 1909 no more dies were made for that institution. In June 1909 *Mehl's Numismatic Monthly* included this unsigned article (excerpted):

Guarding the Mint's Millions

At the Philadelphia Mint the vaults are sunken deeply below the ground and are approached only by descending several flights of stairs and by passing by doors guarded at night by a body of picked guards armed with Mauser and Remington rifles. The vaults have double Remington-Sherman combinations. That to the outer door is known only to the cashier, Joseph E. Murphy; that to the inner door is known only to the custodian, F. Lambertson. To enter a vault both men, of course, have to be together. . . .

Once inside one of these huge vaults, one finds himself now in a vault which the average mind pictures, but rather in an immense room partitioned off into 15 apartments by steel gratings of the strongest kind. A straight passageway extends the length of the room between the rows of cages. . . .

In each compartment devoted to the storage of gold there is $40,000,000. It is stored in bags which contain $5,000 each in half eagles, eagles, and double eagles, and then these bags are arranged 20 on a shelf. Each shelf thus has $100,000 on it. The shelves are arranged in rows downward, 10 from top to bottom of the compartment, thus each row having $1,000,000 of gold in it. . . .

The Numismatist, June 1909, reported that two different varieties of $50 patterns, both dated 1877 and in gold, were acquired by William H. Woodin of New York City, an industrialist who specialized in gold coins and patterns and who was currently preparing a manuscript on early $5 gold coins from 1795 to 1834. The sellers were John W. Haseltine and Stephen K. Nagy of Philadelphia. The pieces "are probably from the Idler Collection, which, piece by piece is being offered at private sale, and in many instances they are specimens not previously recorded." Theories were aplenty. Mint Director A. Piatt Andrew, seemingly unaware that Mint officials had made and privately sold tens of thousands of pattern coins from spring 1859 to the summer of 1885, stated that, as no Mint records could be found for nearly all of the patterns in collectors' hands, they were illegally held. Federal agents started seizing patterns, creating a furor. Finally, the witch hunt was dropped. The two $50 coins that precipitated the matter had come not from Idler but from former Mint Superintendent A. Loudon Snowden. The coins were returned to Snowden, who gave Woodin thousands of other pattern coins in exchange. Snowden then turned them over to the Mint Cabinet.[72]

The Lincoln cent made its debut in August 1909. In an order issued on May 23, 1910, Mint Director A. Piatt Andrew stated:

> As a safeguard against possible irregularities in the issue of coins in future years the following instructions should be carried out at the earliest possible date:
>
> There should be defaced and destroyed in the presence of the Superintendent and Assayer, first, the dated hubs of past years. Second, the hubs and dies of superseded design. Third, all dies, including mother dies, obverse and reverse, dated and undated, except the working dies of the current year and one mother die of each obverse and reverse of each current design upon which is based the manufacture of the hub for the working dies. Fourth, all hubs, dies and mother dies of whatever character, except those used for the production of dies for the current coins of the United States should be destroyed at the end of each calendar year.

Unfortunately, no accurate inventory was kept, nor were any impressions taken from the dies, some of which dated back to when Christian Gobrecht was engraver.

In late 1910 the Philadelphia Mint was working around the clock to produce cents and nickels, which were in great demand for coin-operated machines and for admission to amusements. In 1911 the Denver Mint began producing cents.

CURATOR COMPARETTE

From the 1790s onward the Assay Commission met at the Philadelphia Mint annually to analyze sample silver and gold coins taken from coinage at the several mints the preceding year. On December 12, 1911, Dr. T. Louis Comparette, curator of the Mint Collection since 1905, wrote to State Librarian George S. Goddard in Hartford, Connecticut: "That examination takes place in February, and if you wish me to look out for specimens for you then, just let me know." The library housed the extensive Joseph Mitchelson Collection of coins that had been willed to it, and additions were made each year to keep up with current issues.

In 1912 a book by Comparette, *Catalogue of Coins, Tokens, and Medals in the Numismatic Collection of the Mint of the United States at Philadelphia, Pa.*, was published and was widely sold to numismatists and others who expected to learn much from it. However, it was virtually useless to the coin collector. Sample listings (preceded by Mint catalog number and lightly edited) include:

> 98. Double-eagle, 1855. Similar to No. 92, but with S (San Francisco Mint) beneath the eagle, on reverse, and dated 1855.

> 126. Double-eagle, 1883. Similar to No. 120 except date.

> 276. Half-eagle, 1822. Similar to No. 269 except date.

> 307. Half-eagle, 1852. Similar to No. 294, but with D (Dahlonega Mint) beneath the eagle, on reverse.

> 369. Three-dollars, 1855. Similar to No. 367, but with S (San Francisco Mint) beneath the wreath, on reverse, and dated 1855.

Except for basic introductory information describing the physical appearance of the designs (without reference to the engraver, legislation, etc.), little information was transmitted. No indication was given as to the style of coin (circulation strike or Proof), present grade, rarity, etc. Comparette, whose specialty was ancient numismatics, was a lightweight with regard to American coins, and among other actions, he caused large quantities of 1907 With-Periods Indian Head $10 coins by Saint-Gaudens to be melted down for bullion, when numismatists would have gladly paid a strong premium to buy them.

In 1912 the Denver and San Francisco mints struck nickels for the first time. The dies were made in Philadelphia. On February 17, 1913, mintage of Indian Head or Buffalo nickels began on a small press at the rate of 120 per minute.

THE WAR IN EUROPE

What became known as the Great War and later as the first World War began on July 29, 1914, one month after the assassination of the heir to the Austrian throne in Bosnia. Riding in a 1912 Graf und Stift motorcar at Sarajevo, Archduke Franz Ferdinand, 51, and his wife, Sophie, were killed with two Browning revolver shots fired by high-school student Gavrilo Princip, 16, who was hired by Serbian terrorists to kill the archduke, the nephew of Austrian emperor Franz Josef. One diplomatic and military reaction led to another, and within a month several countries were involved. In time this led to the United States becoming the most important source for war materials for the principal Allied Powers, including England and France. Factories ramped up production, the national economy strengthened, and for the next several years America's mints produced record numbers of coins.

Gold arrived in quantity. *Mehl's Numismatic Monthly*, September–October 1915, included this:

Gold from England

English Sovereigns are Being Melted into U.S. Dollars at Philadelphia Mint. New York, Sept. 25—

You have heard a lot lately about the dollar ruling the world, and some undoubtedly have been puzzled by all this talk of drop in foreign exchange, balance of trade and the like. But if you had been down in the big United States Assay Office last week you could have seen something that a child could understand—£2,000,000 of English gold disappearing and rough, dirty gold bricks taking their places. The bricks will soon become good American money.

New York has recently got three gold shipments from England, each of about $20,000,000, brought by a fast British cruiser to Halifax and then carried overland with elaborate precautions.

The first two consignments were in United States coin—some we have had to ship over to Europe in other years when the trade balance was the other way. Then the Bank of England evidently ran out of eagles, for in the last shipments were $7,850,000 in United States coins and $11,615,000 in sovereigns. The British treat our eagles most shamefully. They ship them by weight. To bring each bag up to the required heft, they chop coins into many triangular bits and add these piece by piece until the scales tip.

The Sub-Treasury here credits the importer only with the whole coins, and the chips are turned over to the Assay Office, which melts them up into bricks. These chips amount to several thousand dollars in the recent three big shipments. As to the sovereigns, the Assay Office found itself with 2,550,000 on its hands. The importers were anxious to get the benefit of the amount.

A law has been passed allowing them to be credited with the sovereigns as bullion without melting them up, but there has been a dispute over the deductions to be made for wear on the coins, dirt and other considerations, which could not be settled, and this has made the law a dead letter. Uncle Sam charges $1 per 1,000 ounces for melting the gold. This does not cover the cost by a lot. It takes one's breath away at first to look around the furnace room of the assay office.

COINAGE AND OTHER MATTERS

In 1916 the mintage of Proof coins for collectors was discontinued. A set contained only the cent and nickel. Gold Sand-Blast Proofs had been very unpopular and were last made in 1915. In 1916 new designs for silver coins were struck. Winged Liberty Head (Mercury) dimes were released in October and garnered widely favorable reviews. Although Standing Liberty quarters and Liberty Walking half dollars were struck late in the year, they were not released until January 1917.

Charles E. Barber, chief engraver since 1880, passed away on February 18, 1917, and was succeeded in the post by George T. Morgan, who had come from England to work at the Mint in 1876. On April 6, 1917, the United States declared war on Germany. In that year no gold coins were struck except for about 5,000 McKinley Memorial commemorative gold dollars. England was in dire need of silver bullion to be shipped to India. Congress passed the Pittman Act on April 13, 1918, with the result that 270,232,722 long-stored, unwanted silver dollars were melted and the resulting ingots exported. The legislation stated in part:

> To conserve the gold supply of the United States; to permit the settlement in silver of trade balances adverse to the United States; to provide silver for subsidiary coinage and for commercial use; to assist foreign governments at war with the enemies of the United States; and for the above purposes to stabilize the price and encourage the production of silver.
>
> Be it enacted by the Senate and House of Representatives of the United States of America in Congress assembled, That the secretary of the Treasury is hereby authorized from time to time to melt or break up and to sell as bullion not in excess of 350,000,000 standard silver dollars now or hereafter held in the Treasury of the United States.

No account was kept of the dates and mints of the coins destroyed, prompting Frank Duffield, editor of *The Numismatist*, to write:

> With such a wholesale melting process going on, the Mint reports of dollar coinage will cease to be of any value in estimating the rarity of any date between 1878 and 1904, and time alone will tell the story. The dollars coined since 1878 have not been popular with collectors, compared with other United States issues, and the number in the hands of collectors or dealers is probably small. None, or very few, are in circulation, and probably few of those that will finally be left in the Treasury will ever be in circulation again. It would seem as if these considerations might tend to popularize this coin with collectors, with the possibility that in the future some very great rarities may be determined.[73]

No gold coins were made in 1918 or 1919. They became scarce in circulation, and most banks stopped paying them out. The public hoarded gold, expecting that, with the current inflation, problems in Europe, and other factors, the government would raise its buying price for gold.

The war in Europe ended with the signing of an armistice on November 11, 1918. The American economy was going full speed ahead, with every expectation that the pace would continue. The hoarding of gold coins eased after peace was restored and the Treasury failed to raise the gold price. The coins became available once again. Regardless, except for in the city of San Francisco, few gold coins were used in circulation. The public found paper money to be more convenient and knew if could be converted into gold coins at any time. Quarter eagles were an exception, and from this time onward they were only available at banks and elsewhere by paying a small premium for them. Silver dollars continued to be popular in commerce in the Rocky Mountain states.

INTO THE 1920S

In 1920 the international price of silver dropped. The Treasury began buying the metal in quantity. In 1921, mainly to provide backing for Silver Certificates, over 92,000,000 silver dollars were struck.

In October 1921 *The Numismatist* quoted Director of the Mint Freas M. Styer regarding the subject:

> The Philadelphia Mint is cutting down the interest-bearing debt of the United States $5 million a month. Each day the Mint is cutting $5,000 from the annual interest on paper held against the United States. All of that comes about because of the concentration of the work of the three United States mints on the coinage of silver dollars to replace 350 million dollars that were melted down during the war to sell to the English as bullion. When all those dollars were melted the United States had to call in all the silver certificates—the $1, $2, and $5 "bills," to speak in common lingo—representing the dollars that were deposited in the vaults of the mints. Under the law of the land the Treasury must hold a silver dollar for each dollar silver certificate issued. So with the melting of the silver dollars the silver certificates had to be recalled. To cover that loss in currency, the government issued short-term certificates of indebtedness bearing 2% interest. The silver dollars now being coined allow for the issuance of new silver certificates which are being used in calling in those certificates of indebtedness.
>
> There are three mints—Philadelphia, Denver and San Francisco. The Philadelphia Mint is equal in output to those in Denver and San Francisco combined. Last April all three plants were started on the making of the silver dollars. They were put on 24-hour working days for six days of the week. Two shifts of 12 hours each are now working in the Philadelphia Mint. Until a few weeks ago there were three shifts of eight hours each. But when Freas Styer succeeded Adam Joyce as superintendent of the mint the third shift was put to work counting the money in the vaults in the making of an audit due to the change in administration. In the four months since April, 20 million silver dollars have been coined. There remains on hand to be pressed into coin of the land 30 million ounces of silver, which will make approximately 35 million silver dollars.
>
> Robert Clark, superintendent of the coinage in the Philadelphia Mint, says that the greatest production in the history of the plant is now being obtained. The daily

average in production for the last month has been 260,000 silver dollars. In some days it has run as high as 275,000. That rate will be maintained until the present supply of silver is exhausted, and then the Mint for a time will go back to the coining of the smaller coins. . . .

The economy lapsed into the doldrums, and mintages of coins other than silver dollars and double eagles were reduced in 1921 and 1922.

On April 3, 1923, the Treasury Department made the announcement that the Mint Collection, earlier known as the Mint Cabinet, would be transferred (not given) from Philadelphia to the Smithsonian Institution. Not stated was the fact that certain officers in Philadelphia found visitors and their constant questions to be annoying. The immediate reason was that the daring robbery at the Denver Mint on December 18, 1922, in which a Mint guard was killed and $200,000 was stolen, precipitated a closing of all mints to visitors. On May 28 over 18,000 coins and other numismatic items were moved. Theodore L. Belote, the Smithsonian's curator of the Division of History, was in charge of the coins, which remained in their shipping containers. It was not until November 26 that unpacking commenced, which was completed on January 21 of the next year.

Meanwhile the national economy was in a rapid growth period. In coming years prices rose on stock markets, many large buildings were erected in cities, luxury items sold well, and there was a general air of prosperity. Nationwide Prohibition had been in effect since 1920, catalyzing a crime wave involving liquor smugglers and illicit distillers.

In 1925 Chief Engraver George T. Morgan passed away and was replaced in the post by John R. Sinnock, who had been an assistant to him at the Mint for two years.

The Annual Report of the Director of the Mint for the fiscal year of 1928 included this:

> *Additions and Improvements:* At the Philadelphia Mint a chromium plating plant has been installed and is being used for greatly improving the wearing qualities of dies, coin collars, machinery parts, and models. A new type of reducing machine has greatly facilitated the preparation of more perfect coin and medal dies. Improvements have been made whereby the time required for finishing medals by sandblast has been reduced by one-half.

In 1929 quarter eagles were minted for the last time. All during the 1920s they had been available from banks only by paying a premium. They were officially discontinued by the Act of April 11, 1930.

Storm clouds gathered on the economic horizon, and in autumn 1929 prices crashed on the New York Stock Exchange. Bankers, brokers, and others said the setback was temporary.

THE 1930s

Despite President Herbert C. Hoover's comment that "prosperity is just around the corner," the depression, later to be known as the Great Depression, deepened. In 1932 Congress established the Reconstruction Finance Corporation in an effort to stimulate the economy. Bonus marchers, as they were called, marched in Washington, claiming bonuses for service in the World War. Most left after Congress rejected their plea, and troops led by Douglas MacArthur forcibly removed the others. Unemployment was at a record 24.1 percent. Radio City Music Hall opened in New York City and featured the Rockettes chorus line and the largest Wurlitzer theatre pipe organ ever made. Not far away the Empire State Building was nearing completion, but renters for space were scarce. The Washington quarter, first intended as a commemorative and then viewed as a regular issue, was first struck in 1932, the 200th anniversary of the first president's birth. Coinage production at the three mints was reduced during this era.

The November 8 election drew more interest than any other such contest in American history. Incumbent Republican Herbert Hoover sought reelection on his record (which he thought was excellent) and the promise that good economic times were in the offing. Democrat Franklin D. Roosevelt saw no such possibility and offered a "New Deal" for Americans, with specific plans for new programs. On September 9, 1932, with the election in the offing, the Dow-Jones Industrial Average closed at 41.63, down 91 percent from its level on the same day in 1929. Unemployment increased, factory production decreased, and many banks, securities firms, and others became bankrupt. Over 5,000 banks had failed by this time. Roosevelt won in a landslide. Wooden nickels were introduced in Tenino, Washington, for use in local commerce after the town's only bank closed. "Brother, can you spare a dime," became a sad but popular question on the street.

The End of Gold Coinage

On March 9, 1933, an act titled "An Act to Provide Relief in the Existing National Emergency in Banking, and for Other Purposes" was passed by Congress. By the time of Roosevelt's inauguration in April 1933 there was further widespread concern for the security of the American monetary system.

On April 5 the president demanded that gold coins, Gold Certificates, and bullion held by the general public were to be turned over to banks and the Federal Reserve System by May 1. Amounts of coins and Gold Certificates less than $100 were exempted, as were coins of numismatic value. This seems to have been done with the advice and consent of his secretary of the Treasury, William H. Woodin, years earlier a familiar figure in numismatic circles.

By that time coinage for the year was already substantial. In January alone 50,000 $10 pieces bearing the 1933 date had been minted; 200,000 additional pieces were minted in February, and 62,500 were struck in March. Of the 1933-dated $20 piece, 100,000 were struck in March, 200,000 in April, and 145,500 in May. As the Roosevelt proclamation was made on April 5, 1933, it is reasonable to conclude that anyone visiting the Treasury Department in March 1933 asking Woodin to obtain a specimen by exchanging another coin could have legally acquired a double eagle bearing that date. However, in later years the Treasury Department took the strong position that, while 1933 $10 pieces could be legally held by collectors, the 1933-dated $20 pieces were illegal to own and were subject to confiscation, as indeed happened on several occasions.

Woodin, in very poor health, turned his duties over to acting secretary Henry Morgenthau Jr., who on December 28, 1933, issued the Order for the Return of All Gold, emphasizing that banks and individuals must surrender all gold coins and certificates in their possession (revoking the $100 exemption) excepting numismatic coins and quarter eagles, for which they would be paid $20.67 an ounce. All quarter eagles, a denomination discontinued in 1929, were automatically considered to be of numismatic value. Jewelry manufacturers and certain others could obtain licenses to handle gold.

On December 28 the Treasury estimated that as of November, about $40,000,000 in gold had been turned in by the public since April. At that time there was outstanding $311,044,985 in gold coin and bullion and $217,486,829 in gold certificates in circulation among the general public.

In 1933 mintages of minor and silver coins were reduced, with no production of nickels, dimes, quarters, half dollars, or dollars.

In January 1934 the Treasury raised the price of gold to $35 per ounce. Citizens who had turned in their gold for $20.67 were the losers. Not much was said about gold confiscation at the time. Many people, probably the majority of citizens, felt that whatever the Roosevelt administration did was right—especially after the helpless floundering of his predecessor, Herbert Hoover.

In 1934 signs of an economic recovery began appearing, and with this came a demand for new coins, starting an upward trend. Across the nation hobbies were booming. A lot of people had spare time on their hands. Jigsaw puzzles, crossword puzzles, stamps, and other pursuits drew millions. Coin panels and albums became popular, and hundreds of thousands of people whiled away time looking for coins in pocket change. Numismatics prospered as never before, counter to the still-ill but improving economy.

In 1935 Lee F. Hewitt launched the *Numismatic Scrapbook Magazine*, soon to be the most popular coin periodical ever. Gold coins, including large denominations, attracted many buyers who felt they were a good store of value in an economy that, with Roosevelt innovations, had an uncertain future. Whitman Publishing Company gained a foothold in numismatic supplies by acquiring the business of J.K. Post of Neenah, Wisconsin, and expanding it. That year saw the popularity of the "Penny Board," in which circulation Lincoln cents could be placed, including, hopefully, the highly prized 1909-S V.D.B. and 1914-D.

Commemorative silver coins, which had been produced intermittently since 1892, expanded with new issues in 1935 and a record number in 1936. In the meantime, gold coins continued to draw new collectors. Thomas L. Elder, a New York City dealer since 1904, sent lists of rare-date gold coins to banks and did a strong business buying coins that the public continued to turn in for face value. Countless thousands of scarce issues were thus rescued from the melting pot.

The Philadelphia Mint and the Federal Reserve Banks in the East stored large quantities of surrendered as well as unissued gold. Many employees switched common coins for rare ones and made a beeline to Abe Kosoff, Stack's, Wayte Raymond, and other New York dealers to sell them for a premium—numismatic Robin Hoods, so to speak, for rescuing countless pieces. It worked out just fine for the Treasury, which retained the same face value when wholesale melting began in 1937.

A NEW ERA

By the late 1930s the economy had recovered substantially, the banks that were still in business were operating profitably, and quantity coinage of cents, dimes, quarters, and half dollars was underway. Silver dollars were not minted after 1935. The issuance of commemoratives, which had boomed in 1935 and 1936, was terminated in 1939 due to abuses in distribution, and market prices dropped. No matter—the rest of the coin market was healthy and growing stronger. Hobbies, including numismatics, became more popular in America than ever before.

Proofs of copper, nickel, and silver coins, not made after 1915, were resumed in 1936. Coins could be ordered singly or in sets of five. At the Philadelphia Mint, special coining and packaging facilities were put in place.

In 1937 the Japanese army invaded China and other territories, and in Germany the Nazi movement under dictator Adolph Hitler gained strength. In 1939 the Nazis took over Poland and Czechoslovakia, and in 1940 the Battle of Britain saw the relentless bombing of London. As had been the case in the World War a quarter century earlier, America became the "arsenal of democracy" and furnished military goods to those fighting Germany, including ships and planes. On December 7, 1941, Japanese planes bombed the American Navy base at Pearl Harbor in Hawaii. The United States went to war.

On the home front the economy prospered as factories were kept busy day and night. Consumer goods were scarce and money was common—causing the inflation of prices, including for rare coins. The effect on coinage was significant. Production increased at the Philadelphia, Denver, and San Francisco mints. Nickel, used in five-cent pieces, was in demand for military goods, and in 1942 the alloy was changed to a silver composition, which remained the standard through 1945. Proof coins were struck in 1942 and then discontinued in view of the urgency to mint regular coins for circulation in unprecedented quantities. Copper was in demand for the war effort as well, and in 1943 these coins were made of zinc-coated steel. This did not work out well, the coins became spotted, and copper alloy was resumed in 1944.

THE POSTWAR YEARS

In 1946 the Roosevelt dime was introduced, replacing the Winged Liberty Head (Mercury) type. Commemorative half dollars were made again for the first time since 1939. Chief Engraver John R. Sinnock died on May 14, 1947. He was succeeded in the post by Gilroy Roberts.

In 1948 the Franklin half dollar replaced the Liberty Walking half dollar. In 1950 the production of Proof coins was resumed, this time sold only in sets of five coins—cent, nickel, dime, and quarter—for $2.10, in cellophane sleeves put up in small square gray boxes. Launched on January 17 of that year, these became very popular—so much so that a restriction of five sets per person was made. In Iola, Wisconsin, Chester ("Chet") L. Krause started *Numismatic News*, which two decades later evolved into Krause Publications, an empire with periodicals and books in several categories. In the meantime *The Numismatist* and the *Numismatic Scrapbook Magazine* continued to flourish.

In 1955 at the Philadelphia Mint a press turned out Lincoln cents with blurred or doubled letters on the obverse—the 1955 Doubled Die cent as it was later called. About 24,000 of these were released into circulation. Within a year or two they were in strong numismatic demand. They launched the widespread popularity of die errors, and the study of various die errors dating back to the 1790s became a specialty in the hobby. Coins from proper dies that were misstruck on presses—earlier called freaks—became very popular as well. The supply of these was augmented on occasion by Mint employees who exchanged other coins for them. That did not last long, as all three mints put security checks into effect.

After minting cents and dimes, the presses at the San Francisco Mint were stilled. Soon the name was changed to the San Francisco Assay Office (see chapter 8). Coinage dies for the mints, now with Denver being the only one other than Philadelphia, continued to be made in Philadelphia.

THE 1960s

At the Philadelphia Mint, where dies were made for coinage at all mints, early in 1960 a smaller date was used on the Lincoln cents, then slightly increased. In the spring it became evident that there were Small Date cents, which were scarce, and Large Date cents, which constituted the majority. The Philadelphia Small Date cents garnered the most attention, but there was some notice concerning the more readily available 1960-D Small Date cents. A $50 face-value bag of 1960 Small Date cents rose in value to cross the $10,000 mark, some selling for $12,000. That year *Coin World* was launched by the Sidney Printing & Publishing Company in Sidney, Ohio, the first weekly coin periodical.

A boom was on, investment in coins was the rallying cry for many newcomers, thousands of retail coin shops were opened across the country, excitement prevailed, and prices rose. At one point the circulation of *Coin World* crossed the 150,000 mark.

THE SILVER SENSATION

In November 1962 a vault sealed since 1929 in the Philadelphia Mint was opened to extract bags of old silver dollars, as this denomination, no longer seen in circulation, was popular for holiday gifts. At the time, collecting Morgan and Peace dollars was popular with numismatists but not in the mainstream. The most valuable Morgan dollar was the 1903-O, of which fewer than a dozen Uncirculated pieces were believed to exist and which commanded the top price among listings in the *Guide Book of United States Coins*: $1,500. But lo and behold!: multiple 1,000-coin bags of 1903-O Morgan dollars were found in the vault, as these and other New Orleans coins had been shipped from that long-closed mint in 1929 and not been seen since. Virtual pandemonium took place as collectors and dealers scrambled to obtain the precious coins, soon augmented by two other former rarities, the 1898-O and 1904-O. From all parts of the country, requests were made to Federal Reserve Banks to supply whatever silver dollars could be found, and millions were shipped to different banks to be eagerly bought by dealers, collectors, and investors.

This caused a "run" on silver dollars stored in bank vaults across the country. By March 1964 no more were available for face value anywhere—not even on gaming tables in Las Vegas where they had been used for decades. At the same time the price of silver bullion was rising on international markets. In 1965 it cost more than face value to mint a dime, quarter, or half dollar. To replace silver, a clad cupro-nickel alloy was used for dimes and quarters beginning in 1965, and a composite planchet with outer silver layers was used for half dollars. Millions of older silver coins were purchased for premiums by dealers and were sold to refineries.

Silver coins in circulation disappeared into the hands of citizens, including speculators. Mint Director Eva Adams, appointed in 1961 by President John F. Kennedy, placed the blame squarely on numismatists, and as punishment stopped making Proof sets. As further punishment she decreed that mintmarks were to no longer appear on coins struck at the Denver Mint. Presses at the San Francisco Assay Office were put back in action. All mintmarkless issues of 1965 through 1967 resembled Philadelphia coins. Facilities at the Frankford Arsenal in Philadelphia were used to anneal and clean coin planchets for use at the Philadelphia Mint.

In 1964 Franklin half dollars were replaced by the introduction of the Kennedy half dollar. Chief Engraver Gilroy Roberts designed the obverse and assistant engraver Frank Gasparro the reverse. The General Numismatics Corporation, later known as the Franklin Mint, was expanding its private enterprise. Joseph Segel, founder of that company, made Roberts an offer to become an executive that he could not refuse. Roberts resigned from his Mint post on October 8, 1964, after 16 years of service. During that time he had been responsible for the design and die production of nearly all U.S. coins, foreign coins struck at the Mint, and presidential medals, as well as national and private commemorative medals.

Both the Philadelphia and Denver mints struck all five denominations bearing the 1964 date right through 1965. Philadelphia kept striking 1964 silver half dollars into early 1966. The Treasury had a large supply of silver, and some thought that the market price would come down. It didn't.

The first copper-nickel–clad coins were minted in August 1965, and it was not until November that any of these were actually released into circulation. The quarter dollar appeared first, followed by clad dimes in February of 1966. The silver-clad (.400 fine) half dollars were not minted until the last few days of 1965, though the entire 1965-dated mintage was recorded within 1966 to simplify the report. These coins debuted in circulation the following spring.

The story of Eva Adams, previously mentioned, is most curious. A few years after vacating her directorship in August 1969 she "got religion," seeking and making many friendships in the numismatic community. She even served on the ANA Board of Governors from 1971 to 1974. The ANA awarded her the Medal of Merit, and for some unfathomable reason the Numismatic Literary Guild bestowed its highest award on her, although apart from a reprinted speech, no one could remember anything she had written! The *Annual Reports* with her name on them had been prepared by staff.

In 1968 the mintage of Proof sets, which had been suspended since the coinage of 1964, was resumed, but at the San Francisco Mint rather than the traditional venue of Philadelphia. Henceforth, Proof coins would have an "S" mintmark.

Coinage of the Third Mint, 1901–1969

The third Philadelphia Mint opened with the latest technology. In the years of its operation it produced many coins that are numismatic favorites today.

Mintage totals listed in *A Guide Book of United States Coins* for certain coins from 1964 through 1967 include branch-mint coins struck without mintmarks—when Mint Director Eva Adams decided to punish numismatists for causing a nationwide coin shortage (which in actuality had little to do with coin collectors).

Cents

The production of cents at the third Mint was continuous from 1901 to 1969 with the exception of 1922, when cents were only struck at Denver. None of the Indian Head or Lincoln cent dates are rare, although there are some elusive varieties such as the 1955 Doubled Die. In 1960 the Small Date cent created a market sensation when $50 bags sold for $10,000 or more, launching a boom that lasted through 1964.

Design Types and Dates Minted

Indian Head, Oak Wreath With Shield: *Variety 3,* **Bronze:** 1901, 1902, 1903, 1904, 1905, 1906, 1907, 1908, 1909.

Lincoln, Wheat Ears Reverse: *Variety 1, Bronze:* 1909 (with and without V.D.B.), 1910, 1911, 1912, 1913, 1914, 1915, 1916, 1917, 1918, 1919, 1920, 1921, 1923, 1924, 1925, 1926, 1927, 1928, 1929, 1930, 1931, 1932, 1933, 1934, 1935, 1936, 1937, 1938, 1939, 1940, 1941, 1942, 1944, 1945, 1946, 1947, 1948, 1949, 1950, 1951, 1952, 1953, 1954, 1955, 1956, 1957, 1958. *Variety 2, Zinc-Coated Steel:* 1943.

Lincoln, Memorial Reverse: *Copper Alloy:* 1959, 1960, 1961, 1962, 1963, 1964, 1965, 1966, 1967, 1968, 1969.

Nickel Five-Cent Pieces

Liberty Head nickels were produced continuously from the opening of the Mint in 1901 to 1912. In 1913 five Liberty Head nickels were struck under unrecorded circumstances. James Earle Fraser's Indian Head or Buffalo design was made through 1937, except for 1931 to 1933. The Jefferson design was minted from 1938 to 1969 with a variation in metal content during World War II.

Design Types and Dates Minted

Liberty Head: 1901, 1902, 1903, 1904, 1905, 1906, 1907, 1908, 1909, 1910, 1911, 1912, 1913.

Indian Head or Buffalo, Variety 1, FIVE CENTS on Raised Ground: 1913.

Indian Head or Buffalo, Variety 2, FIVE CENTS in Recess: 1913; 1914, 4 Over 3; 1914; 1915; 1916; 1917; 1918; 1919; 1920; 1921; 1923; 1924; 1925; 1926; 1927; 1928; 1929; 1930; 1934; 1935; 1936; 1937.

Jefferson: *Nickel:* 1938, 1939, 1940, 1941, 1942, 1946, 1947, 1948, 1949, 1950, 1951, 1952, 1953, 1954, 1955, 1956, 1957, 1958, 1959, 1960, 1961, 1962, 1963, 1964, 1965. *Designer's initials (FS) added:* 1966, 1967. *Wartime Silver Alloy:* 1942-P; 1943-P, 3 Over 2; 1943-P; 1944-P; 1945-P.

Dimes

Dimes were struck continually at the third Mint except for 1922, 1932, and 1933. The Barber design was minted from 1901 to 1916, the Winged Liberty Head (Mercury) from 1916 to 1945, and the Roosevelt design from 1946 to date. The Mercury dime in particular has been a numismatic favorite, although the others are avidly collected as well. Beginning in 1965 all have been struck in clad composition. All years are readily collectible.

Design Types and Dates Minted

Barber or Liberty Head: 1901, 1902, 1903, 1904, 1905, 1906, 1907, 1908, 1909, 1910, 1911, 1912, 1913, 1914, 1915, 1916.

Winged Liberty Head or "Mercury": 1916; 1917; 1918; 1919; 1920; 1921; 1923; 1924; 1925; 1926; 1927; 1928; 1929; 1930; 1931; 1934; 1935; 1936; 1937; 1938; 1939; 1940; 1941; 1942, 42 Over 41; 1942; 1943; 1944; 1945.

Roosevelt: *Silver:* 1946, 1947, 1948, 1949, 1950, 1951, 1952, 1953, 1954, 1955, 1956, 1957, 1958, 1959, 1960, 1961, 1962, 1963, 1964. *Clad:* 1965, 1966, 1967, 1968, 1969.

Quarter Dollars

Barber quarters were minted from 1901 to 1916, including Proofs during the same time. In 1916 the Standing Liberty design made its debut and was continued in production until 1930, except for the recession year of 1922. These are high on the popularity list with numismatists today. Washington quarters were made continuously from 1932 to 1969, except for 1933, and are widely collected.

Design Types and Dates Minted

Barber or Liberty Head: 1901, 1902, 1903, 1904, 1905, 1906, 1907, 1908, 1909, 1910, 1911, 1912, 1913, 1914, 1915, 1916.

Standing Liberty, Variety 1, No Stars Below Eagle: 1916, 1917.

Standing Liberty, Variety 2, Stars Below Eagle, Pedestal Date: 1917, 1918, 1919, 1920, 1921, 1923, 1924.

Standing Liberty, Variety 2, Stars Below Eagle, Recessed Date: 1925, 1926, 1927, 1928, 1929, 1930.

Washington: *Silver:* 1932, 1934, 1935, 1936, 1937, 1938, 1939, 1940, 1941, 1942, 1943, 1944, 1945, 1946, 1947, 1948, 1949, 1950, 1951, 1952, 1953, 1954, 1955, 1956, 1957, 1958, 1959, 1960, 1961, 1962, 1963, 1964. *Clad:* 1965, 1966, 1967, 1968, 1969.

Half Dollars

Half dollars of the Barber design were made continuously from 1901 to 1915. In 1916 the Liberty Walking half dollar, called Liberty *Striding* by the Mint, was introduced. Throughout the series, especially in the early years, there was difficulty in having the design strike fully, with the result that the head of Liberty and her left hand are often weak. Coinage was intermittent, dating at Philadelphia from 1916 to 1921, followed by a large gap to 1934. After that coinage was continuous to the end of the design in 1947.

The Franklin design was first minted in 1948. This was a slow time in the coin market, and collector interest was muted. Coinage continued through 1963. In 1964 the Kennedy design replaced it, made in silver the first year, then in silver clad composition from 1965 to 1969. None of the Kennedy half dollars circulated widely.

Design Types and Dates Minted

Barber or Liberty Head: 1901, 1902, 1903, 1904, 1905, 1906, 1907, 1908, 1909, 1910, 1911, 1912, 1913, 1914, 1915.

Liberty Walking: 1916, 1917, 1918, 1919, 1920, 1921, 1934, 1935, 1936, 1937, 1938, 1939, 1940, 1941, 1942, 1943, 1944, 1945, 1946, 1947.

Franklin: 1948, 1949, 1950, 1951, 1952, 1953, 1954, 1955, 1956, 1957, 1958, 1959, 1960, 1961, 1962, 1963.

Kennedy: *Silver:* 1964. *Silver Clad:* 1965, 1966, 1967.

Silver Dollars

When the new Mint opened for business in 1901, production of Morgan dollars continued. In 1904 the supply of authorized bullion ran out, and mintage was discontinued, only to be resumed in 1921. The first World War ended with the armistice of November 11, 1918. In *The Numismatist* that month, editor Frank Duffield recommend that a "victory coin" would be desirable. More was said on the subject by others. A latecomer was former ANA president Farran Zerbe, who on August 15, 1920, at the ANA convention proposed that to commemorate the late war, a peace coin should be made. This became a reality in 1921 when sculptor Anthony de Francisci was selected to design what became known as the Peace dollar. On the obverse the portrait of Miss Liberty was a close copy of that used by Augustus Saint-Gaudens for the 1907 $10 coin, and the reverse depicted a perched eagle.[74] Peace dollars were struck intermittently through 1935.

Silver dollars of the Morgan and Peace types were available for years afterward, usually upon application to banks. They did not actively circulate in commerce except in a few Rocky Mountain states. In November 1962, when the Philadelphia Mint vault sealed since 1929 was opened to get a supply of dollars to pass out as holiday gifts, hundreds of thousands of mint-state 1903-O dollars were found—up to that time considered to be a great rarity. These listed for $1,500 in the *Guide Book of United States Coins* of that year. Most numismatists had never seen one. A mad rush was precipitated, and from then through March 1964 when the Treasury stopped paying them out (reserving about three million pieces) and bank vaults were emptied, several hundred million went into the hands of the public. Silver dollars, hitherto popular in Nevada gaming casinos, were no longer seen.

Design Types and Dates Minted

Morgan: 1901, 1902, 1903, 1904, 1921.

Peace: *High Relief:* 1921, 1922. *Low Relief:* 1922, 1923, 1924, 1925, 1926, 1927, 1928, 1934, 1935.

$2.50 Quarter Eagles

Liberty Head quarter eagles were made from 1901 to 1907, continuing the same design that had been used without major change since 1840. The Indian Head design replaced it in 1908 and was coined through 1915, then again from 1926 to 1929. Coins of this denomination were difficult to obtain at face value beginning in 1917 and continuing to the end of the series.

Design Types and Dates Minted

Liberty Head: 1901, 1902, 1903, 1904, 1905, 1906, 1907.

Indian Head: 1908, 1909, 1910, 1911, 1912, 1911, 1913, 1914, 1915, 1926, 1927, 1928, 1929.

$5 Half Eagles

Liberty Head half eagles were made from 1901 to 1908, continuing the Liberty Head with motto IN GOD WE TRUST that had been used since 1866. The Indian Head design replaced it in 1908 and was coined through 1915, then again in 1929, the last year being numismatically rare today.

Design Types and Dates Minted

Liberty Head, Variety 2, Motto Above Eagle: 1901, 1902, 1903, 1904, 1905, 1906, 1907, 1908.

Indian Head: 1908, 1909, 1910, 1911, 1912, 1913, 1914, 1915, 1929.

$10 Eagles

Liberty Head eagles were made from 1901 to 1907, continuing the Liberty Head with motto IN GOD WE TRUST that had been used since 1866. In 1907 and 1908 the Indian Head design by Augustus Saint-Gaudens was used without the motto, as President Theodore Roosevelt felt that the use of the Deity's name on coinage was sacrilege. Congress added IN GOD WE TRUST partway through 1908, from which time eagles were coined through 1915 and again in 1926, 1932, and 1933, the latter being rare today.

Design Types and Dates Minted

Liberty Head, Motto Above Eagle: 1901, 1902, 1903, 1904, 1905, 1906, 1907.

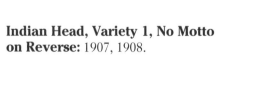

Indian Head, Variety 1, No Motto on Reverse: 1907, 1908.

Indian Head, Variety 2, Motto on Reverse: 1908, 1909, 1910, 1911, 1912, 1913, 1914, 1915, 1926, 1932, 1933.

$20 Double Eagles

Liberty Head eagles were made from 1901 to 1907, continuing the Liberty Head with the denomination lettered as TWENTY DOLLARS, as had been in use since 1877. In December 1907 the innovative double eagle by Augustus Saint-Gaudens, with the date in Roman numerals as MCMVII and with the features in high relief, was released. Coinage from the same dies continued into early 1908, by which time slightly over 12,000 had been coined. These created a sensation and quickly sold for a premium. Probably half or more of the mintage survives today.

In 1907 Chief Engraver William E. Barber modified the design by lowering the relief and using regular or "Arabic" numerals. These were minted from December 1907 through part of 1908 and lacked the motto IN GOD WE TRUST, as President Theodore Roosevelt felt that the use of the Deity's name on coinage was sacrilege. Congress added IN GOD WE TRUST partway through 1908, from which time double eagles were coined through 1915, then again from 1920 to 1933. Those of 1929 and later were mostly stored and melted, making them rare today—the 1933 especially so, with fewer than 20 believed to exist.

Design Types and Dates Minted

Liberty Head, Motto Above Eagle, Value TWENTY DOLLARS: 1901, 1902, 1903, 1904, 1905, 1906, 1907.

Saint-Gaudens, Without Motto IN GOD WE TRUST, High Relief, MCMVII: 1907.

Saint-Gaudens, Arabic Numerals, No Motto: 1907, 1908.

Saint-Gaudens, With Motto IN GOD WE TRUST: 1908; 1909, 9 Over 8; 1909; 1910; 1911; 1912; 1913; 1914; 1915; 1920; 1921; 1922; 1923; 1924; 1925; 1926; 1927; 1928; 1929; 1931; 1932; 1933.

FOURTH PHILADELPHIA MINT (1969 TO DATE)

OPENING OF THE NEW MINT

On Thursday, August 14, 1969, the fourth Philadelphia Mint was officially dedicated. On hand were Treasury Department officials, American Numismatic Association members (the annual convention was in progress in the city), and officials from mints in Austria, Canada, China, England, France, Israel, Italy, Mexico, the Netherlands, Norway, South Africa, Spain, and West Germany.

The tour of the facilities included a demonstration of new "super presses" made by General Motors capable of turning out 10,000 coins per minute. Built at the GM plant in Warren, Michigan, at a cost of $1.5 million, it was expected to revolutionize coinage. This was a "roller press" with multiple obverse and reverse dies mounted around the periphery of two large wheels. At the point where the wheels were closest together, a planchet strip was fed in, and at very high speed coins (in the form of cent-sized bronze coins from fantasy dies) spewed out. Although tests with cent-sized dies bearing fantasy dies were done in GM facilities and enthusiastically publicized (and some Lincoln cents were struck in 1969), in actuality the dies suffered breakage after just a short time of use, and the test coins were defective. The whole matter was a failure. The Mint did not include the number of roller-press cents made in its reports of total production.[75]

FACILITIES AT THE MINT

Although policies were determined by the Treasury Department and at the office of the Bureau of the Mint in Washington, D.C., the Philadelphia Mint remained the central facility for experiments, innovations, and other activities that extended to the branch mints in Denver, San Francisco, and the West Point Bullion Depository. The Engraving Department was there, as were all facilities for making master dies, hubs, working dies, and other aspects of die production. A gallery was open to visitors and permitted views of the coinage operation. A gift shop offered coins and medals for sale. The only other gift shop in the Mint system was at Denver.

Prior to 1985, mintmarks were applied to working dies in the Engraving Department of the Philadelphia Mint. A supply of D and S punches were on hand in a room within the department. An appropriate punch was placed over the proper area of the die and tapped lightly with a hammer. This was a tried and true practice dating back to the making of the first branch-mint dies in 1838. By the mid-1980s the Mint realized that this process sometimes produced varieties that caused excitement among collectors—such as an S mintmark placed over a D. This was done as a matter of economy—to render useful a die that was surplus to the needs of one mint and making it fit for the other branch. Beginning in 1985, the mintmark for Proof dies was placed in the master die used to make working hubs which in turn were used to make working dies. In 1986 the mintmark was placed on original models for commemorative and other special collector-oriented coins, one step earlier than placing it on the master die. In 1990 mintmarks were placed on all master dies for circulating Lincoln cents and Jefferson nickels. In 1991 this practice was extended to the higher denominations. Since that time, the mintmark for a given coin denomination, whether S, D, or for some later a P mintmark for Philadelphia, has not varied in style or position. This resulted of course in Mint efficiency, never mind that no more mintmark errors or oddities take place to delight numismatists.

VISITING THE MINT IN 1988

From May 25 to 27, 1988, I visited the Philadelphia Mint with fellow numismatist Bill Fivaz and photographer Cathy Dumont, arranged by Mint Director Donna Pope. While there we spent time with the employees in all of the departments and were welcomed warmly.

Starting on the production floor, we watched tightly coiled strips of metal being fed into noisy blanking machines. The resulting blank discs were sorted by a mechanical sieve or riddle to remove those that were under or oversized. Later processes included annealing to soften them, cleaning (with a mixture of water, brine, and detergent), drying, and running through one of six milling machines to impart raised rims for easier metal flow when the coins were struck. During our visit, cents, nickels, dimes, and quarters were being made. Kennedy half dollars, by that time made mainly for collectors, were nowhere in sight.

Coiled metal strips from which planchets are punched.

On the production floor were rows of coining presses made by E.W. Bliss, a division of Gulf & Western, a conglomerate that at the time got lots of attention in stock-market news. There were 72 coining presses in all, not all of which were being used. We watched while planchets for copper-coated–zinc cents were fed from a hopper above into a rotating vibrating tray, then fed into two separate chutes, each with two rows of cent planchets that were then fed into the press. The press had two sets of die pairs for a total of four obverse and four reverse dies. From there the coins were mechanically dumped into a bin. Upon examination we saw some cents with prominent striations on their faces, from dies that were not completed finessed. In 1988 the quality of a typical coin made for circulation was lower than it would be in the early 21st century.

From there the cents were taken in bins hoisted by a crane and dumped into riddling machines. Some coins were churned around in the riddling machine for a long time. Cents of the correct diameter passed through holes in the machine, then slid through a device that rejected any that were too thick. Every once in a while an attendant stopped the machine and removed errors that were too large to go through the holes. While to err is divine to numismatists, at the Mint each error is viewed as a mistake, not as a collectible, and every effort is made to catch them.

Bins of finished coins measure about four- to five-feet long, three-feet wide, and three-feet high and are marked with the net weight. One bin of cents weighed 2,791 pounds, and another 3,138 pounds. Nowhere along the line were they visually inspected. The heavy bins were lifted to a platform, a chute was opened on one side, and the coins went into a funnel and then into a hopper. An automatic counting machine then dumped cents into bags of $50 each. Most bags were dated May 26, 1988, but some had the impossible day of May 36, an error. In another area cents were put into large plastic bags by weight, 4,000 pounds, without being counted. These were to be sent to the Federal Reserve System. When the Mint is busy there are three shifts a day, Monday through Friday. The facility is closed on weekends.

We then went to watch nickels being struck at the rate of 100 coins every 25 seconds, again in presses with four dies each. We also watched quarters being made.

A riddling machine sorting planchets.

A bin of freshly minted Jefferson nickels.

After leaving the coin-production area we went to the die shop to see the processes by which master dies, hubs, and working dies are made from sections of steel rods with one end finished in a tapered point. Don Quattrone, a long-time employee, showed us through. A typical die for a cent or dime will strike 600,000 to 1 million coins, we were told. For quarters, 450,000 to 500,000 is normal. Quattrone explained that dies first show wear in the fields, called pitting. At an appropriate time they are replaced. If not they will crack or break. When a small part near the edge breaks it is called a "piece out," creating what collectors call a cud on finished coins. Other parts of the die wear as well, such as the head of Roosevelt on the dime and the less-detailed head of Washington on the quarter. Proper heat treating is a factor in extending die life. If one die in a group of four fails in the press, all four are discarded, on the theory that the other three have had similar, sufficient wear. This was a new procedure at the time of our 1988 visit, as earlier the dies were inspected and just one of the four was thrown away.

We then viewed some dies that had problems and were taken out of service as a result. Some obverse and reverse dies had clashed with each other without a planchet between, causing incusations on both dies. A cent die had a chip out of the field of the die below the word LIBERTY. Such dies were put into a bucket marked "Condemned."

During our days at the Mint we also met with Chief Engraver Elizabeth Jones, by then a fine friend. We visited with engravers Edgar Zell Steever, Michael G. Iacocca, Sherl Joseph Winter, John Mercanti, Chester Morris, and James Charles Licaretz. Mr. Iacocca was in charge of putting mintmarks on coins and even let me punch a D into a Roosevelt-dime die. With a wink, I said that I should have punched an S first, and then a D over it! No chance. We were all on our best behavior. It was a rare treat to see and learn so much. Mr. Morris was retouching the master for the 1989-S quarter, this in advance for the next year. He was strengthening the hair and moving the

initials JF down on the coins. We had a lengthy chat about the Washington portrait, the Jean Antoine Houdon bust from which it was adapted, and of other matters. Miss Jones showed some of her sketches and plasters in her large private office—a familiar room as we had visited it earlier many times, especially when Frank Gasparro was chief engraver. The artists were all tuned into current happenings in numismatics, were readers of *Coin World* and *Numismatic News*, and were sensitive of the wishes and requirements of Representative Frank Annunzio, who was the main contact between Congress (where all of the decisions are made) and the Mint.

Dave Bowers examines plasters being shown to him by Chief Engraver Elizabeth Jones in her office.

We then went to the Medal Department to watch a small-diameter Lincoln medal being struck carefully on an old Ferracute knuckle-action press at the rate of about 70 per minute. Another press was larger, could be adjusted to 350 tons of pressure, and was striking 3-inch medals honoring the late movie actor John Wayne, a popular seller ever since its introduction in 1979. The planchets were purchased by the Mint pre-punched and ready to use. Each medal had to be trimmed on a lathe, then dipped in nitric acid, rinsed, and then sandblasted.

We also watched the later processing, on this day involving the packaging of Lyndon B. Johnson medals.

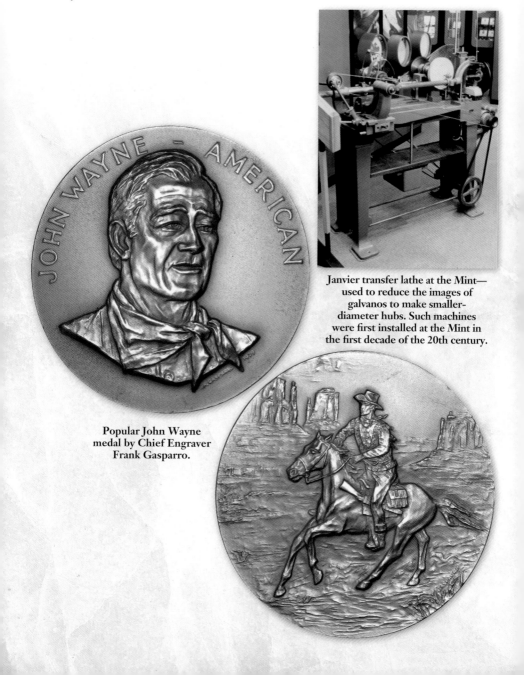

Janvier transfer lathe at the Mint—used to reduce the images of galvanos to make smaller-diameter hubs. Such machines were first installed at the Mint in the first decade of the 20th century.

Popular John Wayne medal by Chief Engraver Frank Gasparro.

Toward the end of our visit we chatted with Eleanor McKelvey and discussed numismatic research. She pointed us to some files she had and also suggested areas of the National Archives to explore. At the end of the third day we were tired and happy, and headed home.

TRANSITIONS

In October 1992 President George H.W. Bush signed authorization to establish an enterprise fund for the Mint, so that profits from the sales of Proof coins, commemoratives, and other numismatic and premium-cost programs would go to the Mint instead of into the general Treasury fund. This permitted the Mint to spend more money on marketing and research for such programs. Most of this was done from the Mint's main office in Washington, D.C.

In 1993 there were changes in Mint positions. The post of chief engraver had been vacant since Elizabeth Jones resigned on December 13, 1990, and now was eliminated, as was the post of assayer at each of the three mints. The title of superintendent for the mints was changed to plant manager. These roles (along with that of the Mint director) had been patronage positions since the 1870s, filled by appointment, with their officeholders serving at the pleasure of the president of the United States. Superintendents often were party officials and fundraisers for whoever won the presidency, with the day-to-day heavy lifting of management falling to more qualified assistants. In the early 1990s Secretary of the Treasury Lloyd Bentsen sought to eliminate the politically filled positions. He brought on board Philip Diehl, chief of staff of the Treasury, who had previously served as staff director of the Senate Finance Committee. Diehl took over as Mint direc-

tor in Washington, and by the time he left office in 2000 he had abolished nine of the Mint's ten patronage positions.

At the Denver Mint on May 13, 1993, a die-making shop went into operation, adding to the facility in place at the present Philadelphia Mint and its predecessors dating back to 1838. Now, up to 150,000 coinage dies could be made each year.

The P mintmark was added to dollars beginning with the Susan B. Anthony 1979-P issue and on nickels, dimes, quarters, and half dollars beginning in 2000.

A master die for a 1994 Washington quarter at the Philadelphia Mint in autumn 1993, getting ready for the 1994 coinage. A master die is in relief and appears the same as a finished coin.

COINAGE OF THE FOURTH MINT, 1969 TO DATE

From 1969 to date the fourth Philadelphia Mint has maintained its traditional status as the home of the Engraving Department and also where many innovations in coinage are made. P mintmarks were added to 1979 Susan B. Anthony dies, followed in 1980 by the other denominations excepting the cent.

All circulation-issue–design coins are readily collectible, including in high grade.

Cents

All are readily collectible.

Design Types and Dates Minted

Lincoln, Memorial Reverse: *Copper Alloy:* 1969, 1970, 1971, 1972, 1973, 1974, 1975, 1976, 1977, 1978, 1979, 1980, 1981, 1982. *Copper-Plated Zinc:* 1982, 1983, 1984, 1985, 1986, 1987, 1988, 1989, 1990, 1991, 1992, 1993, 1994, 1995, 1996, 1997, 1998, 1999, 2000, 2001, 2002, 2003, 2004, 2005, 2006, 2007, 2008.

Lincoln, Bicentennial: *Birth and Early Childhood, Formative Years, Professional Life, Presidency:* 2009.

Lincoln, Shield Reverse: 2010, 2011, 2012, 2013, 2014, 2015, 2016.

Nickel Five-Cent Pieces

All are readily collectible. The Westward Journey nickels of 2004 and 2005 are particularly interesting and include the first designs made by members of the Artistic Infusion Program.

Design Types and Dates Minted

Jefferson: *Nickel:* 1971, 1972, 1973, 1974, 1975, 1976, 1977, 1978, 1979, 1980-P, 1981-P, 1982-P, 1983-P, 1984-P, 1985-P, 1986-P, 1987-P, 1988-P, 1989-P, 1990-P, 1991-P, 1992-P, 1993-P, 1994-P, 1995-P, 1996-P, 1997-P, 1998-P, 1999-P, 2000-P, 2001-P, 2002-P, 2003-P.

Westward Journey: *Peace Medal, Keelboat:* 2004-P.

Westward Journey, Modified Portrait: *American Bison, Ocean in View:* 2005-P.

Jefferson Modified: 2006-P, 2007-P, 2008-P, 2009-P, 2010-P, 2011-P, 2012-P, 2013-P, 2014-P, 2015-P, 2016-P.

Dimes

All are readily collectible.

Design Type and Dates Minted

Roosevelt: *Clad:* 1969, 1970, 1971, 1972, 1973, 1974, 1975, 1976, 1977, 1978, 1979, 1980-P, 1981-P, 1982-P, 1983-P, 1984-P, 1985-P, 1986-P, 1987-P, 1988-P, 1989-P, 1990-P, 1991-P, 1992-P, 1993-P, 1994-P, 1995-P, 1996-P, 1997-P, 1998-P, 1999-P, 2000-P, 2001-P, 2002-P, 2003-P, 2004-P, 2005-P, 2006-P, 2007-P, 2008-P, 2009-P, 2010-P, 2011-P, 2012-P, 2013-P, 2014-P, 2015-P, 2016-P.

Quarter Dollars

All are readily collectible. The Statehood and other designs of 1999 to date are very popular, diverse in their motifs, and are inexpensive.

Design Types and Dates Minted

Washington: *Clad:* 1969, 1970, 1971, 1972, 1973, 1974, 1977, 1978, 1979, 1980-P, 1981-P, 1982-P, 1983-P, 1984-P, 1985-P, 1986-P, 1987-P, 1988-P, 1989-P, 1990-P, 1991-P, 1992-P, 1993-P, 1994-P, 1995-P, 1996-P, 1997-P, 1998-P.

Washington, Bicentennial: 1776–1976 [1975, 1976].

State Quarters (5 designs per year): 1999-P, 2000-P, 2001-P, 2002-P, 2003-P, 2004-P, 2005-P, 2006-P, 2007-P, 2008-P.

District of Columbia and Territories Quarters Series (6 designs): 2009-P.

America the Beautiful (5 designs per year): 2010-P, 2011-P, 2012-P, 2013-P, 2014-P, 2015-P, 2016-P.[76]

Half Dollars

All are readily collectible today. The Kennedy half dollars never circulated in general commerce. In recent times they have been struck for numismatic sale only.

Design Types and Dates Minted

Kennedy: *Clad:* 1971, 1972, 1973, 1974, 1977, 1978, 1979, 1980-P, 1981-P, 1982-P, 1983-P, 1984-P, 1985-P, 1986-P, 1987-P, 1988-P, 1989-P, 1990-P, 1991-P, 1992-P, 1993-P, 1994-P, 1995-P, 1996-P, 1997-P, 1998-P, 1999-P, 2000-P, 2001-P, 2002-P, 2003-P, 2004-P, 2005-P, 2006-P, 2007-P, 2008-P, 2009-P, 2010-P, 2011-P, 2012-P, 2013-P, 2014-P, 2015-P, 2016-P. *Silver:* 2014-P.

Kennedy, Bicentennial: 1776–1976 [1975, 1976].

Dollars

In 1969 a call was made for a new dollar coin. Congress authorized a cupro-nickel–clad coin with the portrait of recently deceased President Dwight D. Eisenhower, designed by Chief Engraver Frank Gasparro. These were produced with dates from 1971 to 1978, except for 1975, when efforts were devoted to pre-striking 1776–1976 dollars for the American Bicentennial. For that occasion a new reverse was used from a design by Dennis R. Williams.

The Eisenhower dollar was used in Nevada casinos for a while in the early 1970s but never caught on with the public. It was hoped that it would replace the dollar bill in circulation, but that never happened. The Treasury considered the matter and felt that citizens found them to be too cumbersome. In 1977 the mini-dollar was proposed, a Liberty Cap design by Gasparro that found favor with the Treasury Department and with the numismatic community. Congress, which had the final say, mandated that the portrait of suffragette Susan B. Anthony be used. Of small diameter, slightly larger than a quarter, the coin would surely replace the paper dollar. After a large coinage the first year (1979), hundreds of millions remained in Treasury vaults. The public confused them with quarters. Coinage continued in lesser amounts in 1980 and in 1981, the last being limited to coins for collectors. In 1999 additional coins were made when the Treasury anticipated that in view of the new millennium, "Y2K," many computer programs would crash as dates beginning with 2 were not programmed. The public might demand hard money in the form of coins, so millions more dollars were struck, even though tens of millions were still on hand.

The Anthony dollars were never popular, so the next idea was to create the "golden dollar" made of manganese-brass of a yellow color and with a plain edge. Surely these would not be confused with quarters. The Sacagawea design made its debut in 2000. While liked by numismatists, the dollars did not gain traction in the marketplace.

In 2009 the eagle reverse of these coins was replaced by a series of Indian–history motifs, the Native American dollar series. In the meantime the Presidential series of golden dollars commenced in 2008 to honor presidents who had been deceased for at least two years. Starting with Washington, the coins were made at the rate of four designs per year, into 2016 when the list of eligible candidates ran out. Living presidents could not be depicted. Native American and Presidential dollars were mostly sold to collectors, with hardly any used in commerce. The Native American and Presidential dollars have P mintmarks on the edge where they are difficult if not impossible to see when they are in albums or holders.

Design Types and Dates Minted

Eisenhower, Eagle Reverse:
1971, 1972, 1973, 1974, 1977, 1978.

Eisenhower, Bicentennial:
1776–1976 [1975, 1976].

Susan B. Anthony: 1979-P, 1980-P, 1981-P, 1999-P.

Sacagawea: 2000-P, 2001-P, 2002-P, 2003-P, 2004-P, 2005-P, 2006-P, 2007-P, 2008-P.

Native American (new reverse yearly):
2009-P, 2010-P, 2011-P, 2012-P, 2013-P, 2014-P, 2015-P, 2016-P.

Presidential (4 designs per year): 2008-P, 2009-P, 2010-P, 2011-P, 2012-P, 2013-P, 2014-P, 2015-P, 2016-P.

THE PHILADELPHIA MINT TODAY

Today the Philadelphia Mint remains the central design and production facility of the U.S. Mint system. Motifs for many of the nation's new coins and medals are developed here, and it is the main source for die-sculpting and engraving services. On the production side, in 30 minutes Philadelphia can produce one million coins—a quantity that took *three years* in the 1790s.

THE PHILADELPHIA MINT'S ARTISTS

In the studio area of the Mint's Engraving Department, sculptor-engraver Don Everhart and each medallic sculptor has his or her own work area, complete with models they have made, sketches, research photographs, and the like—a comfortable creative space. Unlike in olden days the central piece of equipment is a computer. Who would have imagined this a generation ago? In some instances designs are created or "sculpted," so to speak, electronically. In other instances clay models are still made, but translation and fine-tuning for making hubs and dies is by computer. A silent antique Janvier machine has been relegated to the Mint's public tour—a curiosity rather than a workaday piece of equipment.

Each member of the Mint's current staff is classically trained in the fine arts and brings years of experience to the job.

Don Everhart

Sculptor-engraver Don Everhart was born in York, Pennsylvania. He earned a Bachelor of Fine Arts in painting from Kutztown State University in 1972. The following year he joined the Franklin Mint as a designer, eventually attaining the position of staff sculptor, which he held for five years.

Don Everhart.

Q. David Bowers (foreground), David Sundman (background), and John Dannreuther examining galvanos and plasters in the central area of the artists' studio at the Philadelphia Mint, July 2015.

In March 1980 Everhart left the Franklin Mint to pursue a freelance career. During this period, his projects included work with figurines, plates, coins, and medals. His clientele included prominent companies such as Walt Disney Co. and Tiffany, and international mints such as the Royal Norwegian Mint and the British Royal Mint.

Everhart has received numerous commissions. Notable among them is Georgetown University's Sports Hall of Fame, a 24-piece bronze installation. His work has been exhibited internationally and is included in several permanent collections, including those of the Smithsonian Institution, the British Museum, the American Numismatic Society, and the National Sculpture Society. He also received the prestigious American Numismatic Association Sculptor of the Year Award in 1994. In 1997 his submission was chosen as the official inaugural medal for President Bill Clinton's second term.

Everhart is a former president of the American Medallic Sculpture Association (1993–1994) and is a fellow member of the National Sculpture Society.

He joined the Mint's team of sculptor-engravers in January 2004.

Coin Credits:

2005 Westward Journey nickels— American Bison obverse (sculpt); Ocean in View obverse (sculpt)

2005 State quarter—California (sculpt)

2006 State quarter—Nevada (design and sculpt)

2006 American Platinum Eagle reverse (sculpt)

2006 Benjamin Franklin Founding Father silver dollar obverse (design and sculpt)

2006 San Francisco Old Mint gold $5 reverse (sculpt)

2006 San Francisco Old Mint silver dollar obverse (sculpt)

2007 Little Rock High School silver dollar reverse (design and sculpt)

2007 Jamestown silver dollar obverse (sculpt)

2007 State quarters—Montana (sculpt); Idaho (sculpt)

2007 Presidential dollars—James Madison (sculpt); Statue of Liberty common reverse (design and sculpt)

2008 Bald Eagle gold $5 reverse (sculpt); silver dollar obverse (sculpt)

2008 State quarters—Hawaii (design and sculpt); New Mexico (design and sculpt)

2009 D.C./Territories quarter—District of Columbia (sculpt)

2009 Lincoln Bicentennial cent— Professional Life in Illinois reverse (sculpt)

2009 Abraham Lincoln silver dollar obverse (sculpt)

2009 Presidential dollar—Zachary Taylor (design and sculpt)

2010 America the Beautiful quarters— Hot Springs National Park (design); Yellowstone National Park (design and sculpt)

2010 Presidential dollars—Millard Fillmore (design and sculpt); Abraham Lincoln (design and sculpt)

2010 American Veterans Disabled for Life silver dollar obverse (design and sculpt)

2011 U.S. Army silver dollar reverse (sculpt)

2011 Presidential dollars—Andrew Johnson (design and sculpt); Ulysses S. Grant (design and sculpt); Rutherford B. Hayes (design and sculpt)

2012 Star-Spangled Banner silver dollar reverse (sculpt)

2012 Presidential dollars—Chester Arthur (design and sculpt); Grover Cleveland (first term) (design and sculpt); Grover Cleveland (second term) (design and sculpt)

2013 America the Beautiful quarter—Perry's Victory and International Peace Memorial National Park (design and sculpt)

2013 Presidential dollar—Woodrow Wilson (design and sculpt)

2014 National Baseball Hall of Fame silver dollar obverse and reverse (design and sculpt)

2014 America the Beautiful silver bullion coin—Great Sand Dunes National Park (design and sculpt)

2014 America the Beautiful quarter—Great Sand Dunes National Park (design and sculpt)

2015 American Liberty High Relief gold coin (sculpt)

2015 U.S. Marshals Service 225th Anniversary gold $5 reverse (sculpt)

2015 Presidential dollars—Harry S. Truman (design and sculpt); John F. Kennedy (design and sculpt)

2015 March of Dimes silver dollar reverse (design and sculpt)

2016 Presidential dollar—Richard Nixon (design and sculpt)

2016 National Park Service 100th Anniversary gold $5 obverse and reverse (design and sculpt)

2016 Mark Twain commemorative gold $5 obverse (sculpt)

Medal Credits:

2003 Jackie Robinson Congressional Gold Medal reverse (sculpt)

2005 Dalai Lama Congressional Gold Medal obverse (design and sculpt)

2005 Norman E. Borlaug Congressional Gold Medal reverse (design and sculpt)

2005 Dr. Martin Luther King Jr. and Coretta Scott King Congressional Gold Medal obverse (design and sculpt)

2006 Michael Ellis DeBakey, M.D., Congressional Gold Medal obverse and reverse (design and sculpt)

2006 Tuskegee Airmen Congressional Gold Medal reverse (design and sculpt)

2006 Byron Nelson Congressional Gold Medal reverse (design and sculpt)

2006 Henry M. Paulson Jr. Secretary of the Treasury Congressional Gold Medal reverse (design and sculpt)

2006 John Snow Secretary of the Treasury Congressional Gold Medal reverse (design and sculpt)

2007 Code Talkers Recognition Congressional Gold Medals—Crow Nation obverse and reverse (design); Meskwaki Nation reverse (design); Osage Nation reverse (design); Sisseton Wahpeton Oyate Sioux reverse (design)

2007 Daw Aung San Suu Kyi Congressional Gold Medal obverse and reverse (design and sculpt)

2007 President George W. Bush 2nd Term Presidential Medal obverse (sculpt)

2008 Arnold Palmer Congressional Gold Medal obverse (design and sculpt)

2008 Senator Edward William Brooke Congressional Gold Medal obverse (design and sculpt)

2008 Constantino Brumidi Congressional Gold Medal obverse and reverse (design and sculpt)

2008 Code Talkers Recognition Congressional Gold Medals—Rosebud Sioux reverse (design); Cheyenne River Sioux Tribe reverse (design); Ho-Chunk Nation reverse (design and sculpt); Fort Peck Assiniboine and Sioux obverse (design and sculpt); Hopi Tribe obverse (sculpt); Oglala Sioux Tribe reverse (design); Pawnee Nation obverse (design and sculpt); Yankton Sioux Tribe obverse (sculpt); Comanche Nation reverse (design); Choctaw Nation reverse (design)

2009 New Frontier Congressional Gold Medal reverse (sculpt)

2009 Women Airforce Service Pilots Congressional Gold Medal reverse (design and sculpt)

2010 Nisei Soldiers of World War II Congressional Gold Medal reverse (design and sculpt)

2011 Montford Point Marines Congressional Gold Medal reverse (design and sculpt)

2012 Raoul Wallenberg Congressional Gold Medal obverse (design and sculpt)

2014 American Fighter Aces Congressional Gold Medal reverse (design and sculpt)

2014 Doolittle Raiders Congressional Gold Medal obverse (sculpt)

2014 Shimon Peres Congressional Gold Medal obverse (design and sculpt)

2014 Civil Air Patrol Congressional Gold Medal obverse (sculpt)

2015 Jack Nicklaus Congressional Gold Medal obverse (design and sculpt)

2015 High Relief silver medal reverse (sculpt)

High Relief Coin and Medal Credits:
2015 High Relief gold coin and silver medal reverse (sculpt)

First Spouse Gold Coin and Medal Program Credits:
2007 Dolley Madison obverse (sculpt) and reverse (design and sculpt)

2008 Martha Washington reverse (sculpt); Jackson's Liberty reverse (sculpt); Elizabeth Monroe obverse (sculpt)

2009 Julia Tyler obverse and reverse (sculpt)

2010 Jane Pierce obverse (sculpt)

2011 Lucy Hayes obverse (sculpt); Eliza Johnson obverse (sculpt)

2012 Frances Cleveland (first term) obverse (sculpt)

2013 Edith Roosevelt reverse (sculpt); Ellen Wilson reverse (design and sculpt)

2014 Florence Harding reverse (sculpt)

Joe Menna

Medallic sculptor Joe Menna joined the staff of the U.S. Mint following 18 years of professional experience and classical training. He is skilled in all aspects of traditional and digital sculpture and drawing and has helped make the Mint a world leader in digital coin design and production. Menna was the first full-time digitally skilled artist hired by the Mint. He was instrumental in the development of the Mint's first digitally cre-

Joe Menna.

ated coins and continues to help the bureau remain at the forefront of this constantly evolving craft. Menna also maintains an active freelance career and is recognized as one of the world's leading practitioners of digital sculpture. He joined the Mint in 2005.

Coin Credits:

2006 San Francisco Old Mint gold $5 reverse (sculpt)

2007 State quarter—Utah (design and sculpt)

2007 Presidential dollars—George Washington (design and sculpt); Thomas Jefferson (design and sculpt)

2008 Bald Eagle half dollar reverse (sculpt)

2008 State quarter—Arizona (sculpt)

2008 Presidential dollar—James Monroe (design and sculpt)

2009 Louis Braille Bicentennial silver dollar (sculpt)

2009 D.C./Territories quarters—Commonwealth of Puerto Rico (design and sculpt); U.S. Virgin Islands (design and sculpt)

2009 Presidential dollar—William Henry Harrison (design and sculpt)

2009 Lincoln Bicentennial cent—Presidency in D.C. reverse (sculpt)

2010 American Veterans Disabled for Life silver dollar reverse (sculpt)

2010 America the Beautiful quarters—Yosemite National Park (design); Hot Springs National Park (sculpt)

2010 Lincoln Bicentennial cent, Preservation of the Union reverse (sculpt)

2011 U.S. Army $5 gold reverse (sculpt)

2011 U.S. Army half dollar reverse (sculpt)

2011 Medal of Honor gold $5 obverse (design and sculpt)

2011 America the Beautiful quarter—Vicksburg National Military Park (sculpt)

2011 Native American dollar reverse (sculpt)

2012 Star-Spangled Banner $5 gold obverse and reverse (sculpt)

2012 America the Beautiful quarter—Acadia National Park (sculpt)

2013 Girl Scouts of the USA Centennial silver dollar reverse (sculpt)

2013 America the Beautiful quarters—Fort McHenry National Monument and Historic Shrine (design and sculpt); Mount Rushmore National Park (design and sculpt)

2013 Presidential dollar—Theodore Roosevelt (design and sculpt)

2013 5-Star Generals gold $5 reverse (sculpt)

2013 5-Star Generals silver dollar obverse and reverse (sculpt)

2014 American Platinum Eagle Platinum reverse (sculpt)

2014 America the Beautiful silver bullion coin—Everglades National Park (sculpt)

2014 America the Beautiful quarter—Everglades National Park (sculpt)

2014 Presidential dollar—Franklin Roosevelt (design and sculpt)

2014 Native American dollar—Native Hospitality (sculpt)

2015 America the Beautiful quarters—Kisatchie National Forest (sculpt); Blue Ridge Parkway (sculpt)

2015 Presidential dollar—Dwight D. Eisenhower (design and sculpt)

2015 U.S. Marshals Service 225th Anniversary silver dollar reverse (sculpt)

2016 National Park Service 100th Anniversary silver dollar obverse (design and sculpt)

Medal Credits:

2006 Dalai Lama Congressional Gold Medal reverse (design and sculpt)

2008 Arnold Palmer Congressional Gold Medal reverse (sculpt)

2008 Code Talkers Recognition Congressional Gold Medals—Kiowa obverse (design); Kiowa reverse (design); Tonto Apache Tribe obverse (sculpt); Cherokee Nation reverse (design and sculpt); Seminole Nation reverse (design and sculpt); Tlingit Tribe obverse (sculpt); White Mountain Apache Tribe obverse and reverse (sculpt)

2011 September 11 National Medal reverse (sculpt)

2011 Fallen Heroes of 9/11 Congressional Gold Medal–Flight 93 obverse (design and sculpt)

2013 16th Street Baptist Church Bombing Victims Congressional Gold Medal reverse (sculpt)

2014 Shimon Peres Congressional Gold Medal reverse (design and sculpt)

First Spouse Gold Coin and Medal Program Credits:

2007 Martha Washington obverse (design and sculpt)

2007 Abigail Adams obverse (design and sculpt)

2008 Louisa Adams reverse (design and sculpt)

2009 Anna Harrison obverse (sculpt)

2010 Abigail Fillmore reverse (sculpt)

2010 James Buchanan's Liberty reverse (sculpt)

2011 Lucy Hayes reverse (sculpt)

2012 Frances Cleveland (second term) reverse (design and sculpt)

2013 Edith Roosevelt obverse (sculpt)

2014 Florence Harding obverse (sculpt)

2015 Mamie Eisenhower obverse (sculpt)

Phebe Hemphill

Medallic sculptor Phebe Hemphill graduated from the Pennsylvania Academy of the Fine Arts and studied for three years with sculptor EvAngelos Frudakis in Philadelphia. In 1987 she joined the sculpture department at the Franklin Mint, and for the next 15 years she worked on many projects for the firm's porcelain and medallic departments.

Hemphill has displayed her creative talents as a freelance sculptor while working with companies that produce figurines, medallions, dolls, toys, and garden ornaments. She was a staff sculptor for three years with McFarlane Toys in Bloomingdale, New Jersey. Hemphill's extraordinary sculptures have been exhibited by the National Sculpture Society, the American Medallic Sculpture Associa-

Phebe Hemphill.

tion, West Chester University, and the F.A.N. Gallery in Philadelphia. In 2000 she received the Alex J. Ettel Grant from the National Sculpture Society, and in 2001 she earned the Renaissance Sculpture Award from the Franklin Mint.

Hemphill joined the U.S. Mint in 2006.

Coin Credits:

2007 American Platinum Eagle Proof reverse (sculpt)

2008 Presidential dollar—Martin Van Buren (sculpt)

2008 Bald Eagle gold $5 obverse (sculpt)

2008 State quarter—Oklahoma (sculpt)

2009 Abraham Lincoln silver dollar reverse (design and sculpt)

2009 Louis Braille Bicentennial silver dollar obverse (sculpt)

2009 Presidential dollar—John Tyler (design and sculpt)

2009 D.C./Territories quarter—Northern Mariana Islands (sculpt)

2009 American Platinum Eagle reverse (sculpt)

2010 American Platinum Eagle reverse (sculpt)

2010 America the Beautiful quarters—Grand Canyon National Park (design and sculpt); Mount Hood National Forest (design and sculpt); Yosemite National Park (sculpt)

2010 Presidential dollar—James Buchanan (design and sculpt)

2011 American Platinum Eagle reverse (sculpt)

2011 United States Army gold $5 obverse (sculpt)

2011 America the Beautiful quarter—Gettysburg National Military Park (sculpt)

2011 Presidential dollar—James Garfield (design and sculpt)

2011 Medal of Honor silver dollar reverse (sculpt)

2012 Star-Spangled Banner silver dollar obverse (sculpt)

2012 America the Beautiful quarter—Chaco Culture National Historical Park (sculpt)

2012 Native American dollar—Trade Routes in the 17th Century (sculpt)

2012 Presidential dollar—Benjamin Harrison (design and sculpt)

2013 Presidential dollar—William McKinley (design and sculpt)

2013 Native American dollar—Treaty With the Delawares (sculpt)

2013 Girl Scouts of the USA Centennial silver dollar obverse (sculpt)

2013 5-Star Generals half dollar obverse and reverse (design and sculpt)

2013 America the Beautiful quarter—White Mountain National Forest (design and sculpt)

2014 America the Beautiful quarter—Shenandoah National Park (design and sculpt)

2014 America the Beautiful silver bullion coin—Shenandoah National Park (design and sculpt)

2014 Presidential dollars—Calvin Coolidge (design and sculpt); Herbert Hoover (design and sculpt)

2014 Civil Rights Act of 1964 silver dollar obverse (sculpt)

2015 American Platinum Eagle reverse (sculpt)

2015 American Liberty High Relief gold coin obverse (sculpt)

2015 U.S. Marshals Service 225th Anniversary half dollar reverse (sculpt)

2015 America the Beautiful quarter—Bombay Hook National Wildlife Refuge (sculpt)

2015 Native American dollar—Mohawk Ironworkers (sculpt)

Medal Credits:

2008 Code Talkers Recognition bronze medals—Rosebud Sioux obverse (sculpt); Meskwaki Nation obverse (sculpt); Choctaw Nation obverse (design); Crow Creek Sioux Tribe reverse (sculpt); Menominee Nation obverse (sculpt); Muscogee (Creek) Nation obverse (sculpt); Osage Nation obverse (sculpt); Ponca Tribe obverse (sculpt); Santee Sioux Nation obverse (sculpt); Seminole Nation obverse (sculpt); Sisseton Wahpeton Oyate (Sioux) Tribe obverse (sculpt)

2008 Arnold Palmer bronze medal reverse (design)

2008 Senator Edward William Brooke Congressional Gold Medal reverse (design and sculpt)

2008 Tuskegee Airmen Congressional Gold Medal obverse (design and sculpt)

2008 Norman E. Borlaug Congressional Gold Medal obverse (design and sculpt)

2009 New Frontier Congressional Gold Medal obverse (sculpt)

2009 Women Air Force Service Pilots Congressional Gold Medal obverse (sculpt)

2010 Professor Muhammad Yunus bronze medal obverse (sculpt)

2011 Fallen Heroes of 9/11 bronze medals—New York reverse (design and sculpt); Pentagon obverse and reverse (design and sculpt); Flight 93 reverse (design and sculpt)

2011 September 11 national medal obverse (sculpt)

2011 New Frontier bronze medal obverse (sculpt)

2012 Raoul Wallenberg Congressional Gold Medal reverse (design and sculpt)

2014 Monuments Men Congressional Gold Medal obverse (sculpt)

2014 American Fighter Aces Congressional Gold Medal obverse (sculpt)

2015 Jack Nicklaus Congressional Gold Medal (design and sculpt)

High Relief Coin and Medal Credits:
2015 High Relief gold coin and silver medal obverse (sculpt)

First Spouse Gold Coin and Medal Program Credits:

2007 Abigail Adams reverse (sculpt)

2007 Thomas Jefferson's Liberty obverse (sculpt)

2008 Louisa Adams obverse (sculpt)

2009 Margaret Taylor obverse (design)

2009 Letitia Tyler obverse (design and sculpt)

2009 Sarah Polk obverse and reverse (design and sculpt)

2010 Abigail Fillmore obverse (design and sculpt)

2010 Mary Todd Lincoln obverse (design and sculpt)

2010 Mary Todd Lincoln reverse (sculpt)

2011 Lucretia Garfield obverse (sculpt)

2011 Eliza Johnson reverse (sculpt)

2012 Alice Paul obverse (sculpt)

2012 Suffrage Movement reverse (design and sculpt)

2012 Frances Cleveland (second term) obverse (sculpt)

2013 Ida McKinley obverse (sculpt)

2013 Helen Taft obverse (sculpt)

2014 Grace Coolidge obverse (sculpt)

2014 Eleanor Roosevelt obverse (sculpt)

2015 Bess Truman (sculpt)

2015 Jacqueline Kennedy (sculpt)

Jim Licaretz

Medallic sculptor Jim Licaretz graduated from the Pennsylvania Academy of the Fine Arts in Philadelphia. He earned a J. Henry Schiedt Traveling Scholarship, as well as the Edmund Stewardson Prize for Figurative Sculpture and a Philadelphia Board of Education four-year scholarship. He is a former president of the American Medallic Sculpture Association, a fellow member of the National Sculpture Society, and a sculptor member of the Fédération International de la Médaille d'Art.

Licaretz was a member of the U.S. Mint's sculptor-engraver department from 1986 to 1989 before leaving to teach, study abroad, and work for various firms including Franklin Porcelain, the Franklin Mint, and Mattel. He returned to the Philadelphia Mint as a medallic sculptor in 2006.

Licaretz's works are held in numerous private collections and can be seen at the British Museum; the Royal Coin Cabinet in the National Museum of Economy, Stockholm, Sweden; the American Numismatic Society; and the Smithsonian Institution. He has exhibited his sculptures throughout the United States, earning many professional awards including the American Numismatic Association's 2008 Numismatic Art Award for Excellence in Medallic Sculpture.

Jim Licaretz.

Coin Credits:

2008 Presidential dollars—Andrew Jackson (sculpt); Martin Van Buren (sculpt)

2008 Bald Eagle silver dollar reverse (sculpt)

2009 D.C./Territories quarter—Guam (sculpt)

2009 Lincoln Bicentennial cent—Birth and Early Childhood in Kentucky reverse (sculpt)

2010 Boys Scouts of America Centennial silver dollar reverse (sculpt)

2011 America the Beautiful quarter—Chickasaw National Recreation Area (sculpt)

2011 Medal of Honor silver dollar obverse (design and sculpt)

2012 America the Beautiful quarter—Denali National and Preserve (sculpt)

2014 Civil Rights Act of 1964 silver dollar reverse (sculpt)

2015 Presidential dollar—Lyndon B. Johnson (design and sculpt)

2015 U.S. Marshals Service 225th Anniversary half dollar obverse (sculpt)

2015 America the Beautiful quarter—Homestead National Monument of America (sculpt)

2016 National Park Service 100th Anniversary silver dollar reverse (sculpt)

Medal Credits:

1986 Harry Chapin Congressional Gold Medal obverse (sculpt)

1986 Aaron Copland Congressional Gold Medal reverse (design and sculpt)

1986 Natan (Anatoly) and Avital Shcharansky Congressional Gold Medal obverse (design and sculpt) and reverse (sculpt)

1987 Mary Lasker Congressional Gold Medal obverse (design and sculpt)

1988 Young Astronauts silver medal obverse (sculpt)

2006 Treasury Secretary Henry M. Paulson medal obverse (design and sculpt)

2008 Code Talkers Recognition Congressional Gold Medals—Cheyenne River Sioux Tribe reverse (sculpt); Choctaw Nation reverse (sculpt); Crow Nation obverse (sculpt); Fort Peck Assiniboine and Sioux Tribe reverse (sculpt); Standing Rock Sioux Tribe obverse (sculpt); Standing Rock Sioux Tribe reverse (sculpt); Oglala Sioux Tribe reverse (sculpt); Sisseton Wahpeton Oyate Sioux Tribe reverse (sculpt)

2010 Professor Muhammad Yunus Congressional Gold Medal reverse (sculpt)

2011 Fallen Heroes of 9/11 Congressional Gold Medal—New York obverse (sculpt)

2013 16th Street Baptist Church Bombing Victims Congressional Gold Medal obverse (sculpt)

First Spouse Gold Coin and Medal Program Credits:

2008 Martin Van Buren's Liberty (sculpt)

2009 Margaret Taylor reverse (sculpt)

2014 Grace Coolidge reverse (sculpt)

2014 Lou Hoover reverse (sculpt)

2015 Jacqueline Kennedy reverse (sculpt)

Michael Gaudioso

Medallic sculptor Michael Gaudioso graduated from Philadelphia's University of the Arts and earned his Master of Fine Arts from the New York Academy Graduate School of Figurative Art in New York City. From 1995 to 1999 Gaudioso studied sculpture at the prestigious Repin Institute in St. Petersburg, Russia. He is a classically trained draftsman and sculptor and has taught figure drawing at Villanova University. Gaudioso worked as a master painter and designer for America's oldest and largest stained glass studio, Willet Hauser. He joined the Mint's sculpting and engraving department in 2009.

Michael Gaudioso.

Coin Credits:

2008 State quarter—New Mexico (sculpt)

2011 America the Beautiful quarter— Olympic National Park (sculpt)

2011 U.S. Army silver dollar obverse (sculpt)

2011 Medal of Honor gold $5 reverse (sculpt)

2012 America the Beautiful quarters— El Yunque National Forest (sculpt); Olympic National Park (sculpt)

2012 Infantry Soldier silver dollar obverse (sculpt)

2013 Presidential dollar—William Howard Taft (sculpt)

2013 5-Star Generals gold $5 obverse (sculpt)

2013 American Platinum Eagle reverse (sculpt)

2014 Presidential dollar—Warren G. Harding (design and sculpt)

2015 Presidential dollar—Lyndon B. Johnson (design and sculpt)

2015 U.S. Marshals Service 225th Anniversary half dollar obverse (sculpt)

2015 March of Dimes silver dollar obverse (sculpt)

2016 National Park Service 100th Anniversary half dollar obverse (sculpt)

Medal Credits:

2008 Code Talkers Recognition Congressional Gold Medals—Pueblo of Acoma Tribe obverse (sculpt); Cherokee Nation obverse (sculpt); Cheyenne River Sioux Tribe obverse (sculpt); Ho-Chunk Nation obverse (design and sculpt); Oglala Sioux Tribe (design and sculpt)

2011 Montford Point Marines Congressional Gold Medal obverse (design and sculpt)

2014 Civil Air Patrol Congressional Gold Medal reverse (sculpt)

First Spouse Gold Coin and Medal Program Credits:

2011 Julia Grant obverse (sculpt)

2011 Lucretia Garfield reverse (design and sculpt)

2012 Frances Cleveland (first term) reverse (sculpt)

2012 Caroline Harrison obverse (sculpt)

2013 Edith Wilson obverse (sculpt)

2014 Lou Hoover obverse (sculpt)

2015 Lady Bird Johnson obverse (sculpt)

Renata Gordon

Medallic sculptor Renata Gordon earned her Bachelor of Fine Arts in sculpture from the University of the Arts in Philadelphia in 2010. She interned in the U.S. Mint's sculpting/engraving department while in school, studying coin and medal design in addition to traditional and digital sculpture. Her work as an artist includes commissions for public and private murals and portraiture.

Gordon joined the Mint as a full-time employee in March 2011.

Renata Gordon.

Coin Credits:

2013 America the Beautiful silver bullion coin—Great Smoky Mountains National Park (sculpt)

2013 America the Beautiful quarter—Great Basin National Park (sculpt)

2014 America the Beautiful quarter—Great Smoky Mountains National Park (sculpt)

2015 America the Beautiful quarter—Saratoga National Historical Park (sculpt)

Medal Credits:

2008 Code Talkers Recognition Congressional Gold Medal—Hopi Tribe reverse (sculpt); Menominee Nation reverse (sculpt); Meskwaki Nation reverse (sculpt); Muscogee Creek Nation reverse (sculpt); Tlingit Tribe reverse (sculpt); Tonto Apache Tribe reverse (sculpt); Oneida Nation reverse (sculpt); Crow Creek Sioux reverse (sculpt); Crow Nation reverse (sculpt); Osage Nation reverse (sculpt); Ponca Tribe reverse (sculpt); Pueblo of Acoma Tribe reverse (sculpt)

First Spouse Gold Coin and Medal Program Credits:

2013 Ida McKinley reverse (sculpt)

2014 Eleanor Roosevelt obverse (sculpt)

2015 Mamie Eisenhower reverse (sculpt)

2015 Lady Bird Johnson reverse (sculpt)

Charles Vickers

Sculptor-engraver Charles Vickers retired in 2016 after more than 10 years with the U.S. Mint. A student of New York's Art Students League and Frank Reilly School of Art, as well as the Pratt Institute and the School of Visual Arts, he moved to Pennsylvania in 1976 and started a successful career at the Franklin Mint. After leaving to launch his own art studio, he joined the Mint's sculpting-engraving staff in December 2003. Over the course of his Mint career he designed and/or sculpted many coins and medals— commemoratives, State quarters, First Spouse gold coins, Presidential dollars, and others.

Charles Vickers.

THE ARTISTIC INFUSION PROGRAM

In the late 1990s and early 2000s the Mint's design and engraving department was busy with the State quarter program, a constant stream of commemoratives, various medals, and other projects. In November 2003 the Mint launched the Artistic Infusion Program (AIP), "to enrich and invigorate United States coin and medal designs by contracting with a pool of talented, professional American artists representing diverse backgrounds and a variety of interests." Artists in the private sector were invited to contact the Mint, after which more than a dozen were selected to participate in creating coinage designs. AIP artists were to have their initials on coins along with the Mint sculptor-engraver or medallic sculptor who made models from the designs. Mint officials and representatives from the National Endowment for the Arts chose 18 professional artists as "master designers" and 6 graduate-level art students as "associate designers" from 306 interested applicants. The first assignment for AIP participants was the nickel program in the Westward Journey series for the year 2005.

In November 2004 the Mint had a second call for AIP artists and sought to add 16 more names to consist of up to 2 professional artists and up to 14 college- and graduate-level art students. It was announced that those who were selected in the first round would not be employed for continuing work unless they reapplied.

Mint Director Henrietta Holsman Fore reported on the designs already done: "This historic program has produced outstanding results in its first year. For the 2005 nickel, the program's artists have created a striking new obverse portrait of President Thomas Jefferson, and reverse designs of the 'American Bison' and 'Ocean in View!' that revisit and revitalize the honored traditions of American coinage. With this new call to artists, we again seek the best in America, to lend their creativity to our nation's coins and medals."[77]

The AIP has become an ongoing success, with its artists contributing to the America the Beautiful quarters (and their related five-ounce silver bullion coins), the First Spouse gold coins, and other programs.

Paul Balan is one of the current artists in the AIP.

Satin Finish Coins, New Presses, and Other Innovations

In recent decades the U.S. Mint has explored new directions in packaging, design, and production, and has launched many new coinage series. The Philadelphia Mint has been on the forefront of carrying out many of these innovations.

From 2005 through 2010 Uncirculated regular-issue coins made for Mint sets at the Philadelphia and Denver facilities had a special satin finish. On December 13, 2010, the Mint released a statement: "The United States Mint began using the satin finish for its uncirculated coins in 2005 to be consistent with other products in its portfolio, such as commemorative coins. The satin finish, however, highlights surface marks that inherently result from the coin-handling systems. Although the United States Mint modified the process to improve the coin appearance, there is no cost-effective way to completely eliminate the coin-on-coin contact that causes surface nicks. As a result, the United States Mint will revert back to the brilliant finish used on uncirculated coins prior to 2005. This change will result in more aesthetically pleasing coins with a finish that does not highlight surface flaws. It will only be featured on uncirculated-quality one-cent, 5-cent, dime, quarter-dollar, half-dollar and dollar coins."

In 2005 the federal Office of Management and Budget sponsored a competition under which private companies might supply ready-to-use planchets for circulating coins, collectors' coins, and Proof sets. At the time the only planchets for these series purchased outside were for cents, while planchets for bullion and certain other coins were by contract. No changes were made for the series studied, and the mints continued using large rolls of flat metal strips to make planchets and process them for nickels, dimes, quarters, half dollars, and "golden dollars."

The Philadelphia Mint is home to a Gräbener press installed in March 2010 for production of the massive five-ounce, three-inch-diameter America the Beautiful silver bullion coins. These and other modern bullion pieces are cataloged in the book *American Gold and Silver: U.S. Mint Collector and Investor Coins and Medals, Bicentennial to Date*, with many behind-the-scenes photographs from Philadelphia and other U.S. Mint facilities.

The Mint notes in its promotional literature that it's been "green" for more than 200 years. "Since 1792, all excess metal used in the making of coins has been recycled. Even the webbing leftover from blanking is shredded and returned to the metal strip manufacturer to be recycled. Flawed coins and blanks are also recycled. They are crushed between two high pressure rollers which impress a ridged pattern into the metal. 'Waffled' pieces don't have a denomination value and can be returned to the manufacturer for recycling without a United States Mint Police escort."

Starting in 2009, the Philadelphia Mint renovated and dramatically upgraded its public-tour exhibit in a three-year, $3.9 million undertaking. Tom Jurkowsky, director of the Mint's Office of Corporate Communications, was a key supervisor and facilitator throughout the creative design, fabrication, and installation of the Public Tour Upgrade Project—a project for which he and his team earned the Mint's 2012 Rittenhouse Medal for Excellence. Today the Philadelphia public tour attracts some 250,000 visitors every year. The following is an abbreviated account of the tour, as well as some behind-the-scenes reporting, originally published in longer form in *Coin World*:

On Wednesday, June 27, 2012, Mary Counts (president of Whitman Publishing) and I (author Q. David Bowers) joined about a dozen other media people in an invitational preview of the new Visitors' Center at the Philadelphia Mint.

Michael White, a Mint official since 1990 and the only old-timer on the tour, greeted us at the appointed time shortly after 9:00, joined by Mary Lhotsky, deputy director of public affairs, and public-affairs officer Tim Grant. Soon we went upstairs to a meeting room where Director of Public Affairs Tom Jurkowsky gave welcoming remarks. Tom has been in the position for three years and saw a great need for enhanced public-tour facilities as they had not changed since the opening of the Fourth Philadelphia Mint on Independence Square in 1967. The result is a "must see" for coin collectors!

For me this was the latest and most enlightening of the private tours I have had over the past 50-plus years. The first was in the 1950s under Mint superintendent Rae V. Biester in the Third Mint on Spring Garden Street. Later visits at the present Fourth Mint included interviews with Chief Engraver Frank Gasparro, then Chief Engraver Elizabeth Jones, and others.

In those days production at the Mint was in the time-honored manner. Engravers prepared models in clay, sculpting every detail. These were then transferred to plaster impressions from which galvanos were made by electroplating. The galvanos were then placed in a Janvier transfer/reduction lathe which, hours later, turned out a miniature version as a steel hub. Today in 2012 some models and plasters are still made, but that is where tradition ends. The computer is now supreme, guided of course by artists.

Dies were made only in Philadelphia (today Denver makes them as well). In the Engraving Department D and S mintmarks were punched in by hand. During one visit Michael Iacocca let me add a D mintmark to a Roosevelt dime. I joked that I should have placed it upside down!

Coins were struck on presses ranging from old Ferracute knuckle-action machines for Proofs to Bliss and other presses for regular production. It was easy to see the process as a hammer or upper die descended to strike a blank planchet resting on the lower or anvil die. Most equipment was American-made. Today all of the high-tech devices I saw were imported from Europe.

Mr. Jurkowsky realized that so much had changed since the 1960s and also that the story of coin production had not been adequately told even back then. Moreover, he wanted the experiences of the Mint employees to be shared as they have a great passion for their work. Quatrefoil, a company based in Laurel, Maryland, was the prime contractor for the new tour facilities. The result is spectacular.

The invited guests split into two groups. For us the first stop was at the Engraving Department. A half-dozen or so artists were on hand. Mary and I had interviews with two of them before it was time to move on. We could have lingered for an hour. Charles Vickers has been with the Mint since 2003. He designed the Hawaii National Park "Volcanos" quarter which, as luck would have it, we would encounter later in our visit to the production line. The subject of his first medal was Brown vs. the Board of Education. He designed and sculpted the reverse of the 250th Anniversary of the Marine Corps silver dollar. He designed the 2009 cent reverse with Lincoln sitting on a log, as he called it, more formally the "Formative Years" motif. His list of credits includes dozens of works. Charles enjoys his work and is proud and honored to be on the Mint staff. This *esprit de corps* was evident in all other departments during our visit. What a fine asset the Mint has in its *people*.

Next was a visit with Phebe Hemphill, a personable and highly talented graduate of the Pennsylvania Academy of the Fine Arts, who has been with the Mint since 2006. She was the designer or sculptor for three of the first five National Parks quarters—Grand Canyon reverse (designed and sculpted), Mount Hood reverse (designed and sculpted), and Yosemite reverse (sculpted), not to overlook a long list of medals, commemorative coins, First Spouse coins, and more. Of great interest to me was her model of the New Hampshire quarter featuring Mount Chocorua, a famous icon not far from where I live.

We then moved to the area of the Mint where all the dies are manufactured and polished. Dave Puglia told our group that Philadelphia was responsible for all dies used by the Philadelphia and West Point mints and about 20% of those used by San Francisco.

Steel master hubs are created in this division by computer-driven CNC machines with a tiny rotating tool, similar in concept to the old-time Janvier unit. All else is different. It takes about 10 to 12 hours to make a master hub. There is only one master hub per coin design and all are made in Philadelphia. By transfer the hub is used to make master dies and through further transfer the working dies.

Working or coinage dies are made by the thousands, with most work automated, quite a contrast from years ago. Proof dies require special polishing which is about 90% automated, including laser etching to create cameo contrast, after which about 10% of the work, including inspection, is done by trained staff. Certain dies are very tricky and require great talent to finesse. Today a Proof die is good for only about 500 to 600 impressions, a contrast from a century or more ago when many Proof dies were good for over a thousand impressions. The reason seems to be that today the quality of the mirror finish approaches perfection, an elegance of surface not matched years ago. Modern numismatists are very particular about such things! Our forebears were not as finicky—as Liberty Seated silver Proofs with weak features sometimes reveal. A circulating-coin die is another matter entirely. One can handle about 500,000 impressions. There are about 30 workers on staff who create between 800 and 1,000 dies a week

Once a die cracks or has a flaw it is taken out of production and completely defaced so that not a trace of the design remains. I remarked that at one time the Mint simply cut a thin X across the face of some Proof dies and then sold them as collectors' items. "This caused a lot of problems," Mr. Puglia said with a laugh, "so we no longer do that."

Our next stop was at the huge production area or factory—large enough to accommodate several jumbo jetliners, it seemed. Five acres are devoted to planchet preparation and minting. While we were there cents, nickels, and dimes were being made at the rate of about 24 million per day. If need be the full capacity of 50 million per day can be utilized.

For cents the blanks or planchets are purchased from outside sources. For nickel and clad coins long sheets of metal rolled up are placed in a blanking press from which planchets are stamped like a cookie-cutter. The noise, while not deafening, is quite loud. Visitors insert plugs into their ears, place protective covers on their shoes, and put safety glasses on.

From the blanking press the planchets are heated to 1,200 degrees to anneal or soften them, and raised rims are added. While there we saw bins full of raised-rim "golden dollar" planchets, but this denomination was not being made during our visit.

Gone are the days of traditional coining presses. Today the machines, and there are 65 of them, are all made by Schuler. Although in the past multiple-die presses were sometimes used, the Shuler presses use a single pair of dies mounted vertically (instead of horizontally as in older times). The planchets are fed by gravity. Coins are struck at a rate of *12 per second*—faster than the eye can see. That is, if the dies could be seen. Each Shuler press is in a cabinet larger than an SUV and is completely enclosed. There is a small window at each side, but the dies are not in view. At the back of each press the coins come out in a cascade on a conveyor belt.

An announcement board for July 2012 stated that the coining rate would be 28.2 million coins per day. Cents, always called *pennies* during our tour, will be made in the largest quantity, followed by dimes and nickels. Hawaii National Park quarters will be struck as well. One unwelcome (from a numismatic viewpoint!) announcement admonished employees to: "Check pennies for defects. Look for off-center." We all love our error coins, even if the Mint doesn't.

After minting, the coins are put into large bags, sealed, and weighed, after which they are shipped off to Federal Reserve banks which in turn distribute them to member banks, after which they go into commerce.

Our last stop in the production facility was in a special "numismatic area" where three-inch-diameter silver America the Beautiful "quarters" were being struck on a large Gräbener hydraulic press with 577 tons of pressure! The obverse, reverse, and edge of each coin are struck in one process requiring three blows to properly bring up the design. The production rate is 12 per minute—slow enough to easily watch, but fast enough that two attendants, Loyd Paige and Leroy Lockhart, were kept continually busy, one tending to planchet feeding and the other to carefully handling the coins after they were struck.

The Hawaii Volcanoes National Park coins were being made while we were there. Plans call for 25,000 of each current motif to be struck, but public demand falls short of this. Typically, 10,000 or so are made at the outset, after which additional pieces are struck to take care of sales.

Our tour ended with a trip through the new Visitors' Center. This includes a large gift shop on the main floor. From there it is up an escalator to an exhibit area with many rare and unique artifacts, including a hand-powered coining press from the 1790s, "Peter" the Mint eagle who modeled the Gobrecht silver dollar reverse, relics recovered from the First Mint (opened in 1792), and more.

A very interesting series of panels goes year by year and is titled "Reach into your pocket and pull out a coin." Look at your change for recent coins in your collection or keepsake drawer for older ones and see if you can think of what was going on in the world when a particular cent, nickel, or other coin was minted. Samples:

"1946: The ENIAC I computer begins calculations at the University of Pennsylvania. It weighs 30 tons and spans 1,800 square feet."

"1947: Jackie Robinson becomes the first African-American Major League Baseball player, signing with the Brooklyn Dodgers."

Next stop was a darkened room to see a six-minute film narrated by none other than Ben Franklin, Thomas Jefferson, and Alexander Hamilton, shown in silhouette, discussing money and the forming of a national mint.

From there it was up another escalator to the beginning of the walking tour—through long galleries with windows overlooking the coin production floor from 40 feet above. On both sides of the gallery up to about waist level are dozens of pictures, biographical sketches, explanatory notes, and other useful information. A visitor can linger for a half hour or more to take it all in.

On a scale of 1 to 10 I rate the Philadelphia Mint tour a "10"—a job well done.

NEW ORLEANS MINT, 1838–1909

PLANNING THE MINT

Advocated by President Andrew Jackson and vehemently opposed by Senator Henry Clay and his associates, the bill to establish coining facilities in cities other than Philadelphia was hard fought in the Senate. Just about everyone else thought the three proposed branches at Charlotte, Dahlonega, and New Orleans would be a boon to commerce. New Orleans, the leading port in the South, had long been a depot for arriving shipments of silver (in particular) and gold coins minted in other countries. Congressional testimony said the establishment of a mint there would save about 5% of the value of gold as an expense from transporting metal from those regions to the Philadelphia Mint for coining. The Mint Act survived all attacks and was signed into law on March 3, 1835. Not long afterward Martin Gordon, who was rejected by the Senate as collector of New Orleans, was appointed superintendent of the Mint by President Andrew Jackson, who delighted in taking an action that his Senate foes could not prevent.

The New Orleans Mint at the turn of the 20th century.

On April 8, 1835, Mint Director Samuel Moore advertised in the *New Orleans Bee* to solicit bids for the construction of the Branch Mint at New Orleans, as it was designated. Bids were to be received by Martin Gordon, commissioner of the new mint.

On May 9, 1835, the City of New Orleans ceded low-lying property known as Jackson Square, bounded by Esplanade, Front Levee (later changed to North Peters), Barracks, and Levee (later changed to Decatur) streets, to the Treasury Department for use as the Mint site. This arrangement was officially accepted by Martin Gordon on June 19. The construction contract with John Mitchell and Benjamin F. Fox was recorded on August 22. The contract was for $182,000—$40,000 of which was to be retained for the completion of a wing in the future, a provision that became moot, as the entire building was constructed at the same time. The plans were by Philadelphia architect William Strickland, who designed the second Philadelphia Mint (opened in 1833) and also designed the branch mints for Charlotte and Dahlonega.

The total width, including wings on each side (measuring 29 by 81 feet), was 282 feet, and the depth of the main building was 108 feet.

Niles' Weekly Register, September 26, 1835, commented that the building was expected to be ready for operation on May 15, 1836, and that "This edifice will be an ornament to New Orleans; and will equal any public building of its kind in the country, in utility and appearance."

Slavery was the reality in New Orleans at the time, and it was supported by residents of the city with a hard-line attitude. In 1835 a city newspaper carried an advertisement offering an astonishing $50,000 reward to anyone who captured a Northern abolitionist, Arthur Tappan, and turned him over to the Committee of Vigilance of the Parish of East Feliciana.[1] The employment of slave labor was not seriously considered for the Mint at the time of its construction, but it was proposed later.

Not everyone was in favor of the new Mint. The *National Intelligencer* printed this, reprinted in the *Extra Globe*, October 9, 1835, in response to Senator Henry Clay's strong opposition to the idea (a small excerpt):

> It is true, that Mr. Clay opposed the appropriation, and the fact ought to be circulated throughout the whole of the Southwest and West. He also opposed the appropriations for a Branch Mint in Georgia, and one in North Carolina; and he opposed the passage of the Gold Bill, and that legalizing Foreign Silver Coins. He has opposed everything which was calculated in the least to interfere with the Bank of the United States and its interests.

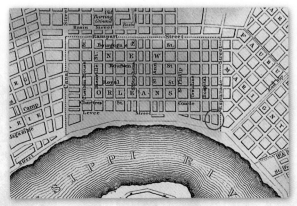

**Approximate location of the Mint, then under
construction, on an 1837 map by T.G. Bradford.**

In relation to the same subject, we find the following in the *Courier* and *Enquirer*:

"We are not among the number of those who would be niggardly in the appropriation of the public funds for national purposes. On the contrary, we wish to see our public buildings so erected as to do honor to the country, and remain to future ages as the monuments of the national munificence; but we dislike these attempts to *deceive* the government into appropriations. If the New Orleans Mint cannot be erected for less than the enormous amount now mentioned, why was Congress asked to appropriate merely a paltry modicum of it—so that every future session must be teased into annual additions to it, on the ground that *what is already expended will be lost, unless we go on with new sums every year?*"

In reply to this, we have only to refer to what is before mentioned in regard to New Orleans and the present [Martin Van Buren] Administration. — But in regard to former appropriations and other cities, we would refer to the origin and progress of the Mint in Philadelphia. . . .[2]

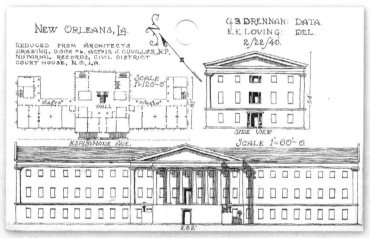

Architectural rendering of the building.

The New Orleans Mint as depicted in *Gibson's Guide and Directory of the State of Louisiana and the Cities of New Orleans and Lafayette*, 1838.

In March 1837 David Bradford was appointed superintendent and Rufus Tyler coiner. The suspension of specie payments by Eastern banks, during the Panic of 1837, that began on May 10 spread to New Orleans, resulting in 14 banks refusing to exchange paper money at par for silver or gold coins. By 1839 normal banking had resumed across the city.

On January 23, 1838, Mint Director Robert Maskell Patterson reported:

> The machinery of the New Orleans branch mint was executed, and the steam engine set in action in May last; and there would have been no difficulty in putting the Mint in full operation, but for the apprehensions from the climate. Two of the officers and all of the workmen were from the middle states, and unacclimated; and I was advised by the resident officers that they would incur great risk in going to New Orleans in the warm season. . . . They are now, however, all at their stations, and making every exertion to commence the operations of the Mint at an early day.

EARLY OPERATIONS AT THE MINT

From *The Annual Report of the Director of the Mint* for 1838:

> The branch mint at New Orleans received its first deposits of bullion the 8th of March, and commenced operations immediately afterwards. The demand for silver change led the officers to confine the coinage to dimes, of which 367,434 were struck before the end of July, when the work was interrupted.
>
> The machinery of the New Orleans branch mint was executed, and the steam engine set in action, in May last; and there would have been no difficulty in putting the mint in full operation, but for the apprehensions from the climate. Two of the officers and all the workmen were from the middle states, and unacclimated; and 1 was advised by the resident officers that they would incur great risk in going to New Orleans in the warm season. The occurrence of a fatal epidemic, soon after, proved that these apprehensions were too well founded, and the officers and men from this place were thereby prevented from reaching New Orleans before the beginning of December.
>
> Two of the officers, and nearly all the workmen of this Mint were from the North, and it was deemed unsafe for them to remain in New Orleans during the sickly season. Accordingly, leave of absence was granted to them on the first of August, the workmen being put on half-pay. They are now, however, all at their stations, and making every exertion to commence the operations of the mint at an early day.
>
> In November, the operations were resumed, but much could not be accomplished before the close of the year. The value of the bullion received at this mint was $40,600 in gold, and $237,000 in silver. The [silver] coinage amounted to $40,243, all in dimes.

On May 7 the first coinage run was made—and 30 1838-O dimes were produced. By year's end a total of 70,000 half dimes and 406,034 dimes had been made, nearly all from melted-down Mexican and related 8 *reales* coins. The Mint turned out a small run of 20 1838-O half dollars as well.[3]

Not all was well. Construction was by local bricklayers and carpenters who seemed to have little experience in building such a large structure on infirm ground—which probably should not have been selected as the location to begin with. The floors were built on groined brick arches, supported by brick pillars. Pressure from the arches caused abutments to fail, and the arches began to sink, threatening collapse to the entire building. Adjustments were made, and on May 17, 1839, the government formally accepted the building. The total cost of the building was $182,000, and close to an additional $118,000 was spent on an iron fence enclosing the facility, metal processing machinery, coining presses, furnaces, fixtures, and other equipment, bringing the cost to about $300,000. Annual expenses for maintenance were estimated at about $52,000.

Soon afterward, irregularities at the Mint were suspected, and Thomas Slidell, the United States district attorney in the city, launched an investigation. Rumors in the press were that the officers had debased the coinage by adding too much alloy, there was infighting, there were irregularities concerning a specific deposit, and expenses were out of control (such as paying out $52,000 in salaries while only $40,243 in coins was produced).

On July 12 Slidell left New Orleans and headed to Washington to deliver his report. On August 17, 1839, coinage operations were suspended.[4] Slidell reported that "disagreements arose among the officers, which ended in mutual denunciations, and rendered an investigation necessary as to the characters of those implicated and the state of the institution generally." In particular, coiner Rufus Tyler was the subject of three depositions by employees, who stated that he wasted a lot of time, handled bullion carelessly, and in essence, was not suited for the job. On the plus side he had cleaned up rusted machinery when it arrived, tried to make some improvements in the presses (one of which was rusted and set aside out of service), and seemingly expended much effort.[5]

Scarcely was the investigation completed when cholera swept through the city, killing coiner Rufus Tyler, melter and refiner James B. Rodgers, and four other employees. In October Superintendent David Bradford was removed from his position. Mint treasurer Edmond Forestall was terminated as well. Joseph M. Kennedy was appointed to replace him. Philos B. Tyler, brother of the late Rufus Tyler, was appointed coiner, and John L. Riddell was named to the post of melter and refiner. "The institution has now an able corps of officers," Director Patterson noted.

In 1839 the New Orleans Mint struck half dimes, dimes, half dollars, and quarter eagles.

INTO THE 1840s

In 1840 New Orleans was the fourth-largest city in the United States and was the busiest port in America for the export of goods, handling more than half of the trade. Coins in circulation there were predominantly silver and gold from the Spanish-American mints. So far, not enough coins had been minted in New Orleans for them to have a noticeable presence in commerce. Most reckoning of accounts was done in Spanish instead of American dollars. Half dimes and Spanish 6-1/4 cent *medios* were interchangeable, as were dimes and Spanish 12-1/2 cent reales or bits.

The port handled large quantities of imports of supplies destined for forwarding to the settlements north on the Mississippi River. Ice packed in straw and hay for insulation and sent by sea from Boston was among the more unusual items.

On July 1, 1843, *Niles' National Register* printed this item from New Orleans:

> Our mint in this city is now in a flourishing condition, and has in deposits at the present time nearly $2,000,000, most of which is gold. It is not generally known that our moneyed men, who are in the habit of receiving quantities of foreign gold, send most of it to this establishment to have it melted and recoined into American money. The premium on this is sufficient to afford them a handsome profit.

THE MINT IN 1845

In 1845 James H. Dakin made structural changes to the Mint building, which continued to have problems with shifting and settling—unsurprisingly, as the foundation was laid on wooden planks. In the same year a steam boiler was put into operation to run coining presses of the general type installed at the Philadelphia Mint in 1836. Previously, screw presses operated by a crew of three men had been used. In early 1845 Mint treasurer H.C. Commack was fired, and in May he was replaced by J.R. McMurdo. There were 13 employees plus a foreman in the Coining Department and 5 in the Melting Department; William P. Hort was the assayer. The average salary of a Mint worker was about $60 per month.

In addition to melting foreign coins into bullion in 1845, gold mined in Alabama was becoming important, and potential was seen in silver from the regions to the west of Lake Superior, precious metals from Arkansas, and gold from mines in Mexico. John Leonard Riddell wrote:

> Gold is presented to us in the form of foreign coin, bars, dust, and old jewelry; the most abundant foreign gold coins being English sovereigns, French Napoleons, patriot doubloons, and the coinage of different German States; while the unwrought gold is principally from the State of Alabama.[6]

This metal was refined, assayed, cast into ingots, rolled into strips, and made available for planchets. Riddell described the minting operation:

The Coining Process

There are four presses in the coining-room, forming a series, in respect to size and strength, adapted to the stamping of the various coins, from the half-dime to the dollar. The mechanical principle brought into play is the same as that in the ordinary printing press—the genicular or elbow power, by which, with sustaining parts of sufficient strength, an almost incalculable degree of pressure may be commanded.

Each operating press requires a man to watch it, to oil the joints occasionally, and to keep a vertical brass tube supplied with the blanks or planchets to be coined. The untiring press goes on, seizing with iron fingers from the tube, a planchet of its own accord, carefully adjusting it to the retracted dies, squeezing it with a degree of force sublime to contemplate, and then quietly and safely depositing it in the box placed to receive it. From 80 to 150 pieces, dependent upon the size, are thus coined in one minute's time.

The obverse, reverse, and indented work upon the edge, are all completed at a single effort of the press. Travel the world over, and you can scarcely meet with a more admirable piece of massive mechanism than the new press in the New Orleans Mint, for the coinage of dollars.

There continued to be many problems at the New Orleans Mint, and one of them was Riddell, dating back to the time he was first employed. He was unpopular with certain of the employees, who accused him of shortages in the amounts of silver and gold bullion on hand and poor skill in the refining and making of ingots (some of which were said to be too brittle to be rolled out into strips). Some allegations had clear merit; the metal shortages were not proven. He and most others who came to the Mint after the purge in 1839 had had little relevant experience and had to learn on the job.

Riddell proposed that money could be saved by employing slaves, thereby spending less for regular work and nothing additional for overtime, stating that such an arrangement had worked well at the Dahlonega Mint in the melting and refining processes. He also argued that slaves could work "over hot furnaces far better than whites who may generally wish to leave during the summer." He proposed that the Mint should provide them with clothing without pockets or places to conceal coins, to prevent theft.[7]

Norman's New Orleans and Environs, a guide published by B.M. Norman in October 1845, illustrated the New Orleans Mint and gave the information quoted below:

> The United States Branch Mint is situated on what was once called Jackson Square, being nearly the former site of fort St. Charles. It is an edifice of the Ionic order, of brick plastered to imitate granite, having a centre building projecting, with two wings; is strongly built, with very thick walls, and well finished.
>
> Our limits will not permit us to go into a detailed description of its interior arrangements; which, however, may be generally spoken of as such as not to discredit the distinguished engineer who planned it. The total length of the edifice is 282 feet, and the depth about 108 — the wings being 29 by 81, and the whole three stories in height. It was begun in September, 1835; and the building was perfectly completed at a cost of $182,000. The machinery is elegant and highly finished, and, when in operation, proves an interesting sight to visitors; which, from the gentlemanly urbanity of the officers of the establishment, may be easily enjoyed. The square is surrounded by a neat iron railing on a granite basement. The coinage of 1844 — gold, $3,010,000 — silver, $1,198,500 — making in all $4,208,500.[8]

THE LATE 1840s

The controversial and widely disliked J.L. Riddell remained at the mint until a letter dated December 4, 1848, from Secretary of the Treasury Robert J. Walker took effect:

> I am directed by the President of the United States to inform you that your services are no longer required as melter and refiner in the Branch of the Mint of the United States at N. Orleans and that you are removed from that office.

Liberty Seated silver dollars were struck for the first time in New Orleans when 59,000 1846-O coins were struck. The Treasury Department envisioned that these would become very popular and compete with Spanish-American 8 reales coins. Accordingly, four pairs of dies were sent from Philadelphia to New Orleans for use in coining 1847-O dollars, but no such coins were ever made. Four more pairs were shipped from Philadelphia to New Orleans for making 1848-O dollars, but these too never materialized. The Treasury also intended that 1849-O dollars be made, and one obverse die bearing that date was shipped to New Orleans, to be mated with one or more reverses left over from earlier die pairs. However, no 1849-O dollars were made. Finally, more Liberty Seated dollars were struck there in 1850.

In August 1847 a telegraph connection was made to Mobile, Alabama. Within a short time after that the city was linked to other states in the East.

THE 1850s

In early 1850 Joseph M. Kennedy and his wife gave a fancy dress ball for Rose and Josephine, his two debutante daughters. At the time the family was an integral part of high society in New Orleans.

In 1850 the double eagle denomination was introduced, and coinage took place at Philadelphia and New Orleans, the latter mint using a press for silver dollars that had been installed in 1845 for the first dollar coinage in 1846. The presses at Charlotte and Dahlonega did not have the capacity to strike coins larger than the diameter of a half eagle. Gold from California arrived in quantity. In early 1850 the SS *Falcon* docked in New Orleans with $500,000 in treasure (as gold dust and nuggets were often called) and the SS *Alabama* landed with $350,000.[9]

In June 1850 this appeared in the *New Orleans Bulletin:*

> The Mint in this city is now in full operation, turning out double eagles with such rapidity, that they will soon work up the immense heap of 'the root of all evil,' which was so long locked up in its vaults, to the great inconvenience of many parties. As a small piece requires as much time to coin as a larger one, the plan adopted by the Mint is judicious, of applying their force to the double eagle, as it will enable them to work up the accumulated stock of gold in a very short period. Our California friends may continue to forward their gold to this city, as it can now be coined here with the greatest promptness.

In 1851 the New Orleans had a record year with coinage totaling $10,122,000, mostly from California gold that arrived by ship. In 1852 gold arrivals from California by steamship totaled $470,783. In 1853 a yellow fever epidemic claimed 8,000 lives in the city.

The Mint continued to have structural problems. On April 17, 1854, Mint Superintendent Charles Bienvenu wrote to Mint Director James Ross Snowden in Philadelphia, stating that conditions were so dangerous that repairs were needed in order to preserve the building. On May 31, 1854, the Senate passed a resolution to direct the secretary of the Treasury to determine whether it was better to try to repair the original building or to construct an entirely new one. The work for repairs was variously estimated to be $25,000 to $37,000 and to take about six months' time.

Repairs commenced and were expected to be routine. However, the opposite proved to be the case. The project was not finished until September 30, 1858, by which time an incredible $588,812.70 had been spent on the project. Government reports chronicled poor planning, waste, inefficiency, and disarray as reasons for the huge cost override. Along the way the coining presses remained in service, but production quantities were lower than in earlier times.

In its issue of September 11, 1858, *Ballou's Pictorial Drawing Room Companion* featured the Mint this story:

The United States Branch Mint at New Orleans

This building, which is shown in the engraving on this page from an engraving made expressly for us by Mr. Kilburn, is a fine substantial structure of the Ionic order of architecture and consists of a centre building and two wings. It is built of brick with a mastic coating to represent granite. The internal arrangement is very creditable to the architect who planned it. The total length of the building is 282 feet and the depth 108, the wings being 27 by 81 feet, and is three stories in height. The erection of the building commenced in 1835, and when completed cost $182,000.

The Mint is situated on what was once called Jackson Square, being nearly the former site of Fort St. Charles. Our view is from Esplanade Street, one of the finest avenues of New Orleans. The machinery of the Mint is substantial and highly finished, which from the gentlemanly urbanity of the officers of the establishment can be easily enjoyed.[10]

A large amount of coin is produced at this Mint, both of gold and silver. The Mint is of course an object of great interest, and an especial mystery to those who have fewer dealings with current coin than they desire—a sort of Aladdin's cave which their imaginations invest with the untold wealth to which the enchanter's lamp gave access to the adventurous Oriental. In reality, however, the Mint is accessible, and its treasures are by no means untold. . . .[11]

There is no such thing as gold dust or gold dollars "lying around loose" there; every grain is accounted for. . . . The bars or ingots of gold or silver, after having been cast, are taken out of the mould, and their surfaces cleaned or *blanched* by being placed in a bath of hot diluted acid, after which they are immediately dried. They are then flattened by rollers and reduced to the proper thickness to suit the species of money about to be coined. To render the plates more uniform they are sometimes drawn by passing them through narrow holes in a small device. The plates, whether of gold, silver, or copper [*sic*], when reduced to their proper thickness, are next cut out into round pieces called *blanks* or *planchets*.

THE CIVIL WAR

On January 26, 1861, Louisiana seceded from the Union. State troops seized the Mint and held it until March 31, when control was shifted to Confederate men. From then until May 14, $1,356,136 was coined in half dollars and double eagles. Four pattern half dollars were made by combining an 1861 Liberty Seated obverse with a newly made Confederate-design reverse. On May 31 the uncoined bullion was turned over to the Confederate Treasury. A resolution of the Congress of the Seceding States, introduced by T.R.R. Cobb of Georgia, that the Dahlonega and New Orleans mints remain open and produce coins on a continuing basis for the Confederacy was signed by President Jefferson Davis on March 14. The president of the Congress of the Seceding States was Howell Cobb, who had earlier been secretary of the Treasury in Washington during the James Buchanan administration.

The *Springfield* (Massachusetts) *Republican* included this exchange item from Philadelphia on March 26, 1861:

A Want of Confidence

We understand that a considerable amount of gold, of foreign coinage, has recently been forwarded by a bank in New Orleans to a Philadelphia Bank, for the purpose of having it converted into American coin at the United States Mint in this city— the reason assigned for transmitting it here, instead of having it coined at the New

Orleans Mint, being a want of confidence in the latter institution since it has fallen into the hands of the secessionists. This incident is quite a suggestive one, and shows how much distrust of the disunionists is felt in the financial circles of the South.—*Philadelphia Press*, 23d.

The Mint Report for the fiscal year ending June 30, 1861, included this:

No reports have been received from the branches at New Orleans, Dahlonega, or Charlotte. Although New Orleans is now, and has been for some months, in the possession of the Union forces [who retook the city from the Confederates], yet the operations of the Branch Mint in that city have not been resumed; nor is it expedient or necessary that they should be. . . .

At the Branch Mint the amount of deposits received up to the 31st day of January, 1861, was $1,243,449.01; of which the sum of $334,410.77 was in gold, and $909,038.24 in silver; coined during the same period, $244,000 in gold, and $809,000 in silver; silver bars stamped, value, $16,818.33; total coinage, $1,069,818.33; number of pieces, 1,237,800. Since the 31st day of January, 1861, no report has been received from this branch.

No consideration of public or private interest would, under the most favorable circumstances justify the re-opening [after the war] of the branches at Dahlonega or Charlotte. They ought not to have been established; and having been the source of useless expenditure, they should not, even in the event of the states in which they are respectively located, returning to their allegiance, be again employed for Minting purposes. Whether gold or silver coins were struck at any of the defected branches of the Mint during the past year, I have not been able to ascertain with certainty; if any, the amount was small.

Prior to the defection of the branch at New Orleans, the dies in that institution were defaced, or destroyed by some of the local employees, under the direction of one of the officers who remained true to his duty and to his country.[12] This destruction of the dies must have delayed, if not altogether prevented any coinage at that branch.

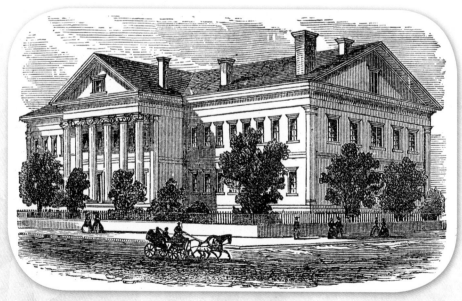

The New Orleans Mint in 1861.

A booklet, *History of the Mint*, published in 1895, told this:

> Between January 26 and May 31, 1861, the State of Louisiana had possession of the New Orleans branch mint and coined during that period $195,000 in double eagles, and a coinage of $59,820 in double eagles was accomplished by the Confederates, during that time. This was during the sequestration of the Mint, and a total gold coinage of $254,820 was reached. In February or March of the same year the State of Louisiana coined 1,240,000 silver halves; by the Confederacy in April and May, 962,633 silver halves; of total silver coinage, in half dollars, by the State of Louisiana in the Confederate States of $1,101,316.50. All of this coinage was done with the regular dies of the United States, supplied late in 1860 for the following year for government use. For obvious reasons none of the coinage executed at the United States Mint at New Orleans, while out of the control of the United States government, has ever been taken up into statements of the coinage of the United States.

The Confederate Congress passed an act on January 27, 1862, to establish an assay office at the Mint. On May 1, 1862, Union forces repatriated New Orleans, and former Superintendent M.F. Bonzano was soon ordered by Secretary of the Treasury Salmon P. Chase to take possession of the Mint and its contents. General Benjamin F. Butler was put in charge of the city. He earned the nicknames "Spoons Butler" and "Beast Butler" from Union troops ransacking homes in the city and for his treatment of the residents.

The *Annual Report of the Director of the Mint* for the fiscal year ended June 30, 1862, noted:

> Although New Orleans is now, and has been for some months, in the possession of the Union forces, yet the operations of the Branch Mint in that city have not been resumed; nor is it expedient or necessary that they should be. . . .
>
> Prior to the defection of the branch at New Orleans, the dies in that institution were defaced, or destroyed by some of the local employees, under the direction of one of the officers who remained true to his duty and to his country.[13] This destruction of the dies must have delayed, if not altogether prevented any coinage at that branch.

THE SECOND ERA OF THE MINT

The *Report of the Director of the Mint* for the fiscal year ended June 30, 1865 included this:

> The suppression of the rebellion and the anticipated early return of the recusant States to their allegiance present the question, what shall be done with the branch mints at New Orleans, Louisiana; Charlotte, North Carolina; and Dahlonega, Georgia? In my annual report of 1862 it was suggested that the branch mint at New Orleans, after the re-establishment of law and order in Louisiana, might be successfully operated, and that the branch mints at Charlotte and Dahlonega ought not to be employed again for minting purposes. My opinions on this subject are unchanged. The commercial importance of New Orleans, and the relations of that city to every portion of our country, justified the establishment there of a branch mint; and the amount coined in that institution from its organization, in 1838, to January, 1861, confirmed the propriety of its location at that place. During the period of its active operations, the total coinage was over seventy millions of dollars, as follows: $40,381,615 in gold, and $29,890,037 in silver. The deposits of silver at this branch have always been large; and it is worthy of consideration whether the coinage there should not, for the present, at least, be confined to silver.

This did not happen. The New Orleans Mint was essentially an unoccupied building from 1862 into the 1870s. On January 15, 1865, 32 pairs of dies left over from earlier days were destroyed.[14] An inventory performed by M.F. Bonzano reported the equipment on hand and its condition. This included three milling machines and five coining presses, the smallest suitable for half dimes, dimes, and gold dollars; the next for dimes, quarter eagles, and half eagles; the next for quarters, half dollars, half eagles, and eagles; and the largest suitable for half dollars, silver dollars, eagles, and double eagles.

By 1867 continuing repairs amounted to $26,013.10. Many carpetbaggers— opportunists who came from the North during the Reconstruction period, took up residence in the Mint. Unpopular with local residents, there was a revolt against them in 1874. Federal troops came to their defense and stayed on duty until 1877. In that year there was an economic recession in the United States.

In 1878 the Treasury Department, faced with the challenge of coining millions of silver dollars each year, investigated the possibility of reopening the New Orleans Mint. At the second session of the 45th Congress provision was made to put the facility back in service to resume coinage, last done in 1861. Michael Hahn was appointed superintendent and began studying the situation. The City of New Orleans intimated that as it had, in effect, been abandoned by the Treasury Department, it now held title to the land.

It would seem that per the original agreement in which the city ceded the land on May 11, 1835, this was true:

> Resolved, That the use of the Square of ground now enclosed and known as Jackson Square, situated in the City of New-Orleans, and bounded as follows, to wit: by Esplanade, Garrison and Levee streets, and the Public road, be ceded to the United States for the express and only purpose of erecting thereon a branch Mint of the United States together with the necessary appendages; and that the Mayor be, and he is, hereby authorized to convey by notarial act to Martin Gordon, Esq., the Commissioner appointed by authority of the said United States, to superintend the erection and building of said Mint, the use and occupation of said Square for the purposes aforesaid.
>
> Be it further resolved, that should it hereafter be deemed necessary by the Government of the United States to remove the Mint contemplated to be established as aforesaid, or to cease to occupy it for such purposes, then the said act to be null and void.

Incensed by this and apparently unfamiliar with the 1835 provision, Hahn demanded that the deed be cleared. On July 18 a quitclaim was signed. Plans were put into effect to refit and refurbish certain areas and to install machinery. Most of the equipment had been damaged, rusted, or had other problems from the long period of disuse. Not long afterward yellow fever broke out and caused great distress, eventually claiming about 4,500 lives. Sick people fleeing New Orleans carried it to areas north on the Mississippi River. Eventually there were over 100,000 cases with tens of thousands of fatalities.

This delayed the activity at the Mint, and experienced mechanics from other mints were told to stay at home until the epidemic cleared. Eventually it did, and everything was in place. Mint Director Henry R. Linderman stated in his *Annual Report:*

> The various operative rooms in the Mint have been placed in good condition, the melting and annealing furnaces restored, the engine and machinery repaired, and such additional machinery as is required to render this mint effective for coinage has been procured, and it is believed that by the end of the current calendar year everything pertaining to the mint will be in a condition to commence coinage.

The New Orleans Mint will add to the coining capacity of the mints about $1,000,000 in silver dollars per month, but this capacity would be somewhat reduced if called upon to execute any considerable amount of gold or fractional silver coinage.

On December 28, 1878, Henry S. Foote was named superintendent, replacing Hahn. In the same month Mint Director Linderman left his post. The Mint officers were allowed to reside in apartments there, a practice that continued until 1888.

COINAGE BEGINS AT THE MINT

Delays occurred during this interregnum. In February 1879 Horatio C. Burchard was appointed Mint Director. On the 20th of that month coinage operations began. Silver bullion was in short supply in the district at the time as it cost more to deliver the metal to New Orleans than to the Mint in Philadelphia, and the Mint worked at less than capacity.[15] Eventually 2,887,000 silver dollars were made that year plus just 1,500 $10 and 2,325 $20 gold coins. This would be the only year that double eagles were made at New Orleans in the new era.

The *Annual Report* for the fiscal year ending June 30, 1880, included this:

New Orleans Mint

In preference to purchasing bullion for delivery at the New Orleans Mint at figures regarded as in excess of the market price, it was at first deemed advisable to transfer from the New York assay office 1,798,167.82 standard ounces purchased prior to June 30, 1879. This was insufficient to supply that mint with an amount of bullion equal to its capacity and the demand upon it for silver coinage; it therefore became necessary to purchase additional bullion at such rates as were offered or to suspend coinage at that mint.

The rates, though at first above the New York price, were less than the cost to the department to purchase and transfer from Philadelphia or New York. Offers were accepted during the year for the delivery at New Orleans of 1,684,158 standard ounces at the lowest rates attainable, but above the New York price.

The difficulty continued through to obtain silver bullion for the mint. However, demand in the gold states for silver coins necessitated allotment at the mint for the coinage of 400,000 silver dollars. Following the death of Superintendent Henry S. Foote on May 19, 1880, an audit was taken to transfer the administration to incoming Superintendent M.V. Davis, on June 11, 1880 (Davis was previously coiner at the mint), and a sack of 1,000 silver dollars was found to be missing. The discrepancy was apparently a surprise to the cashier and officers of the mint. There had previously been frequent urgent demands for the shipment of standard dollars, and the error was supposed to have occurred from an undiscovered mistake in the report or count of the sacks or some delivery for distribution.

CONTINUING OPERATIONS

The silver problem eased when the director of the Mint negotiated with suppliers, including Western refineries. Silver that was earlier shipped to New York City and then to New Orleans was in many instances now sent there directly. Shipment by rail became increasingly efficient. In his *Annual Report* for fiscal year 1883 Director Horatio Burchard included this about New Orleans:

The coinage of gold at this mint was inconsiderable, amounting to only $43,000. Standard silver dollars were struck of the value of $8,040,000, being over $3,000,000 in excess of the coinage of the fiscal year 1882. "This institution is of little local advantage, except that it furnishes a large amount of silver coin for circulation in the Southern and Southwestern states. The bullion used in its coinage is obtained from distant refineries.

Dr. Andrew W. Smythe, a surgeon, was appointed superintended and began in the post on August 21, 1882. In February 1883 the Assay Commission, evaluating sample coins coined in 1882 at the various mints, found some $10 gold coins from New Orleans to be of correct weight, but to have lower gold content replaced with more alloy. Newspapers nationwide reported the incident and suggested that a clean sweep might be made of all at the branch mint. In July 1885 Smythe was routinely succeeded by Gabriel Montegut.

Gold $5 and $10 coins were struck intermittently in New Orleans. Liberty Seated dimes and quarters were struck in 1891. In 1891 the average cost of minting coins at the New Orleans Mint was $0.0203+ or about two cents. Most efforts were concentrated on silver dollars which were made every year through 1904, when the supply of silver authorized by Congress for coinage of this denomination was exhausted.

In 1892 the Barber design made its debut on the silver dime, quarter, and half dollar. The New Orleans Mint struck these pieces continually until 1909, when coinage operations stopped forever. No 1904-O dimes were made. In July 1892 Superintendent Smythe made the news by firing 21 employees, including some with long-time service. No reason was given for the action. In an era before Civil Service it was common for federal employees to be terminated without explanation.

Punching planchets from metal strips at the New Orleans Mint in 1897. Silver dollars were being produced when these photographs were made.

Women adjusting overweight planchets by filing. Underweight planchets are set aside to be melted.

Putting a raised rim on blank planchets with the milling machine.

"Stamping Room" with coin presses.

SCANDAL AT THE MINT

When the annual settlement of accounts was taken by auditors on June 30, 1893, in the counting of cash in the form of paper money in the cashier James M. Dowling's vault, a deficiency of $25,000 was found. The cashier claimed that a fire had occurred there between the closing of the Mint on Saturday, June 24, and the reopening on Monday morning. A lighted oil lamp in the vault had tipped over, he said. The superintendent wired the Treasury Department in Washington, and an expert in evaluating charred currency was sent to the Mint. Traces of only $1,182 in cash were found, leaving a shortage of $23,818. A.R. Barrett, a Secret Service operative, was called in to investigate. Incriminating evidence was found and the cashier was arrested on July 20. President Grover Cleveland fired the superintendent, assayer, melter and refiner, and the coiner of the Mint, and Overton Cade was appointed as the new superintendent, replacing Dr. Andrew W. Smythe. These men began service on July 23, 1893.

Dowling was tried in court in December and acquitted. Suit was then filed against former Superintendent Smythe, who was responsible for security, for recovery of the amount. This dragged on in the courts into the early 20th century.

In fiscal year 1896 in Philadelphia it cost $0.009, or slightly less than a cent for the expense, metal not included, of coining a dollar. In San Francisco it was $0.011527, or slightly more than a cent. In New Orleans the cost was $0.036564, or nearly four cents. Year in and year out, various Treasury officials complained about the New Orleans facility. In 1899 the coinage of silver dollars caused all three mints to run overtime in some months.

FURTHER COINAGE

From 1902 to 1904 there was discussion within the Treasury Department of closing the New Orleans Mint. It was found that the Philadelphia and San Francisco mints operated more efficiently, with lower cost for each coin struck, and work on a new mint in Denver was underway. The stock of bullion for coining dollars under the Act of July 14, 1890, was scheduled to run out, after which the New Orleans Mint would not be needed unless work was cut back in Philadelphia and San Francisco, an unacceptable prospect. Coinage of gold in New Orleans was not large, nor was there a demand for gold coins anywhere. The *Annual Report* for fiscal year 1902 stated:

Weighing and counting room with 1897-O Morgan dollars being evaluated. March and April figures are on a blackboard.

It is opportune here to call attention to the fact that the gold coinage of the country is now entering almost entirely into storage and that the cost of coinage is an unnecessary expense. The Treasury holds now about $500,000,000 of coined gold, which is doubtless more than will be called for in a generation to come. Practically all of the current coinage is being deposited in the Treasury for certificates. When gold is required for export, it is wanted in bars, while for domestic circulation the public prefers the Treasury certificates, which, with some modification of the statutes, might as well be issued against bars. . . .

The cost of operating the New Orleans Mint last year was $259,158.98. The estimates for Philadelphia and San Francisco are not increased, but it will be possible for those institutions to do the entire coinage for the year 1903–04 within the appropriations that are asked for them.

In February 1904 there were 219 people employed at the Mint. By June the number had been reduced to just 97, with more cuts planned. From November 1904 into the summer of 1905 the Mint was idle as the coinage of dollars had ended. At the time $32,000,000 in coins was stored in its vaults, including $29,000,000 in silver dollars, the production of which ended in 1904. In 1918 $12,400,000 in silver dollars was shipped to the San Francisco Mint to be melted under the terms of the Pittman Act. Many other dollars remained in storage in New Orleans for a long time and in 1929 were removed to a vault at the Philadelphia Mint, which was sealed so that it would not need to be audited.

In 1906 there was a shortage of silver, coinage was suspended, and most of the employees, except those needed for maintenance, were furloughed without pay, a situation that lasted for four months. In that year some employees were sent to Denver to help with the opening of the new mint there.

In 1909 there was also a coinage of 1909-O half eagles. On April 1 of that year the last coin was minted. In 1910 the coining presses and related machinery were crated and shipped to Philadelphia. Records show that since 1838 the New Orleans Mint had produced $298,660,707.60 in coins of which only $48,704,172.50 was in gold.

Assaying was continued at the Mint after coinage ended. In later years the building was used by various government agencies and for storage. At present the Louisiana State Museum operates exhibits for visitors.

Rear view of the New Orleans Mint, an unusual perspective.

COINAGE OF THE NEW ORLEANS MINT, 1838–1909

Three-Cent Pieces

The 1851-O three-cent piece or trime is the only silver coin of less than five-cent denomination to have been struck at a branch mint. The O mintmark is a tiny perfect circle, not an ellipse.

DESIGN TYPE AND DATE MINTED
Silver, Variety 1: 1851-O.

Half Dimes

Half dimes were struck at the New Orleans Mint from 1838 to 1859, with the exceptions of 1843 and 1845 to 1847. Most were made in quantity and are easily enough found today, the 1853-O, Without Arrows, being an exception. Most half dimes range from scarce to rare in high Mint State levels.

The New Orleans Mint building
in 1963. (Dan Leyer survey)

The New Orleans Mint circa 1909, the last year of coinage.

Design Types and Dates Minted

Liberty Seated, Variety 1, No Stars on Obverse: 1838-O.

Liberty Seated, Variety 2, Stars on Obverse: *No Drapery From Elbow:* 1839-O, 1840-O. *Drapery From Elbow:* 1840-O, 1841-O, 1842-O, 1844-O, 1848-O, 1849-O, 1850-O, 1851-O, 1852-O, 1853-O. *Variety 3, Arrows at Date, Reduced Weight:* 1853-O, 1854-O, 1855-O. *Arrows Removed, Weight Remains:* 1856-O, 1857-O, 1858-O, 1859-O.

Liberty Seated, Variety 4, Legend on Obverse: 1860-O.

Dimes

Liberty Seated dimes were made at the New Orleans Mint from 1838 to 1860 except for 1846 to 1848. Nearly all range from scarce to very rare in high Mint State levels.

Design Types and Dates Minted

Liberty Seated, Variety 1, No Stars on Obverse: 1838-O.

Liberty Seated, Variety 2, Stars on Obverse: 1839-O, 1840-O, 1841-O, 1842-O, 1843-O, 1845-O, 1849-O, 1850-O, 1851-O, 1852-O. *Variety 3, Arrows at Date, Reduced Weight:* 1853-O, 1854-O. *Arrows Removed, Weight Remains:* 1856-O, 1857-O, 1858-O, 1859-O.

Liberty Seated, Variety 4, Legend on Obverse: 1860-O.

Quarter Dollars

Quarters were made continuously at the New Orleans Mint from 1840 to 1860 and were popular in commerce. Circa 1841 a large quantity of these was buried in the ground. Discovered generations later during when they were unearthed by a bulldozer, these coins are known as the New Orleans Bank Find. The coins are believed to have been a secret reserve for some long-forgotten merchant or bank.

DESIGN TYPES AND DATES MINTED

Liberty Seated, Variety 1, No Motto Above Eagle: *No Drapery From Elbow:* 1840-O. *Drapery From Elbow:* 1840-O, 1841-O, 1842-O, 1843-O, 1844-O, 1847-O, 1849-O, 1850-O, 1851-O, 1852-O.

Liberty Seated, Arrows at Date: *Variety 2, Rays on Reverse:* 1853-O. *Variety 3, No Rays:* 1854-O, 1855-O. *Arrows Removed, Weight Remains:* 1856-O, 1857-O, 1858-O, 1859-O, 1860-O.

Half Dollars

The first issue, 1838-O, is a rarity with just 20 coined, according to a note attributed to coiner Rufus Tyler. The later issues were made in quantity, except for 1853-O, No Arrows, of which just three are known today. Certain of the 1861-O coins were struck by occupying forces after New Orleans was captured in early 1861.

DESIGN TYPES AND DATES MINTED

Capped Bust, Reeded Edge, Reverse HALF DOL.: 1838-O, 1839-O.

Liberty Seated, Variety 1, No Motto Above Eagle: 1840-O, 1841-O, 1842-O, 1843-O, 1844-O, 1845-O, 1846-O, 1847-O, 1848-O, 1849-O, 1850-O, 1851-O, 1852-O, 1853-O.

Liberty Seated, Arrows at Date: *Variety 2, Rays Around Eagle:* 1853-O. *Variety 3, No Rays:* 1854-O, 1855-O. *Arrows Removed, Weight Remains:* 1856-O, 1857-O, 1858-O, 1859-O, 1860-O, 1861-O.

Barber or Liberty Head: 1892-O, 1893-O, 1894-O, 1895-O, 1896-O, 1897-O, 1898-O, 1899-O, 1900-O, 1901-O. 1902-O. 1903-O, 1904-O, 1905-O, 1906-O, 1907-O, 1908-O, 1909-O.

Silver Dollars

Four different dates of Liberty Seated dollars were minted. The 1846-O was circulated regionally, with the result that most seen today are worn. The three later issues were mostly exported. A small group of Mint State 1859-O and 1860-O dollars was marketed in the 1960s. Morgan dollars were struck continuously from 1879 to 1904.

The Annual Report of the Director of the Mint for the fiscal year dated 1885 told of the distribution of silver dollars as of June 30, 1885. New Orleans had 16,221,999 on hand and had distributed 5,193,639.

DESIGN TYPES AND DATES MINTED

Liberty Seated, No Motto: 1846-O, 1850-O, 1859-O, 1860-O.

Morgan: 1879-O; 1880-O, 80 Over 79; 1880-O; 1881-O; 1882-O; 1883-O; 1884-O; 1885-O; 1886-O; 1887-O, 7 Over 6; 1887-O; 1888-O; 1889-O; 1890-O; 1891-O; 1892-O; 1893-O; 1894-O; 1895-O; 1896-O; 1897-O; 1898-O; 1899-O; 1900-O; 1901-O; 1902-O; 1903-O; 1904-O.

Gold Dollars

Gold dollars were minted from 1849 to 1855. These are seen with some frequency. The 1855-O is scarce, and when seen it is usually lightly struck.

DESIGN TYPES AND DATES MINTED

Liberty Head (Type I): 1849-O, 1850-O, 1851-O, 1852-O, 1853-O.

Indian Princess Head, Small Head (Type II): 1855-O.

$2.50 Quarter Eagles

All of the New Orleans quarter eagles are collectible. The 1839-O is in special demand as the only year of its type. All are rare in correctly graded Mint States.

DESIGN TYPES AND DATES MINTED
Classic Head: 1839-O.

Liberty Head: 1840-O, 1842-O, 1843-O, 1845-O, 1846-O, 1847-O, 1850-O, 1851-O, 1852-O, 1854-O, 1856-O, 1857-O.

$3 Gold Coins

Three-dollar gold coins were struck at the New Orleans and Dahlonega mints only in 1854. When seen the typical 1854-O shows wear.

DESIGN TYPE AND DATE MINTED
Indian Princess Head: 1854-O.

$5 Half Eagles

Half eagles were mostly coined in the earlier years with two Liberty Head issues after the Mint reopened after the Civil War, and for 1909-O a lone Indian Head coin, the last being scarce today.

DESIGN TYPES AND DATES MINTED
Liberty Head, Variety 1, No Motto Above Eagle: 1840-O, 1841-O, 1842-O, 1843-O, 1844-O, 1845-O, 1846-O, 1847-O, 1851-O, 1854-O, 1855-O, 1856-O, 1857-O.

Liberty Head, Variety 2, Motto Above Eagle: 1892-O, 1893-O, 1894-O.

Indian Head: 1909-O.

$10 Eagles

Most eagles were minted from 1841-O to 1860-O, followed by lower coinages into the early 20th century.

DESIGN TYPES AND DATES MINTED

Liberty Head, No Motto Above Eagle:
1841-O; 1842-O; 1843-O; 1844-O; 1845-O;
1846-O, 6 Over 5; 1846-O; 1847-O; 1848-O;
1849-O; 1850-O; 1851-O; 1852-O; 1853-O;
1854-O; 1855-O; 1856-O; 1857-O; 1858-O;
1859-O; 1860-O.

Liberty Head, Motto Above Eagle:
1879-O, 1880-O, 1881-O, 1882-O, 1883-O,
1888-O, 1892-O, 1893-O, 1894-O, 1895-O,
1897-O, 1899-O, 1901-O, 1903-O, 1904-O,
1906-O.

$20 Double Eagles

Double eagles were the workhorse denomination for the New Orleans Mint in the early days. 1854-O and 1856-O are major rarities coined while the Mint was undergoing repairs and production was low. The stray 1879-O is very elusive.

DESIGN TYPES AND DATES MINTED

**Liberty Head, Without Motto on
Reverse:** 1850-O, 1851-O, 1852-O,
1853-O, 1854-O, 1855-O, 1856-O,
1857-O, 1858-O, 1859-O, 1860-O,
1861-O.

**Liberty Head, Motto Above Eagle,
Value TWENTY DOLLARS:**
1879-O.

CHARLOTTE MINT, 1838–1861

GOLD IN NORTH CAROLINA

In 1799 in Cabarrus County, North Carolina, 12-year-old Conrad Reed found a 17-pound "glittering stone" in Little Meadow Creek on his father's farm.[1] Not knowing what it was, he took it home where it was used as a doorstop. Not long afterward it was purchased by a local jeweler, who recognized it as gold. This discovery in various iterations became a twice-told tale in the annals of 19th-century Americana. By 1802 the news of gold in the area drew many fortune seekers. In 1804 John Pfifer, a miner in the area, sent a shipment of dust and nuggets to the Philadelphia Mint, and others did as well, creating the first notice of native North Carolina gold in Mint records. That year about $11,000 was received. From then through 1824, shipments were regular, according to reports, but never reached the 1804 total. From the *Annual Report* for 1829:

The Charlotte Mint (drawn by George Osborn).

In 1824, the amount received was $5,000; in 1825, it had increased to $17,000; in 1826, it was $20,000; in 1827, about $21,000; and in 1828, nearly $46,000. In 1829, it was $128,000. This remarkable increase in the amount of gold received from North Carolina, during the years following 1824, has been considered of sufficient interest to be noted in the *Annual Reports* from the Mint, since that period. The circumstance will attract additional attention, from the fact now ascertained, that the gold region of the United States extends far beyond the locality to which it has heretofore appeared to be limited. Gold bullion had not been received from Virginia, or South Carolina, until within the last year; or, if at all received, it has been in quantities too inconsiderable to have been specially noticed. The gold from all these localities is found, in its native state, to be, on an average, nearly of the same fineness as the standard of our gold coins.

Of the amount of gold bullion, deposited at the Mint, within the last year, about $131,000 were received from Mexico, South America, and the West Indies; $22,000 from Africa; about $12,000 from sources not ascertained; and the residue, about $134,000, from North Carolina, and the adjacent States of South Carolina and Virginia. The proportion from North Carolina may be stated at $128,000; that from South Carolina, at $3,500; and that from Virginia, at $2,500.

Such news inspired action, including this from the records of Congress, Monday, December 28, 1829:

Resolved, That a select committee be appointed to inquire into the expediency of establishing a branch of the United States' Mint in the gold region of North Carolina. And on the question, Will the House reconsider the said vote? It passed in the affirmative.

Mr. Cabson rose and said that he had been induced to offer the resolution from various considerations, one of the most important of which was, the highly interesting information he was induced to believe would be elicited by such an inquiry. That it will prove necessary to establish a mint in North Carolina, to the extent to which such an establishment now exists in Philadelphia, he was by no means prepared to say; and were he to hazard an opinion, as at present advised, he would say that it would not be necessary. A branch, however, of the Mint might be found necessary. For instance, [said he] an office, under national authority, connected with the mother institution, to assay our metals, and show us their correct value—to stamp our bars of gold, and prepare them for a circulating medium, or as an article of deposit, upon which circulating medium might issue. This would also prevent frauds from being practiced: for, while it would show the owner the real value of the metal, it would also secure the purchaser from frauds, such as mixing alloy with the gold, which otherwise would be difficult to detect. In a word, [Mr. C. said] the inquiry would do no injury, while there was a probability of its doing good, for any report made by the committee will be subject to the future action and control of the House.

From the *Annual Report of the director of the Mint* for 1832:

Of the amount of gold coined within the past year, about $80,000 were derived from Mexico, South America, and the West Indies; $28,900 from Africa; $678,000 from the gold region of the United States; and about $12,000 from sources not ascertained.

Of the amount of gold of the United States, above mentioned, about $34,000 may be stated to have been received from Virginia; $458,000 from North Carolina; $45,000 from South Carolina; $140,000 from Georgia; and about $1,000 from Tennessee.

Mint Director Samuel Moore warmly welcomed these deposits, but was against suggestions for branch mints, feeling that the capacity in Philadelphia was more than sufficient to handle all of the gold deposited. This was true, but for miners to send gold from Charlotte or from Dahlonega to the south (see the next chapter) and receive gold coins in return involved about two months of time and 5% in insurance and transportation fees. How nice it would be if gold coins could be minted regionally was the prevailing opinion of many in North Carolina and Georgia, usually seconded by their representatives in Congress.

PLANS FOR THE BRANCH MINT AT CHARLOTTE

The Act of March 3, 1835, provided for the establishment of branch mints in New Orleans, Charlotte, and Dahlonega. New Orleans was the busiest port in the South and received large quantities of foreign gold and silver coins ideal for melting and converting into federal coins. The other two were in mining areas, as noted. In March 1835 General R.M. Saunders was appointed by President Andrew Jackson to be superintendent of the Charlotte Mint, this at the inception of planning.

Philadelphia architect William Strickland was selected to draw up plans for which he charged $150. These were similar to those for the mint intended for Dahlonega and of the same general style as the much larger building for New Orleans.

That summer Director Moore resigned his post and made a smooth transition with his successor, Jackson appointee Robert Maskell Patterson. In September 1835 Christian Gobrecht was added to the Philadelphia Mint staff as "second" (not assistant) engraver to William Kneass, who had been at the Mint since 1834. It was anticipated that the three branch mints would require many more dies to be made.

James M. Hutchison was selected as commissioner in North Carolina to oversee the project, but on May 6, 1835, he declined the appointment. On May 12 Samuel McComb was selected and took the post. He was a man of proven ability, with the result that all proceeded smoothly for the most part. He traveled to Philadelphia, and on June 9 and 10 met with Mint Director Samuel Moore, who was transitioning smoothly out of the post as Robert Maskell Patterson was coming in. Not long afterward Director Patterson suggested that construction be budgeted at $33,000, and that $17,000, later revised to $15,000, be allowed for machinery and its installation.

On July 22 McComb placed advertisement soliciting bids from contractors in the *Charlotte Journal*, which ran the ad the same day, and in the *Washington Globe, Richmond Enquirer, Raleigh Standard,* and *Fayetteville Journal,* to run continuously until September 1st. He carefully studied each qualified proposal. It was not until October 15, 1835, that a contract was awarded. From *Niles' Weekly Register*, November 7, 1835:

> Charlotte, N.C. October 23. We understand that Messrs. Perry & Ligon, of Raleigh, have contracted for the erection of our branch mint building, for the sum of $29,800.[2] The building is to be 125 feet in front by 33-1/2 feet in depth, with a projection, in the rear of the center, of 33 by 36 feet, and to be two stories in height, with a basement story of 5 feet above the surface of the ground. The basement, in front, to be of hammer-dressed range work—the principal and attic stories of brick. The sills and heads of the windows to be of stone; the cornice of brick, the roof covered with zinc. The basement and principal stories are to be arched, with groin arches, throughout the front building. To be completed by the 6th January, 1837.

Thomas Ligon went to the Philadelphia Mint after being awarded the contract, met with Director Patterson, and found many errors and flaws in the plans, most if not all of which were subsequently corrected by drawing new plans for both the Charlotte and Dahlonega mints.

McComb arranged to purchase from William Carson and F.L. Smith a nearly square parcel of 3.8 acres on Trade Street. This was done on November 2, with the deeds recorded on the 25th.

Coleman, Sellers & Sons, Philadelphia makers of heavy machinery, furnished the steam engine, shafting, and related items for $8,297. Merrick, Agnew & Tyler of the same city provided the coining presses, draw benches, planchet punch, and related items for $6,690.

CONSTRUCTION AND OUTFITTING

On January 8, 1836, the cornerstone was laid with due notice that this was the 21st anniversary of General Andrew Jackson's victory over the British in the Battle of New Orleans. Construction proceeded mainly with local material. At the time the population of Mecklenburg County was close to 20,000, and about one thousand people lived in Charlotte. Various mills and factories were available in the region to supply some needed items, in contrast with Dahlonega, where most supplies for that Mint had to be brought from a distance. This took all of the year 1836, into the next. On January 19, 1837, John H. Wheeler was appointed superintendent, succeeding Saunders. This was a political plum—perhaps the best federal job in the State of North Carolina. Often such appointments were given to men of questionable business ability, but who had supported the president and his party. In the case of Wheeler, he had recently become a widower, and his son, his only child, was terminally ill. Though faced with these personal challenges, he set about his Mint directorship with determination and with a bright outlook for his new beginnings in life.

John R. Bolton, who worked on presses at the Philadelphia Mint, was appointed coiner, and Dr. John Gibbon was named as assayer. Both of these positions were essential to the integrity of coinage quality. As Charlotte was a bustling city at the time (quite unlike Dahlonega), labor, supplies, and other requirements were usually met with ease.

In April 1837 the coining equipment, steam engine, and other machinery were shipped to Charlotte, first via sea to Charleston on the *Langdon Cheves* (named for a president of the late Bank of the United States), then inland by a steamship, then overland. Transporting these heavy items, packed in wooden crates, was difficult, and it was not until August that these were in place and set up.

PEALE VISITS CHARLOTTE

In August 1837 in connection with reported problems in Charlotte and Dahlonega, Director Patterson wrote to Treasury Secretary Levi Woodbury proposing to send Franklin Peale, skilled mechanic and chief coiner at the Philadelphia Mint, to the South to view the situation and give instructions on remedying any problems.[3]

Peale left Philadelphia, accompanied by his daughter Anna, and arrived in Charlotte on August 23.[4] There he found that the minting equipment had not arrived yet. The town was devoid of sophistication, he reported to Patterson, "the only active being are the hogs," and proposed that he go to Dahlonega and then to New Orleans.

The director told Peale that he was needed in Philadelphia, and to make the trip as quickly as possible.

There were some annoyances upon his arrival in Charlotte, however. Castings and copper vessels for the melting furnace sent eight days earlier from the port of Charleston had yet to arrive. He dispatched Superintendent Wheeler to take the road to Charleston and try to locate them. Workmen were idled in the meantime. Peale and his daughter killed time by visiting regional mining sites and watching the operations. The locally made bricks were below the quality of those available in Philadelphia. Some of the windows were poorly installed, and it took two men to tug them to open or close. Finally, the strayed shipment from Charleston was tracked down in Columbia, South Carolina. Two wagons were sent to collect them, returning on October 27.

He reported that except for the preceding troubles all was well during his first stop:

> The buildings in Charlotte have been faithfully executed according to the designs and contracts made and furnished for that purpose. The commissioner Maj. McComb has been and still continues in assiduous attendance and has carefully watched over the interests of the government. The materials employed have been the best that the country supplies, and the execution all that could be expected, with the necessary allowances for the resources and habits of the neighborhood.

The problems at Dahlonega were severe, and Peale left for there on November 10.

EARLY OPERATIONS

From *The Annual Report of the Director of the Mint* for 1838:

> The machinery for the branch mint at Charlotte, North Carolina, was shipped to Charleston in April; and, in the following month, competent workmen were sent forward to put it in operation. Great delay and difficulty occurred, however, in transporting the heavy machinery to Charlotte, so that the steam engine was not set in action until the middle of August.
>
> To put up the furnaces and refineries of these mints, masons were sent from Philadelphia; and, in order that this important part of the work might be properly executed, as well as for the purpose of giving advice and aid in other essential points, one of the officers of this Mint was requested to go to Charlotte and Dahlonega; and I have reason to believe that his visit was of great importance.
>
> The mint at Charlotte is entirely finished, and has commenced operations, so far as to be receiving deposits and assaying. The certificates of this branch are, for the present, paid here; so that it may, in this way, retain gold to furnish a fund, for the purpose of enabling it hereafter to make prompt payments on the spot, for the bullion that may be furnished for coinage.

THE FIRST COINAGE

In November 1837 dies for quarter eagles and half eagles were shipped from Philadelphia. Each had a C mintmark above the date 1838. On December 4, the first day that Superintendent John H. Wheeler said he could receive bullion, a deposit of about 111 ounces of native gold was made. Coinage might have progressed, but acids essential for the parting and assaying process had not yet arrived. In addition, Superintendent Wheeler desired to adhere to the letter of the law and not strike coins in advance of the year on the dies. On the same day Wheeler published this:

Regulations

With regard to visitors to the Mint

1. The Mint will only be opened to visitors from 9 o'clock A.M. to 12 M, during which time they will be conducted by an attendant who will exhibit to them the different operations of the Mint and the various machinery.

2. They cannot be admitted in wet weather.

3. Visitors are requested not to handle any of the coin, precious metals, apparatus, or the machinery, unless in the presence and with the consent of an officer.

4. Each visitor will write his name and residence in the register kept for the purpose.

5. Visitors are requested not to offer any fee to the attendants or workmen, as no person employed in the Mint is permitted to receive such fee, under pain of dismissal from his situation.

By March all was in order, the necessary chemicals were on hand, and on the 28th the first coins were struck—the initial coinage at any of the three branch mints.

The *Financial Register*, January 17, 1838, included this misinformation:

> The Branch Mint in North Carolina has commenced coining in gold and silver, and it is expected that the branches at New Orleans and Georgia will commence operations soon. The principal coinage is of pieces most proper for common circulation—such as quarter eagles in gold, and twenty-five, ten cent and five-cent pieces in silver.

Although the possibility of striking silver coins was discussed over the years, no such coinage ever materialized.

The *Annual Report of the Director of the Mint* for 1838 told this:

> The branch mint at Charlotte commenced its operations in December 1837, and has received deposits of gold to the value of $130,600. The amount of coinage has been $84,165, composed of 12,886 half eagles, and 7,894 quarter eagles.

In the meantime Christopher Bechtler and his family were refining and coining gold at their private mint in Rutherfordton, in the same state (see chapter 12). This unwanted competition annoyed Superintendent Wheeler, and he investigated the possibility of legal actions. However, the Treasury Department, although it was aware of the Bechtler coinage, assayed pieces now and again, and mentioned the operation in some reports, never decided to take action to close down the operation. The Bechtler mint expired on its own account in 1852, by which time Augustus Bechtler was in charge.

INFIGHTING AT THE MINT

In 1839 the deposits of gold were about double that of the preceding year. All seemed to be going well. By 1839 the Charlotte Mint had coined quarter eagles and half eagles. By resolution of the General Assembly of North Carolina this was issued on January 4:

> Whereas, it is believed that a great deficiency of specie change now exists in the United States, and that at points remote from the mints authorized to coin this kind of currency, there exists a scarcity perplexing and obstructive to the interests not only of the trading and manufacturing classes, but to the great injury of all the productive branches of trade and industry, so much so as to induce the legislatures of many of the states to seek relief in the issue of treasury notes: And whereas, it is known that the coinage of the branch mint in this state is confined to gold bullion alone, and that said mint is located in a region suffering peculiarly from the want of small coin as change:

Be it, therefore, resolved, That our Senators and Representatives in Congress be requested to use their best endeavors to have a law passed, directing the coinage of small change at the Branch Mint in Charlotte, in this State, and that the governor forward to our senators and representatives a copy of this resolution.

Read three times, and ratified, in General Assembly, this the 4th day of January, 1839.

"Small change" probably referred to silver coins. Silver, while not important on its own in regional mining, was a byproduct of gold refining.

In 1840 rumors began to circulate that newly minted coins were short of the proper gold content. Whether these unfounded comments were started or furthered by those seeking positions at the Mint or by those associated with the Bechtlers is not known. By late spring 1841 the situation caused a sharp drop in deposits. This precipitated enmity between Superintendent Wheeler and assayer Gibbon. Eventually, the director accused Gibbon of poor assays and also remonstrated that he bothered to show up at the Mint only one or two days a week, instead of adhering to the proper six-day schedule. In the assayer's defense it could be said that there were now slack times with reduced deposits on hand.

The brouhaha was outlined to Director Robert Maskell Patterson in Philadelphia, who responded that as assayer Gibbon had routinely sent slivers of gold from his various ingots, and because these as well as finished coins were found to be of proper content, there was no problem here. However, Patterson reported to his superiors in the Treasury Department that Gibbon was insubordinate to the superintendent and was often truant. The situation was that for any given situation at any mint at the time, the superintendent's political job was very secure, but assayers, coiners, and other personnel could be replaced without difficulty.

BURGESS GAITHER AS SUPERINTENDENT

On March 4, 1841, William Henry Harrison of the Whig Party was inaugurated president, ending the Democratic rule of the Martin Van Buren administration. Harrison caught cold on inauguration day, died a few weeks later, and was succeeded as chief executive by Vice President John Tyler. The controversy between Wheeler and Gibbon at the Mint was solved when on July 3, Tyler dismissed Wheeler and named Burgess Gaither, a loyal Whig Party member, superintendent. Harmony did not last long, as Gaither demanded that Gibbon fire a blacksmith, Robertson, whom he had hired or else pay his salary personally. Robertson was terminated. Gibbon changed from working only one or two days a week to not showing up at all.

This amounted to a reprise of the earlier contretemps. Director Patterson became involved. He viewed Gibbon as the main source of the problem. The assayer seemed to have friends in high places. The result was that in the spring of 1843 charges against Superintendent Gaither were fabricated by a friend of Gibbon, and in May, Gaither was shown the door.

FIRE!

Green (sometimes in print as Greene) Washington Caldwell, generally referred to as Green W. Caldwell, was appointed in June as Gaither's successor. As a representative in Congress from 1841 to 1843, Caldwell had supported the maintenance of the branch mints in the gold district when yet another attempt was made to close them down.[5]

On July 26, 1844, Superintendent Caldwell was away from the Mint, taking some time in the fresh air of the nearby mountains. Between three and four o'clock in the early morning of July 27, 1844, fire broke out in the Coining Room, spread, and left the building in ruins. Fortunately, the records and other valuable items were stored in a vault and remained intact. Caldwell reported it was the work of a thief and arsonist, as his apartment at the Mint had been entered and items stolen.

Public speculations included spontaneous combustion in the assaying room and that a tinner was repairing the roof and started the fire by accident. College students smoking cigars on the roof of the Mint was suggested, but ruled out when no traces of ashes were found on the remaining roof. Calvin, a slave owned by former Superintendent Burgess S. Gaither, had been discharged, after which he was heard to mutter that he would burn the building down to ashes. He went to trial, but was absolved when it was proved that he spent that night not with his wife, but with a girlfriend named Rose. Such are the footnotes in Charlotte Mint history!

Only the steam engine and a planchet-cutting press were salvageable. The disaster had a tremendous impact. On December 14, Director Patterson wrote to the Secretary of the Treasury, describing the loss:

> Of the main building it may be assumed that there is nothing left which can be made available, except a portion of the material, and perhaps of the old foundation. The outbuildings are all saved. In the department of the superintendent and treasurer, the coin, bullion, scale beams, furniture, books, and papers were saved. In the assay room and in the melting room, but little damage was done. In the separating room the destruction was more considerable; but all the losses of the apparatus and material can be replaced without resort to any new appropriation. In the coiner's department the steam engine was slightly injured.
>
> The draw-bench is so much injured that it will be expedient to replace it. Of the cutting presses, one can be repaired, but the other must be replaced. The coining presses are past repair. The milling machine and the rolls are destroyed.

James Knox Polk, victor in the presidential election of 1844, took office on March 4, 1845. With ties to the state he was a strong advocate for rebuilding the Mint, although by that time there had been many calls in Congress to abolish both the Charlotte and Dahlonega mints because the cost per coin of production was far higher than in Philadelphia. D.M. Barringer, who represented the district in Congress, was instrumental in having an appropriation passed. The result was the Act of March 3, 1845, which provided for the building to be replaced using an adaptation of the earlier plans, but now only one story in height above the foundation, rather than two as before, the new plans being drawn under the supervision of Franklin Peale.

On April 1 the local *Jeffersonian* paper commented:

> The Superintendent of our mint, Green Caldwell, is a great fellow—a real business man. He received on this day [a week ago] from the Director of the Mint his instructions for putting up a new building, and [the] on Monday after he made a contract for the whole job at less cost than the Government appropriated. Our enterprising fellow townsman, H.C. Owens, Esq., took the contract for $20,000, the building to be completed by the 1st of January, next.

Caldwell stated that just one coining press was needed. Construction was completed at a cost of $31,572.97 and the new building was occupied the following year. Coinage resumed in October 1846. Accordingly, there was no coinage in 1845. To prevent a recurrence of theft Matthew "Devil" Wallace, a large and strong man, was hired to guard the Mint in off hours.

In the spring of 1847 Caldwell resigned as superintendent to join the army as a captain of a company of dragoons in the War with Mexico, with E.C. Davidson, J.K. Harrison, and Alfred A. Norman as his lieutenants. The unit did not see much action, and most of them returned home.

LATER YEARS

William Alexander was appointed superintendent. Not long afterward the first coins of a new denomination, the gold dollar, were struck at Charlotte. In June 1849 President Zachary Taylor named James Osborne as superintendent. Alexander stayed on until August.

On March 4, 1853, the month after Franklin Pierce was inaugurated president, he appointed Green W. Caldwell to a second term as superintendent.

In 1859 many regional miners joined the Gold Rush and headed west. Many if not most returned later, often bringing gold dust and nuggets with them. In his reminiscences, Ebenezer Locke Mason Jr., Philadelphia coin dealer, told this:

> While making a coin speculating tour of the chief cities and towns of North Carolina, during the year 1859, we visited the promising little village of Charlotte, and its neighboring gold mines. After satisfying our curiosity by a personal examination of the different methods of extracting the precious metal, and witnessing the slow and laborious process of mining, as conducted by Negro slaves, with the aid of mule power, we accepted an invitation from the landlord of the prominent hotel in Charlotte, to visit the U.S. Branch Mint, then in operation at this town.
>
> During our peregrinations through the different departments of the Mint, we accidentally met Dr. Andrews, a very prominent citizen of the place, and one who had devoted many years to scientific subjects, including geology and numismatology. The Dr. very kindly invited us to his residence, and while there we had an opportunity to inspect a fine cabinet of coins and minerals. . . .
>
> Dr. Edwards, the former superintendent [*sic*] of the Charlotte Mint, exhibited to us in 1859 a handsome numismatic cabinet, arranged around the reception room of the mint; and this collection attracted many visitors. We do not know what became of this cabinet of coins after the government changed it from a mint to an assay office.[6]

THE CIVIL WAR

On April 21, 1861, with the Civil War underway for about a week, the Charlotte Mint was taken over by Colonel Bryce and his troops. "No resistance was made, or the slightest disturbance occurred. Several military companies are awaiting searching orders."[7] Operations continued and gold on hand was used to strike 2,044 half eagles in April and 887 in May and pay them to depositors. By the end of May coinage had stopped. Superintendent Caldwell and assayer Dr. John H. Gibbon were still at their posts.

The *Report of the Director of the Mint* for the fiscal year ending June 30, 1861, included this:

> The deposits at the branch mint at Charlotte, up to the 31st day of March, 1861, were $65,558 30; coinage, $70,580; and number of pieces 14,116. The deposits at this branch and Dahlonega are exclusively of gold. No report has been received from this institution since the day last named.

The Confederate Assay Office was opened in the Mint under an act of the Confederate Congress passed on August 24, 1861, and Gibbon was appointed assayer. In May 1862 the building was turned over to the Navy Department.

The *Report of the Director of the Mint* for the fiscal year ended June 30, 1865, included this:

> In my annual report of 1862 it was suggested that the branch mint at New Orleans, after the re-establishment of law and order in Louisiana, might be successfully operated, and that the branch mints at Charlotte and Dahlonega ought not to be employed again for minting purposes. My opinions on this subject are unchanged. . . .
>
> Reasons for re-opening the branches at Charlotte and Dahlonega do not exist. They are away from the commercial centres, inland, and of little commercial importance in themselves. The existence of gold mines in their respective localities may be a reason for re-opening them as assay offices, but not for minting purposes. The results of their operations from their commencement, in 1838, to February, 1861, do not sustain the policy of their original establishment. The coinage of both these branches is limited, by act of Congress, to gold. At Charlotte the total coinage during the twenty-three years of the existence of this branch was only $5,048,641.50; and at Dahlonega for the same period, $86,121,919; an average annual coinage of about $250,000, declining at Dahlonega, from 1857 to 1861, to an annual coinage of about $70,000; and at Charlotte, for the same period, of less than $150,000. These facts seem to be conclusive on the question of re-opening these branches for minting purposes, and particularly when there is no great probability of a large increase in the gold production of those localities.
>
> To meet every commercial want of those places, and also the interests of the miners of gold, the re-opening of these branches for melting, refining, assaying, and stamping gold bullion would be amply sufficient; giving to the superintendent or treasurer of each branch authority to issue, in payment for gold-dust, bullion, or bars deposited for assay, drafts or certificates of deposit, payable in specie at the treasury, or any sub-treasury of the United States, to any depositor electing to receive payment in that form. This provision would wholly supersede the necessity of coining at these branches, or any imaginary benefits resulting therefrom.
>
> On the subject of assay offices for our gold-mining regions, and the impolicy of multiplying branch mints, my sentiments were fully expressed in my last annual report, to which you are respectfully referred.

The Charlotte Mint in 1931.

After the war John H. Gibbon set up his own assaying business in the Mint building. On March 19, 1867, the Treasury Department authorized its Assay Office there and commanded that Gibbon leave the premises. The facility opened early the next year. Isaac W. Jones was appointed as assayer. This business continued until July 1, 1875, when it closed for a time. It reopened on October 16, 1876, and continued until June 30, 1913, after which the Treasury Department closed it forever.

In later years it was used for a federal courtroom and Post Office, American Red Cross (during World War I), and for meetings, including those of the Charlotte Woman's Club.

In 1932 the building was slated for demolishing to make way for an expansion of the Post Office. Civil pride came to the fore, the Mint was dismantled, moved to a park, and on October 22, 1936, opened as the Mint Museum of Art, with a gilded federal eagle over the doorway. Today the museum is a rich cultural center with many activities that are detailed on its Web site at www.mintmuseum.org.

COINAGE OF THE CHARLOTTE MINT, 1838–1861

During its operation from 1838 to 1861 the Charlotte Mint struck slightly more than 1.2 million gold coins with a face value of $5,059,188. Gold sources included North Carolina $4,520,730.79, South Carolina—$460,523.34, and from miners returning from California $87,321.01.

Gold dollars were struck from 1849 to 1859 except for 1854, 1856, and 1858. The 1849 was made in two reverse varieties, Open Wreath (a rarity) and Close Wreath. Coins of 1855 onward are usually miserably struck. Quarter eagles were made intermittently from 1838 to 1860, except for 1845 (year of rebuilding the Mint), 1853, 1859, and 1859. On June 1, 1854, a pair of dies for an 1854-C $3 gold piece coinage was sent to Charlotte, but they were not used. Half eagles, the most important denomination, were made continuously from 1838 to 1861 except 1845.

The Charlotte Mint Museum of Art.

Gold Dollars

Gold dollars were struck at the Charlotte Mint from 1849 to 1853 and in 1855, 1857, and 1859. The surface quality of the gold dollars 1855-C and later is usually rough, probably caused by poor planchet preparation. Their appearance is quite rustic. For many years a mintage for the 1854-C gold dollar was listed, but in modern times it was discovered that this was a bookkeeping error.

DESIGN TYPES AND DATES MINTED

Liberty Head (Type I): 1849-C, 1850-C, 1851-C, 1852-C, 1853-C.

Indian Princess Head, Small Head (Type II): 1855-C.

Indian Princess Head, Large Head (Type III): 1857-C, 1859-C.

$2.50 Quarter Eagles

Charlotte quarter eagles are easily enough found in grades such as Very Fine and Extremely Fine, but their popularity has made them expensive. In conservatively graded Mint State all range from scarce to rare.

DESIGN TYPES AND DATES MINTED

Classic Head: 1838-C, 1839-C.

Liberty Head: 1840-C, 1841-C, 1842-C, 1843-C, 1844-C, 1846-C, 1847-C, 1848-C, 1849-C, 1850-C, 1851-C, 1852-C, 1854-C, 1855-C, 1856-C, 1858-C, 1860-C.

$5 Half Eagles

Similar to Charlotte quarter eagles, half eagle from this mint are easily enough found in grades such as Very Fine and Extremely Fine, but their popularity has made them expensive. In conservatively graded Mint State all range from scarce to rare.

DESIGN TYPES AND DATES MINTED

Classic Head: 1838-C.

Liberty Head, Variety 1, No Motto Above Eagle:
1839-C, 1840-C, 1841-C, 1842-C, 1843-C, 1844-C, 1846-C, 1847-C, 1848-C, 1849-C, 1850-C, 1851-C, 1852-C, 1853-C, 1854-C, 1855-C, 1856-C, 1857-C, 1858-C, 1859-C, 1860-C, 1861-C.

DAHLONEGA MINT, 1838-1861

GOLD DISCOVERED IN GEORGIA

In 1804 the first deposit of native gold from North Carolina was made at the Philadelphia Mint.[1] Into the 1820s the deposits from that region increased and became important. In 1828 an important discovery of gold was made in adjacent Georgia in the area known, as of December 3, 1832, as Lumpkin County, at a spot about two miles from the Chestatee River. The finder, 26-year-old Benjamin Parks, who came from North Carolina, was hunting deer when he glimpsed a flash of gold in a lump of quartz. The land was owned by Reverend Robert Obarr (also spelled Obar in some accounts), a Baptist minister. Parks told of his find to Obarr, who was skeptical, but agreed to give Parks a 40-year lease in exchange for a quarter interest in any gold that might be found.

Parks signed a partner, and both of them found gold in quantity panning in streams. Obarr tried to rescind the lease, but was not successful. He then sold his land to Judge Underwood, who then sold it to Senator John C. Calhoun of South Carolina, one of the most powerful political figures of the era (who at one time threatened that his state would withdraw from the Union if the Import Tariff of 1828 took effect). Imbued with the prospect of riches, Calhoun then negotiated with Parks and bought his lease. He subsequently made some of his slaves work the mining operation.

Sketch of the Dahlonega Mint in the 1860s.

The news of gold could not be contained, and within a year or two an estimated 10 to 15 thousand fortune seekers poured into the district. Governor George R. Gilmer of Georgia sent troops to halt the arrivals, but with little success. Some of his men decided to join the miners, many of whom took up residence in a ramshackle collection of buildings known as Nuckollsville, later Auraria, set up about six miles southwest of the later-established town of Dahlonega. By 1830 the district of northeast Georgia was swarming with prospectors. Templeton Reid, an assayer, set up a coining business in Milledgeville, the state capital at the time, then moved it to Gainesville, where mining activity was intense (see chapter 12). In 1830 the Philadelphia Mint received its first shipment of Georgia gold in the amount of $212,000.

THE DAHLONEGA DISTRICT

In 1830 the State of Georgia took over much of the land in the Auraria area and had a corps of surveyors divide it into segments, including 60-acre farms and 40-acre gold-bearing parcels. Lumpkin County was laid out, named after the current governor. Auraria was selected as the county seat, but there were difficulties with land titles.[2] An uninhabited area six miles distant, Lot 930, 12th District, 1st Section, was selected. Buildings were erected. The first session of the Superior Court was held there, and it was decided to name the place Talonega—a Cherokee word for "yellow," after the gold, Spellings varied, and Dahlonega became official in 1833.

In 1835 the Bank of Darien, headquartered in the Georgia seacoast town of the same name, opened a branch in Dahlonega. It also had branches in Augusta, Auraria, Macon, Milledgeville (the state capital at the time), and Savannah. The earliest known notes of the Dahlonega branch are dated November 1, 1835, and are signed by Ebenezer S. Rees as cashier and Anson Kimberley as president. By May 20, 1837, the signatories were Rus and Jacob Wood, the latter being a former president of the state senate. By 1838 A. Mitchel replaced Wood. The latest known note of the branch is dated December 12, 1838. The branch closed by the summer of 1841.[3]

In the meantime, federal coins were common as money, as were gold coins privately made by the Bechtler family (see chapter 12) in Rutherfordton, North Carolina. Other operations in the area included the Cain Creek Placer Mining Company, the Finley Ridge Mine, the Barlow Mine, and a host of others, mostly small enterprises. Much

$20 note of the Dahlonega branch of the Bank of Darien signed by E.S. Rus and Anson Kimberly, November 1, 1835. "Dahlonega" is inked in at the lower left.

gold was sent to the Philadelphia Mint, which involved a 5% fee for insurance and transit and a delay of over two months in receiving compensation. As an example, John N. Rose sent a shipment of gold to Philadelphia on September 3, 1835, via Augusta, then downriver to Savannah, then north by sea, reaching the Mint on October 7.[4]

PLANNING THE DAHLONEGA MINT

The Act of March 3, 1835, provided for three branch mints, including one "at or near Dahlonega in the State of Georgia," with coinage to be limited to gold. As might be expected, Senator John C. Calhoun was among the 24 yea votes, as was Senator Thomas Hart Benton of Missouri, nicknamed "Old Bullion" for his strong stance on gold coins as opposed to paper money. Nays in the Senate numbered 19. The bill authorized $50,000 for construction. A further $15,000 was earmarked for salaries of the superintendent, chief coiner, melter, assayer, refiner, and up to five employees.

Colonel Ignatius Alphonso Few, a Methodist preacher with an L.L.D. from Princeton as well as a lawyer of Columbus, Georgia, who also received mail in Athens and Covington, was appointed as commissioner to supervise construction. He was expected to go to Philadelphia to discuss the program with Mint Director Samuel Moore. It seems that Few, who had been a colonel in the War of 1812, was well connected politically. On April 16 the director sent him questions to research before coming. These included the cost of land for the Mint, the possibility of having water on the site to drive an engine of 10 to 15 horsepower, the cost of transporting a 10 horsepower engine from the nearest port (Savannah) to Dahlonega, the availability of building materials, and costs for labor.

Apparently Colonel Few was in marginal health and also unmotivated. He did not go to Philadelphia, and, as it developed, did little actual oversight and visited Dahlonega only occasionally. Complicating matters, Director Moore delayed in sending plans, and it seems that communications with Few were incomplete.

Meanwhile, Philadelphia architect William Strickland was selected to draw up plans. These were similar to the mint intended in Charlotte and of the same general style as the much larger building for New Orleans. The plans called for the building to be 125 feet wide and 33-1/2 feet deep, but with the center section having an extension of 53 by 35 feet to the rear. The building was two stories high, built of brick with a white stucco facing, and set on a granite foundation of five feet high and more than three feet thick. The basement was 10 feet from floor to ceiling, and the two floors above each had 12-foot ceilings. Across the front were 17 windows. Two round stone pillars framed the front entrance.

Colonel Few visited Dahlonega on May 11, investigated possible building sites, and identified six that offered possibilities.

CONSTRUCTION DELAYED

Delays occurred, and by the summer little had happened. On July 30, 1835, regional citizens held a meeting to raise their concerns. A committee appointed at the gathering sent a petition to President Andrew Jackson and a strongly worded complaint to Secretary of the Treasury Levi Woodbury to begin construction, stating that Few's health was so poor that he could not possibly fill his duties, and some business-minded person must be appointed to replace him.

After delays and correspondence a 10-acre parcel owned by William Worley was selected, the Treasury Department agreed, and in August the land was purchased for $1,050. Strickland's plans finally arrived in Dahlonega the same month, and the project was put out for bids to be received by September 22, 1835.

On September 18, Few wrote to Mint Director Robert Maskell Patterson, who had entered that post earlier in the summer. He stated that Director Samuel Moore had delayed in sending plans and had not responded to certain of Few's questions and concerns—this being the reason work had not started. Throughout this period there were complaints and accusations to and from Few and the director.

Eight proposals were received, six from within the state and one each from North Carolina and Tennessee. Benjamin Towns of nearby Athens had the lowest, $33,450, and was chosen on September 22. The agreement with him specified that construction was to be completed by March 22, 1837.

Towns was an acquaintance of Dr. James Tinsley, a jack of all trades whose credits include publishing a journal, running a tavern, practicing surgery without surgical instruments, and assisting in the construction of buildings in Athens. Mint Director Patterson asked Towns to come to Philadelphia to review plans for the Mint and to learn more about operations. In January 1836 Dr. Tinsley went in his stead. It seems that he absorbed little. Later, when additional plans for the Mint arrived in Dahlonega, neither he nor Few could understand them. Tinsley wrote to the director asking what several mysterious characters were, to which Patterson replied, "The square black spots, as in all architectural drawings, represent flues."

CONSTRUCTION BEGINS

Workers were hired and construction began, but it began very slowly. Lumber was cut locally, granite was quarried, and other materials were used by the crew of workers. Some workers were inexperienced, and several walls collapsed and had to be rebuilt. Little was available locally in the way of hardware, tools, and other supplies. The necessity of digging and connecting a well to supply water had been forgotten. By the due date, work was still unfinished. Congress appropriated additional funds to correct past problems and to outfit the facility. In January 1836 a contract for the machinery was awarded to Merrick & Agnew, Philadelphia machinists who were in the process of setting up the Philadelphia Mint's first steam press at (which was inaugurated in a ceremony on March 23). For Dahlonega two small-size steam presses were ordered, the idea of using water power having been dismissed. Work continued, and by April most of the massive foundation was in place, and construction of the upper part began.

The Cherokees of the district had recently fallen victim to the terms of the Treaty of New Echota, a Trojan-horse offering from the United States government that initially had provided for the fulfillment of the Indian Removal Act of 1830 with the key distinction that individual Cherokees could decide to remain in the district as long as they agreed to live under federal and state sovereignty as U.S. citizens rather than under the sovereignty of the Cherokee Nation. However, President Jackson had removed this key provision from the treaty during negotiations. Moreover, these negotiations were not made with the governing body of the Cherokees, but by a self-appointed council of 20 people without legislative authority who believed that removal was unavoidable. Chief John Ross refused to sign the treaty, and most Cherokees supported John Ross in his efforts to lobby Congress not to ratify the treaty. Despite the fact that Chief John Ross gathered signatures from nearly every person living within the Cherokee Nation, and despite the fact that the treaty was not made with the Cherokee government, Congress ratified the Treaty by a single vote on March 1, 1836, meaning that all Cherokees would be forced to move West to what was known as the Indian Territories. Tensions ran high in Dahlonega.

There were torrential rains in late summer, flooding the site. By September the upper walls had been raised only to the bottom of the windows on the first floor. Letters flew back and forth, Colonel Few was accused of dereliction of his duties, and he in turn blamed the Treasury Department for the delays. It later developed that Few was busy with other matters that he felt were more important.

SUPERINTENDENT SINGLETON

In autumn 1836 it was announced that President Andrew Jackson had appointed Dr. Joseph James Singleton, a plantation owner from Athens, Georgia, to be superintendent of the Dahlonega Mint. Singleton had served in various offices, including as judge and as a representative and senator in the Georgia legislature. In 1834 he had lost in the election for a representative at large to the United States Congress. The loss was due at least in part to a caper pulled by William E. Jones, editor of the *Athens Whig*. Before the election he created and published Singleton's obituary! This deception caused a furor, and Jones resigned and scuttled off to Texas. Singleton entered service in or about January 1837.[5]

David H. Mason, an engraver in the private sector and a maker of dies at the Philadelphia Mint, was named coiner, and Dr. Joseph W. Farnum, a professor of chemistry and natural science at Washington College in Lexington, Virginia, was named assayer and melter. The three officers went to Philadelphia to study Mint procedures and returned to Georgia in March. Colonel Few wrote that the Mint would probably be finished my May 15, to the extent that machinery could be installed.

Singleton learned that nowhere in the construction had provision been made for accommodating the steam-powered equipment. Moreover, the arches supporting the floor inside the building created lateral thrust that the exterior walls had not been built to withstand, causing some arches to fail.

Two medium-size, steam-powered presses made in Philadelphia by Merrick & Agnew and other machinery and equipment were packed into 15 crates and shipped by sea aboard the brig *New Hanover* to Savannah, then hauled upriver to Augusta, and then taken on May 11 by ox-hauled wagons over a rough road to Dahlonega, arriving on May 29.

Niles' Weekly Register, June 10, 1837, had an update:

> *Branch mint at Dahlonega, Ga.* The machinery for the branch mint of the U. States has arrived at Dahlonega. It weighs upwards of fifty thousand pounds, and cost the government upwards of one thousand dollars for its transportation. The building intended for its reception will not be ready before the middle of the summer. The whole amount of gold which is extracted from the gold mines in the gold region of Georgia, during the last year, is estimated at upwards of two hundred thousand penny-weights, and it is believed that the labor of the present year will yield as much as the last.

William Baker and his men from Philadelphia, having completed their work installing equipment in Charlotte, arrived in Dahlonega on August 30. David H. Mason traveled from Baltimore and arrived in town 13 days later. Three days later he sent a letter to Mint Director Robert Maskell Patterson detailing some ideas about improving the structure. Mason, Baker, and their men left the town on November 10 and headed back to Philadelphia.

Patterson later gave this overview:

> The difficulties naturally incident to any new undertaking have been fully presented at the branch mints, and it would, therefore, not be just to form conclusions as to their importance and efficiency from the operations of the past year. They are now, however, in good condition, and the officers and men have acquired the necessary experience. It is to be hoped, therefore, that the labors at the branch mints during the present year, will be such as to satisfy the expectations which led to their establishment.
>
> The machinery for the branch mint at Dahlonega, in Georgia, was sent to Savannah in May, and difficulty and delay also occurred in its transportation thence by land. After the workmen employed at Charlotte had finished their task, they proceeded to Dahlonega, to erect the machinery there; and they completed this work early in November. The mint at Dahlonega is nearly completed, except as to the enclosure and out-buildings.[6]

PEALE VISITS DAHLONEGA

After inspecting and reporting on the Charlotte Mint (see preceding chapter), chief coiner Franklin Peale and his daughter Anna departed on November 10, 1837, and arrived in Dahlonega on the afternoon of Wednesday, November 15.[7] By that time Baker, who had helped set up the equipment, had left to return to Philadelphia. David H. Mason stayed on as coiner for the new Mint.

Franklin Peale.

Upon arriving in Dahlonega, Peale learned that Superintendent Joseph Singleton was not to be found, nor were any other Mint officials. At first glance the construction problems of the building seemed to have been solved in a satisfactory manner, but after a couple of days Peale concluded that the construction was crude, the bricks were poorly laid with insufficient mortar to bond them together, and some of the workmen were drunk and displayed other bad habits. Peale wrote to Patterson that he hoped to return home soon. However, the more that Peale examined the building, the worse the conditions appeared, per this in a letter to Patterson:

> The workmanship of the Mint edifice is abominable, a letter might be three times filled with the details or errors and intentional mal- constructions, the first and greatest of which may fairly be traced to Philadelphia, in *ordering* a brick building in a country where there is no *clay*, the material employed for the brick making being the *red soil* of the Gold region, *a decomposed granite* . . . put into brick by men who certainly deserve diplomas for *botching*. . . . All this in an area full of granite, which would have been an ideal building material for the Mint.

Peale placed the responsibility on Colonel Few, who was supposed to oversee the construction. He concluded that if the Mint were to be put in proper shape, an additional congressional appropriation would be needed, an unlikely possibility. The only realistic answer was to accept the structure in its present condition and work with it, otherwise there would be no Dahlonega Mint.

Peale and his daughter Anna, who had accompanied him, left Dahlonega on November 27 and traveled north by stagecoach, then by train. In Virginia the locomotive was derailed and crashed, injuring 20 people. Franklin Peale was unharmed, but his daughter had a slight back injury when their car left the tracks and was thrown to the side. Peale stopped in Washington in mid-December and was back at the Philadelphia Mint by December 23. In the meantime, Few had not been in Dahlonega since the summer.

FEATHERSTONEHAUGH'S VISIT

Pursuing his own interests and not related to the Mint, geologist and adventurer George W. Featherstonehaugh visited Dahlonega on September 2, 1837, and wrote this:

> The situation of Dahlonega is a very good one, being built on a fine knoll sloping to the east, with the ground rising agreeably to the west. It possessed a rather imposing looking granitic building, called the Mint, with a brick Court-house in the centre of the square, and a few tolerable looking private buildings, all [of] which gave to the town an air of pretension that was indifferently borne out by the inhabitants. There were two excellent springs of water at the village. The Court-house was built on a broad vein of hornblende slate, and the soil of the public square was impregnated with small specks of gold.[8]

THE MINT IN OPERATION

David H. Mason and his family moved into rooms on the second floor of the Mint, as did bachelor Joseph W. Farnum—this arrangement was convenient for them and also an addition to security. Director Singleton lived locally.

On April 8, 1838, President Martin Van Buren implemented the Indian Removal Act of 1830 with brutal force. He ordered Major General Winfield Scott to go to the land of the Cherokees in North Carolina and Georgia and to remove them by whatever means were necessary. This took place in the "Trail of Tears" displacement in 1838 and 1839 when they were forced to settle in Indian Territory west of the Mississippi River. Close to 4,000 Cherokees, almost a quarter of the total population, died on the forced march. They were killed by disease, exacerbated by the fact that they were given blankets and supplies from medical waste at plague hospitals; starvation; and exposure, as they were forced by their drovers to avoid white cities and settlements—many of the 1,000 miles marched in the dead of an unusually harsh winter were unnecessary, added only to circumvent white dwelling places. The white locals they did interact with often charged exploitative prices for the barest necessities; many froze to death by the Ohio River while waiting for a ferry that charged each individual Cherokee octuple the fare normally charged to transport an entire Conestoga wagon. After crossing the river and arriving in Golconda, Illinois, many Cherokees were also murdered by locals before they could resume their journey West. In Georgia and North Carolina the land formerly belonging to the Cherokee was distributed to farmers and gold-seekers via lottery.

The Dahlonega Mint was just starting to strike coins. In his 1838 *Report* Director Patterson related:

> The branch mint at Dahlonega commenced its operations in February, and has received deposits of gold to the value of $141,800. The amount of its coinage has been $102,915, composed of 20,583 half-eagles.

The first coinage, in the form of half eagles, took place on April 21, 1838, when 80 pieces were struck. Most gold deposited was from local mining, but foreign gold coins were also included. In 1838 bullion received included 7,619 ounces of native gold and 327 ounces in the form of British sovereigns. In the meantime, Superintendent Singleton was also working a mining property by means of cuts and tunnels, processing the rock in a wooden 12-stamp mill powered by an undershot wheel driven by water in Yahoola Creek.[9]

Colonel Few was named president of Emory College in Covington, Georgia, in 1838. His absence from Dahlonega was explained by his activity in forming this college, which would later move to Atlanta, becoming Emory University, but still maintaining a campus known as Oxford College of Emory University in the same location.

There was concern in Dahlonega that regional gold output might not be sufficient enough to sustain minting operations. In the meantime, continuing for years afterward, there were constant complaints in Congress and elsewhere that the Southern mints were a waste of time and money, and that minting could be done more efficiently in Philadelphia. This was true, but the reason for establishing the branch mints was to enable holders of gold to receive newly minted coins without long delays. David H. Mason acquitted his duties as coiner by turning out very attractive pieces, despite some complaints about the quality of the dies (in 1838) and problems with the strip-rolling machinery. In the meantime, the roof of the building leaked, and there were difficulties with the structure.

Within the Mint there were many quarrels among the officers, many complaints about Singleton's abilities or lack thereof, problems with lazy as well as intoxicated workers, difficulties with operations, excessive expenses, missing mail, adversarial communications to and from Philadelphia, reports that regional mines were becoming depleted, and more. Affairs did not run smoothly. All the while there were many comments in Congress and elsewhere that the Dahlonega Mint was costly and unnecessary. By one estimate it cost 12.5% of face value to coin gold there.

ROSSIGNOL APPOINTED AS SUPERINTENDENT

Many believed Superintendent Joseph J. Singleton had performed admirably in the difficult first several years of the Mint's operation. But did he? Opinions differed widely. The Whig victory in November 1840 ended 12 years of the combined Jackson and Van Buren administrations. In March 1841 newly inaugurated President William Henry Harrison named Paul Rossignol to replace Singleton as Mint superintendent, a traditional political "plum," and widely considered to be the best political patronage job in the state. Rossignol had served as a teller at the Bank of Augusta for 15 years prior to this appointment.

When the new superintendent took office he soon learned that there were multiple problems with die breakage, with dies failing after 500 coins instead of lasting until about 12,000 were struck—a more typical duration. He dealt with frequent complaints from depositors about the time it took to process gold. In Congress many continued to call for the Dahlonega Mint to be abolished. In the autumn of 1841 he filled out a long questionnaire sent by Congress, demanding information on all aspects of the business.

Ex-director Singleton remained nearby and continued his mining business in the region until his death on April 9, 1854. He was a frequent visitor to the Mint, not always to the satisfaction of the officers, who sometimes thought he interfered with operations.

Quarrels among Mint employees continued to be the rule, not the exception. It was anything but a happy family. There were squabbles about hiring relatives, the amenities or lack thereof in the residential apartments within the Mint, and the absence of some personnel who were attending to outside interests.

Matters looked up in May 1842 when an important new gold vein was found in the mine leased by Senator C. Calhoun. The senator was too busy in Washington to come to Georgia, so he sent his son-in-law Thomas Green Clemson, a highly qualified engineer. Clemson erected a four-stamp mill and ran it for two or three years. He found that many mounds of processed ore could yield even more gold—apparently, certain early extractions were done hastily. By that year it is said that the mine had yielded at least $100,000, a considerable sum at the time.[10] After Calhoun's death in 1850 the works were idle for a time, but by 1858 they were operated by a William G. Lawrence.

Complaints from depositors continued at the Mint. Coins were not available until their gold had been melted and assayed. Mint certificates were issued and could be traded regionally, but only at a discount in comparison to coins. Disappointed office seekers, disgruntled employees, and others accused Rossignol of various malefactions. The accusations were not much different from what Singleton had endured. Some of these complaints reached President John Tyler, who suggested an investigation be conducted. Mint Director Robert Maskell Patterson and others in Philadelphia assured him that operations were satisfactory.

COOPER REPLACES ROSSIGNOL

On May 1, 1843, Secretary of the Treasury John A. Spencer, new in the post, told Mint Director Patterson that Rossignol had been replaced. The new appointee was James Fairlie Cooper, a West Point graduate. The news caught Rossignol and everyone else in Dahlonega by surprise.

Cooper took up residence in an apartment in the Mint, the first superintendent to do so, after persuading David H. Mason and his family to move out. Employee dissatisfaction and infighting continued and resulted in some changes of personnel early in his administration, but conditions improved greatly by the end of his tenure. Former Superintendent Joseph J. Singleton, first in the position, continued to hang around the Mint, on occasion writing to state and government officials that he would be interested in being reappointed, and at times mounting an intense campaign to achieve this goal.

The fineness of native gold from the region was outstanding, often .950 to .980, in comparison to the .900 standard for federal coins. Accordingly, ingots could be made simply by adding alloy. Bullion deposits rose to new highs, allaying fears that the Mint would run out of metal. Coinage in 1843 reached an all-time high.

The time lag from deposit to coin delivery was shortened to an average of four business days. The Mint was in operation daily except for Sunday. Deposits slowed in 1845. There were continuing problems with maintaining the presses and other machinery. Valentine Hodgson, a skilled machinist, arrived from Philadelphia in October 1845 to join the staff.

Coiner David H. Mason, one of the finest coiners in any of the mints, became ill, and on August 28, 1848, he died. Dies for the 1849 coinage were sent from Philadelphia on September 20, 1848, a very early date for the succeeding year's coinage. They arrived on October 3, carried with other items from Philadelphia, after stopping in Washington with Hodgson, who was visiting the north to seek appointment as coiner in Dahlonega.[11] However, John D. Field Jr., son-in-law of the late David Mason, was chosen instead, probably due to his endorsement by Superintendent Cooper.

In 1849 the news attending the California Gold Rush reached Dahlonega, and over a thousand miners in Georgia decided to head west. Dr. Matthew F. Stephenson, assayer at the Mint, pleaded with some of them to remain, but to no avail. In the same year a new denomination, the gold dollar, was coined on a press that had previously been used to coin half dimes in Philadelphia. On July 12 Superintendent Cooper sent two gold dollars to Director Robert Maskell Patterson in Philadelphia.

REDDING AS SUPERINTENDENT

Zachary Taylor won the presidential election of 1848 and was inaugurated on March 4, 1849. In July Cooper learned through rumors that a new superintendent had been appointed, but facts were scarce. The appointee was Colonel Anderson W. Redding, keeper of the State Penitentiary. Redding arrived in Atlanta in August, before his appointment had been officially confirmed. Cooper, resisting all the while, continued with his duties until the official change on October 1. This created the unusual situation of two overlapping superintendencies.

Beginning in 1850, much gold from California was turned into coins in Dahlonega. In 1852 at the annual Assay Commission meeting in Philadelphia, samples of coins from the four mints in 1851 were analyzed. Gold from Dahlonega was found to be .8993 fine, minutely short of the .900 standard. This prompted the director to take more oversight with the Dahlonega processes. In that year the Mint director ordered the Dahlonega coiners to refine gold so that it contained no more than 8% silver.

In 1851 rumors were afloat that Redding would vacate the Mint and resume his position as head of the State Penitentiary, much to the pleasure of Singleton, who was, as ever, waiting in the wings. This did not happen, however. In 1853 Franklin Pierce was inaugurated as president. Howell Cobb, who in time would become federal secretary of the Treasury, had been governor for two years. He requested that Pierce oust Redding and name Jacob R. Davis as superintendent. This was done, and the nomination was confirmed by the Senate on April 7, 1853. This caused much consternation at the Mint, and various charges against Davis were sent to Washington—these including drunkenness (in the back of the tailor shop he owned), not properly observing the Sabbath, gambling (at cock-fighting, pitching coins, and a game called bluff), and deserting his wife (after exhausting the wealth she had brought to the marriage). Meanwhile Pearce had to deal with over 700 appointments around the nation and thousands of office-seekers. On May 11 Davis was removed—another curious footnote in Dahlonega Mint history: a confirmed superintendent who never served!

In 1853 gold from California comprised $269,617.78 as compared to only $18,293.94 made up of regional gold. Some of the California gold was brought back to Georgia by miners who had been part of the Gold Rush. Around 1,800 men fell into this category according to an estimate by Redding. The Californian deposits contained silver as an "impurity," which was not unusual as much Georgia native gold also contained silver.

Depositor's statement for gold bullion deposited on July 1, 1850, and subsequently assayed. (Birdsall)

PATTON REPLACES REDDING AS SUPERINTENDENT

In July 1853 in the new Democratic administration of President Franklin Pierce a new superintendent, Julius M. Patton, replaced Redding (or by some accounts, replaced Jacob R. Davis). In the meantime Davis was in Washington and in various places in Georgia petitioning and suing to be installed as superintendent, a *cause célèbre* that echoed widely. Patton, who securely remained in his new position, was an attorney from Cassville and had earlier served as the Georgia state treasurer. Robert H. Moore replaced John D. Field Jr. as coiner. Field went into the mining industry. He later returned to his former Mint position.

In early 1854 the assay of 1853 coins showed Dahlonega gold at an unacceptable .8986. In the spring of 1854 the San Francisco Branch Mint opened for business. The utility of sending gold to the branch mints in the East was diminished, and in 1855 only $47,429 was processed in Dahlonega.

On November 24, 1854, a shipment, comprising 3 pairs of each of the gold $1, $2.50, and $5 denominations, was sent in one box and forwarded by Adams Express Company from Philadelphia, but was lost for a time in Savannah, Georgia; after it was found it was forwarded to Dahlonega and arrived there on December 27, 1854. The Assay Commission of 1857 found gold of 1856 to be slightly light. It was concluded the person weighing gold in Dahlonega may have had defective balance scales. In 1858 the 1857 coinage was found to be acceptable.

Mint Director James Ross Snowden, in that post since 1853, used the Adams Express Company to send dies, which took from 10 days to a month. In contrast, by the late 1850s rail service had improved to the extent that sending dies through the Post Office took only four days. Snowden's preference for the slower method caused problems when dies were needed urgently.

The November 1856 election and March 4, 1857, inauguration of James Buchanan as president resulted in the Democrats keeping the White House. Accordingly, Buchanan had less than usual of the customary appointment-reshuffling to oversee, and Patton's position was not threatened. Deposits of regional gold had been declining, but there was a sharp uptick with new discoveries in 1857.

In 1858 deposits of gold touched a record low as little other than regional metal was brought to the Mint. However, deposits of Georgia gold on their own were the best since 1852.

Hydraulic mining, popular in California, came to the district when the Boston-owned Yahoola River and Cane Creek Hydraulic Hose Mining Company incorporated in Georgia on December 11, 1858, and began constructing aqueducts. The year 1859 began and ended without much taking place, at least as reflected in Mint deposits.[12] On December 20, 1860, the Nachoochee Hydraulic Mining Company, incorporated on December 22, 1857, took over the previous company's interests.

On July 7, 1860, Patton resigned his superintendency without stating a reason. His resignation went into effect on October 1. He moved to Marietta, Georgia, to retire.

KELLOGG AS SUPERINTENDENT

In October 1860 George Kellogg, a businessman and former member of the State Legislature who resided in nearby Athens and had been an original settler of the town of Forsyth, became the latest and, as it turned out, *final* superintendent of the

Dahlonega Mint. Kellogg was postmaster of Coal Mountain, Georgia, and conducted a general store there from the 1830s through the early 1850s. In 1853 he had been under consideration when Patton was chosen. On February 14, 1854, he was an incorporator of the Forsyth & Lumpkin Rail Road Company, chartered on that date.

The appointment was made by President James Buchanan, who as chief executive was betwixt and between—trying to please the North and South at the same time, and doing neither. His secretary of the Treasury was Howell Cobb, who after Abraham Lincoln's election in November would decamp to the South.

Kellogg's son Henry C. Kellogg was appointed clerk at the Mint and handled much of the correspondence. In 1860 there were important deposits of gold from the new discoveries in and near Denver, in the Colorado Territory.

On January 7, 1861, dies for the year's coinage arrived from the Philadelphia Mint.

THE CIVIL WAR

On January 19, 1861, Georgia seceded from the Union and joined the Confederate States of America. *The New York Times* published this on January 27, 1861:

> Richmond, Va., Sunday, Jan. 27.
>
> Intelligence has reached here that previous to the passage of the Georgia ordinance of secession, HARRISON W. RILEY, a prominent citizen of the mountain region of that State, proclaimed in another portion of Georgia that he was on his way home, and intended to raise a party and take possession of the United States Mint at Dahlonega for the United States Government, in case Georgia passed a secession ordinance. His declaration caused considerable excitement at Milledgeville, and the Governor was requested to send a military force to Dahlonega, but had not done so at the date of the advices.
>
> On the same authority as the above we learn that there is a considerable Union element in the mountains of Georgia, and an anti-secession meeting was held in Pickens County, when they heard of the passage of the secession ordinance. The old stars and stripes were run up, and the demonstrations of resistance to secession were emphatic and unequivocal.
>
> The impression, however, was, at Milledgeville, that the people of Georgia will generally sustain the ordinance of secession, and cooperate in the formation of a Southern Cotton States Republic.

After the secession, under a new ordinance Governor Joseph E. Brown assumed control of federal assets including the Mint. Kellogg kept up his duties on behalf of the Treasury Department, and on January 31 sent a report of the month's operations to Director Snowden in Philadelphia. On March 1 he sent the next report to Snowden, with sample coins from 1860 to be reviewed by the Assay Commission.

Lincoln was inaugurated president on March 4. On March 19 Director Snowden wrote to Salmon P. Chase, new secretary of the Treasury, noting that "notwithstanding the revolutionary proceedings in the State of Georgia, the Branch of the Mint at Dahlonega continues to recognize itself as a branch of the Mint of the United States." Snowden suggested that "some effort be made to retrieve the coining dies."

Meanwhile, Kellogg had written to Howell Cobb, now at the Confederate capitol in Montgomery, Alabama, stating that he was willing to "resign at any time and be commissioned under the Southern Confederacy."

Another man entered the scenario. Benjamin Hamilton wrote to Confederate Treasurer C.G. Memminger on March 1 stating that for $5,000 plus $1,000 for contingent expenses he would assume control of the Mint and run it.

On March 14, 1861, President Jefferson Davis signed a resolution introduced by T.R.R. Cobb of Georgia calling for the New Orleans and Dahlonega mints to remain open during the impending crisis with the Union, and to be used to strike coins for the South. It was noted that there had been rumors of the closing of the Dahlonega Mint.

On April 8 the Confederacy took over the Mint. On April 17, under the title of "Office of the Mint," Kellogg sent a report of the first three months of operation for the year to Treasury Secretary Memminger.

On April 25 Kellogg sent his resignation to the United States government, effective May 15, and inferred that he might retain his post under the Confederacy. This did not happen. In early May, Kellogg received orders not to strike any more coins from federal dies, but to await the receipt of Confederate dies.[13] This never happened, despite newspaper reports that such dies had been received.

The *Report of the Director of the Mint* for the fiscal year ending June 30, 1861, included this:

> At the branch mint at Dahlonega, the deposits received up to the 28th day of February, A. D. 1861, were $62,193.05; the coinage, $60,946; and the number of pieces, 13,442. No report has been received from this branch since the day last named.

The Confederacy officially closed the Mint on June 1. Kellogg stayed on to tidy accounts. On June 14 he submitted an inventory of the Mint's assets. A final report was sent to Memminger on June 27.

It is thought that $8,760 in gold coins from federal dies was struck in early 1861, including $2,209 in April, although estimates vary. The *Guide Book of United States Coins* places the mintage figures as 1,597 half eagles and about 1,250 gold dollars, the latter all coined by the Confederacy.

The Kelloggs, father and son, went to Coal Mountain in Forsyth County, to which address they requested that their mail be forwarded. Henry C. Kellogg became a captain in the military.

LATER YEARS

An Act of the Confederate Congress, approved August 24, 1861, authorized the openings of assay offices in Charlotte and Dahlonega. For the latter Lewis W. Quillian was appointed assayer. The Confederate States of America Assay Office activity continued into the first quarter of 1864. The business was low volume and comprised 345 assays, including 216 for the Confederate government. These were valued at $67,202.38. The Assay Office issued bars stamped with weight, fineness, and value. None seem to have survived.

After the Civil War ended Georgia's Congressional representatives were urged to have the Mint reopened, but they did not take any such action, nor did they ask to operate an assay office there. The coining machinery was removed. Federal troops set up in the Mint in January 1867 and stayed until mid-May of 1869. During that time the building seems to have been used as a detention area. A guard house was set up, and a facility was erected for the draconian punishment of hanging soldiers by the thumbs for short periods–a sentence for having become drunk or for other offenses. Following this, the Mint was used as a school for black children until a separate building was erected for this purpose by the Freedmen's Bureau. In February 1871, the federal government offered the building for sale, but was unable to attract a high offer of much more than $1,300, which was deemed to be unacceptable. Thus, on April 20, 1871, Congress authorized the secretary of the Treasury to convey the title to the building and land to the State of Georgia for use for educational purposes. The Academy at Dahlonega was planned, but it never materialized.

The North Georgia Agricultural College, a division of Georgia State University, opened its doors on January 1, 1873. Despite the specialty in its title, it served as a general high school equivalent to many regional people who had not had school in upper grades. Tuition was free. In 1873 there were 209 students enrolled in courses including reading, geography, arithmetic, grammar, book-keeping, history, rhetoric, natural philosophy, chemistry, physiology, botany, geology, mental philosophy, algebra, geometry, surveying, and the Latin and Greek languages. Reverend A.A. Lipscomb, D.D., L.L.D., was chancellor, David W. Lewis, A.M., was president, and there were four assistants on the staff.[14]

On February 28, 1877, the Georgia State Legislature acted to transition the school to train students for military service under the university system. Four years later it transitioned to become a military college without a change in name. On December 20, 1878, the building caught fire and burned down to the cellar. A new structure was built on top of the foundation and remains to this day.

COINAGE OF THE DAHLONEGA MINT, 1838–1861

After the first few years the striking of coins seems to have been a rather rustic operation at Dahlonega, which, in terms of the coins produced, resulted in pieces that had a very distinctive appearance. Whereas gold coins struck in Philadelphia in 1838 and later years were apt to be much alike, later Dahlonega products were often crudely struck, with weak areas in the dentils or devices, and they were sometimes indistinct,

An early advertisement for the North Georgia Agricultural College.

The new college building erected on the Dahlonega Mint foundation.

The Dahlonega Mint as photographed in 1878.

particularly at the centers. Of course, it is precisely this rusticity that makes such pieces appealing to numismatists today. The same naive or rustic characteristic applies to Charlotte gold, but not so much as with Dahlonega coins.

The initial Dahlonega coinage of 1838 consisted of a single denomination, the half eagle or $5 gold piece, which was struck to the extent of 20,583 pieces, representing slightly more than $100,000 in face value. The design was of the Classic Head style which had been in use at the Philadelphia Mint since 1834 and had been designed by Chief Engraver William Kneass. To distinguish the branch mint products, a D mint-mark was added above the date on the obverse. As it developed, in 1839 the half eagle type was changed to the Liberty Head or Coronet type, thus isolating the 1838-D as the only year of its design produced at Dahlonega.

After the first year the mintmark was transferred to the reverse below the eagle. Half eagles were generally made in small quantities, in relation to those of the Phila-delphia Mint in particular. However, among Dahlonega gold denominations, mint-ages of half eagles were sharply higher than of other denominations. This was simply a matter of practicality; local miners and holders of gold dust, ore, and nuggets, wanted to convert their metal to coinage form, and the most efficient way to do it was to use the largest denomination then in effect at Dahlonega, the $5. It took approximately twice as much effort to coin two quarter eagles as one half eagle.

Gold Dollars

For the coinage of gold dollars one pair of dies was sent from Philadelphia on June 6, 1849, and another pair on June 7; the four dies arrived in Dahlonega on June 16, 1849. Apparently, it was felt that there would be little call for large quantities of this new federal denomination in that district, although Bechtler gold dollars, which were privately minted in North Carolina, circulated extensively in the area. The hesitancy to plan for greater coinage of the denomination was unfounded, and gold dollars continued to 1860. In antic-ipation for the 1861 coinage, two obverse and two reverse dies were sent to Dahlonega. However, it was not until the Mint was under control of Confederate forces that 1861-D gold dollars were struck. Accordingly, these can be considered as official CSA coinage.

DESIGN TYPES AND DATES MINTED

Liberty Head (Type I): 1849-D, 1850-D, 1851-D, 1852-D, 1853-D, 1854-D.

Indian Princess Head, Small Head (Type II): 1855-D.

Indian Princess Head, Large Head (Type III): 1856-D, 1857-D, 1858-D, 1859-D, 1860-D, 1861-D (CSA issue).

$2.50 Quarter Eagles

Quarter eagles were a staple denomination of the Mint from 1839 onward, but they were not as important as half eagles. Today, while all dates are scarce, especially in AU and higher grades, there are no "impossible" rarities.

DESIGN TYPES AND DATES MINTED

Classic Head: 1839-D.

Liberty Head: 1840-D, 1841-D, 1842-D, 1843-D, 1844-D, 1845-D, 1846-D, 1847-D, 1848-D, 1849-D, 1850-D, 1851-D, 1852-D, 1853-D, 1854-D, 1855-D, 1856-D, 1857-D, 1859-D.

$3 Gold Coins

On June 1, 1854, one obverse and one reverse die for the new $3 denomination were sent from Philadelphia, and they arrived in Dahlonega on June 10, 1854. This represented the extent of $3 dies for Dahlonega; no later dies were made.

DESIGN TYPE AND DATE MINTED

Indian Princess Head:1854-D.

$5 Half Eagles

For the initial 1838-D coinage two obverse and two reverse dies were taken from Philadelphia to Dahlonega by David Hastings, arriving at Dahlonega on January 26, 1838. Coinage of half eagles was continuous from that point, through early 1861.

DESIGN TYPES AND DATES MINTED

Classic Head: 1838-D.

Liberty Head, Variety 1, No Motto Above Eagle: 1839-D, 1840-D, 1841-D, 1842-D, 1843-D, 1844-D, 1845-D, 1846-D, 1847-D, 1848-D, 1849-D, 1850-D, 1851-D, 1852-D, 1853-D, 1854-D, 1855-D, 1856-D, 1857-D, 1858-D, 1859-D, 1860-D, 1861-D.

SAN FRANCISCO MINT, 1854 TO DATE

FIRST SAN FRANCISCO MINT (1854–1874)

MINTING IN SAN FRANCISCO

On December 4, 1853, the United States Assay Office of Gold ceased operations (see chapter 12) and the building was sold to the Treasury Department. Remodeling was planned, including an expansion of the building. Machinery and equipment were transferred to the new San Francisco Mint. The construction contract was given to a Mr. Butler at $239,000. This was subsequently acquired by Curtis, Perry & Ward, former owners of the facility, who negotiated a new contract with the secretary of the Treasury. The partnership undertook to provide both the building and machinery. Curtis supervised construction details while Perry tended to the arrangements for the machinery. The San Francisco Mint, as finished, consisted of a structure 60 feet square and 3 stories high—20 feet wider on the western side than the earlier private mint, which measured 40 by 60 feet.

From the 1853 *Mint Report:*

> The branch mint at San Francisco, California, it is expected, will be ready to receive deposits and commence operations about the 1st of March next. In consequence of a change in the grade of the street on which the building is being erected, more time will be consumed in its completion than was anticipated. The machinery, which was constructed in Philadelphia, arrived there in good condition on the 12th of December last; but a portion of the fixtures and apparatus had not arrived on the 30th of December, the date of my last advices, the vessel containing them having then been out one hundred and forty-five days. These circumstances will probably delay the commencement of coining operations until the time above stated. The coins to be issued by this branch of the Mint will be designated by the letter S on the reverse. It is proper to remark, that the coins of the other branches are designated as follows: New Orleans by the letter O; Dahlonega, D; Charlotte, C. The coins of the principal mint are not marked by any letter.

Under a revised contract the cramped and poorly ventilated three-story Moffat building was refurbished and extended 20 feet on the western side, now measuring 80 by 40 feet, leading some to suggest that there was a cozy arrangement between the

Moffat interests and the Treasury Department. Whatever the circumstances, the "new" Mint was hardly new. Coining presses and other equipment were ordered from the East and were shipped around Cape Horn by clipper ships, including in the cargo holds of the *Trade Wind* and *Southern Cross*.

On March 15, 1854, a supply of 1854-dated coinage dies arrived from Philadelphia, where all dies were made. Some additional machinery came as well. Anticipation had run high, and when notices of employment positions were posted, the response was overwhelming. Soon, the Mint was well staffed, including over a dozen ladies assigned to weighing gold planchets and filing them slightly to remove excess metal in order to bring them down to precisely the authorized standard. Nearly 70 people were on the payroll being trained, setting up equipment, and attending to other duties. The refurbished old building of Moffat & Co. was given an outside coat of white cement. Published reports lauded it to the skies. The facility, despite a slight expansion, remained cramped and somewhat stuffy. In reality it was hardly a "new" mint, but was an updated old one.

The San Francisco Mint opened for business on April 3, 1854, and on April 15 the first coins, $20 pieces, were made. The first delivery of coins took place on April 19 and consisted of $2.50, $5, $10, and $20 denominations.[1] Gold dollars were delivered later in that month. Thus, by the end of April all current gold denominations had been made, save for those of the new and somewhat curious $3 denomination, which that year were coined only in Philadelphia, as well as limited production in New Orleans and Dahlonega. San Francisco would not strike $3 pieces until 1855. In various contemporary records the facility was usually referred to as the Branch Mint at San Francisco.

The first San Francisco Mint was on Commercial Street in the heart of the business district in 1854. Purchased by the government in 1853, the building had been used earlier by Moffat & Co. and related companies (see chapter 12). It was expanded and began business in March 1854. The facility was occupied until November 1874.

The San Francisco Mint in 1856

Hutchings' Illustrated California Magazine, October 1856, included "Coining Money," here slightly edited:

On the north side of Commercial Street, between Montgomery and Kearney, there stands a dark, heavy looking building, with heavy iron bars and heavy iron shutters, to windows and doors; and high above, standing on, and just peering over a heavy cornice, there is a large American eagle; looking down into the building, as if he meant to see, and take notes, of all that is going on within, "and print'em too." At his back there is a small forest of chimney stacks, from which various kinds of smoke, and different colored fumes, are issuing. This building is the Branch Mint of San Francisco.

On the pavement, in front, stands a number of odd looking, square boxes, containing bottles with glass necks rising above the top, and in which are the various kinds of acid used in the manufacture of gold and silver coin within.

In the street can be seen drays and wagons with men unloading supplies of various kinds for the Mint; express wagons with packages of the precious metal from all parts of the mines; men going up with carpet sacks hanging heavily on their hand, all desirous of having their gold dust converted into coin.

At the entrance door a man is sitting whose business it is to inquire your business whenever you present yourself for admission; and, if it is tolerably clear to him that you have no intention of obtaining a hatful of gold without a proper certificate; and more, that you have business dealings with Uncle Samuel; or, at least, wish to see how gold and silver is made into coin; why, it is probable that you may be allowed to pass.

By the kindness of Mr. Lott, the superintendent of the Mint, and the courtesy of the officers of the different departments, every facility was offered us for obtaining sketches, and all the necessary information concerning the *modus operandi* of coining, cheerfully given in all its branches.

To make the subject as plain as possible, we will suppose that the reader has just placed a bag of gold at the treasurer's counter, for the purpose of having it coined. Here the receiving clerk takes it, and after accurately weighing it, hands to the depositor a certificate for the gross weight of gold dust received, before melting. It is then sent to the *Melting Room*, where it is put into a black-lead crucible, melted, (each deposit is melted by itself,) and run into a "bar." A "chip," weighing about a tenth of an ounce, is then taken from each end of the bar, at opposite corners, one from the top, the other from the bottom side.

These chips are then taken to the *Assay Room* where they are carefully analyzed, by chemical process, and the exact amount of gold, silver, and other metals contained in each chip, accurately ascertained. The Assayer then reports to the Treasurer the exact proportion of gold, silver, and other metals, found in the chips. The standard fineness of the whole bar is then determined, and the value of

the deposit ascertained; it then awaits, in the Treasurer's Office, the orders of the depositor. When it is withdrawn, the depositor presents his certificate to the Superintendent's Clerk, who issues a warrant upon the Treasurer for the net value of the deposit; and, upon the payment of this warrant, in coin, or bar, the Treasurer delivers the Mint memorandum, which contains the weight of the deposit before and after melting, fineness, net value, &c., &c.

To facilitate business and prevent delay, a large amount of coin is always kept on hand, so that depositors are not required to wait until the gold dust taken in, is coined; but the moment its value is ascertained from the Assayer, the value is promptly paid the depositor: this is a great public convenience.

Now with the reader's permission let us see the gold bars accurately weighed in the Treasurer's Office; and let us carefully watch the many and interesting processes through which they must pass while being converted into coin.

On leaving the Treasurer's hands they are first sent into the *Melting Room*, where, as California gold contains from three to twelve per cent of silver, it becomes necessary in order to extract it, to alloy the gold with about twice its weight of silver; and thereby destroy the affinity of the gold for the silver, this enables the acid to act upon the silver. For this purpose, the gold and silver are melted together; and, while in a hot and fluid state, is poured gradually into cold water, where it forms into small thin pieces somewhat resembling the common pop-corn in appearance , and these are called "granulations."

The *Granulations* are then conveyed from the *Melting Room* to the *Refining Room;* where they are placed in porcelain pots that are standing in vats lined with lead. Nitric acid is then poured in upon the granulations, in about the proportion of two and a half pounds of acid, to one of gold; and, after the porcelain pots are thus filled sufficiently, the shutters, by which they are surrounded, are fixed closely down, and the granulations and acids boiled by steam for six hours, by which process the silver and all the base metals are dissolved, while the gold lies upon the bottom untouched. The bright orange colored vapor that we see issuing from the top of one of the chimneys of the Mint is generated from this process.

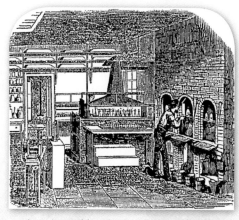

Assaying gold at the San Francisco Mint. When gold is received it is cast into bars or ingots, and from each a chip or piece is taken from two corners. These chips are then assayed to determine the purity of and hence the value of the deposit.

Making the granulations.

After boiling, the solution is drawn out of the pots by means of a gold syphon, (worth over two thousand dollars) into small tubs; it is then carried and emptied into a large tub or vat, twelve feet in diameter and six feet in depth—where a stream of salt water is poured upon it, which precipitates the *nitrate of silver* contained in solution and it becomes chloride of silver. The chloride is then run out of the vat into large filters, where it is washed until the water escaping from the filter is perfectly free from the acid. The *chloride of silver* is then taken out of the filter and placed in a "reducing vat" where it is mixed with *granulated zinc* and water: oil of vitriol is then poured in upon it, where by the action of the *oil of vitriol* upon it, where by the action of the oil of vitriol up on the zinc and the water, hydrogen gas is generated; which, combining with the chlorine of the chloride of silver forms muriatic acid, and leaves pure metallic silver, in fine powder, at the bottom of the reducing vat.

After the silver is thus thoroughly washed, it is placed in a hydraulic press, and subjected to the enormous pressure of twelve thousand pounds to the square inch, and the water nearly all forced out of it, leaving a compact, circular cake of silver, about ten inches in width, by three in thickness. These cakes are then placed on a draying pan, and the remaining moisture dried out. The silver is now ready for melting, and making into coin; or, for use in the granulating process.

Now, if you please, let us return to the porcelain pots and notice what becomes of the gold left in the bottom. This is now subjected to another boiling process of six hours, in fresh nitric in about the same proportion as before, during which time it is frequently stirred, to enable the acid to permeate the whole of the gold in the pot.

After this second boiling the acid is bailed out (and saved for the first boiling process) and the contents of the porcelain pots emptied into a filter, where it is well washed with hot water, prepared expressly for this purpose, and the remaining nitrate of silver is entirely washed out, leaving nothing but pure gold. The water is now pressed out in the same manner as it was from the silver, and the cakes locked up in a drying furnace for about three hours, when they are taken out and are ready for melting.

Let us now go to the *Melting Room*. There we find men moving about among "crucibles," "shoe" and "ingot-moulds," and what not, in front of the furnace, and as they lift back the cover, and the bright light breaks upon the eyes; down in the white heat we can see the crucible, ready to receive the precious metal. The gold is then put into it, with a sufficient amount of copper to reduce the standards of 1000 to 903. The gold is then run off into what are technically called "shoe-moulds." The bar thus run is termed "toughened bar." It is again assayed for the purpose of knowing the exact amount of copper to the added to reduce it to 900–1000, or the United States' standard fineness of coin. It is then again melted and reduced to the above standard; after which it is run into "ingot-moulds," and is again assayed, to determine whether it is now of the fineness required.

These ingots of standards gold, each weighing about sixty ounces, of which there are from thirty-six to forty in one "melt" are then "pickled," which being interpreted, means, to heat them red-hot and immerse them in sulphuric acid

Drawing off the acid from porcelain pots.

water, which cleans and partially anneals them They are then delivered by the Melter and the Refiner to the Treasurer, who weighs them accurately and then delivers them to the Coiner.

The ingots thus delivered, for twenty dollar pieces, are about 12 inches in length, about 1 inch and 7/16th in width, and about 1/2 an inch in thickness; yet for every different sized coin the width varies to suit.

They are now removed to the *Rolling Room* where the ingots pass thirteen consecutive times through the rollers, and at each time decrease in thickness, and increase in length, until they are about three feet six inches long: they are then taken to the *Annealing Room*, enclosed in long copper tubes and securely sealed to prevent oxidation or loss of the metal. They are now placed in the annealing furnace, where, after remaining for about forty-five minutes in sealed tubes, they are taken out and cooled in clear water. The "strips" of gold are now ready for rolling to the finished thickness and are re-taken to the Rolling Room for that purpose; and are afterwards returned to the Annealing Room and subjected again to a red hot heat for forty-five minutes, and again cooled as before.

These "strips" are now carried to the Drawing and Cutting Room, where they are first pointed, then heated, by steam; then "greased," with wax and tallow; and are then ready for the draw-bench. The point of the strip is then inserted in the "draw-jaw" and the whole strip is drawn through the "jaw" which reduces it exactly for the required thickness for coining. The strips thus gauged are then taken to the "cutting press," where, from the end of each strip a "proof piece" is "punched" and accurately weighed; and, if found correct is punched into "blanks" or "planchets" at the rate of about one hundred and eighty per minute. Should any of the strips be found too heavy, they are re-drawn through the "draw-jaw." If too light, they are laid aside to be regulated, by what is technically termed the "doctor"—a process by which the strip is made concave, before the planchets are cut out, and which gives them the required weight. This is an improvement only in use in the San Francisco Branch Mint and is, we believe, the invention of Mr. Eckfeldt, the Coiner; and by which some thirteen thousand dollars in light strips are saved from re-melting every day. Simple as the fact appears; it prevents the melting of about four millions of dollars per annum, and is doubtless, a great saving to the public.

After the blanks or planchets are cut out, the strips are bent in a convenient shape for re-melting, and are sent to the Coiner's Office to be weighed, preparatory to making up his account for the day, and which, with the planchets, must make up the gross amount received in the morning from the Treasurer.

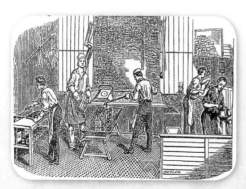

Casting gold into ingots.

Rolling and Cutting Room where ingots of gold and silver are rolled into long strips and planchets cut from them.

They are afterwards delivered to the Treasurer, by whom they are again weighed and then sent to the Melter and Refiner to be again cast into ingots.

The planchets are then carried from the cutting-press to the *Cleaning Room* where they are boiled in very strong soap-suds, from which they are taken and dried in a pan, heated by steam, and then conveyed to the Coiner's Office to be weighed. After which, they are sent to the *Adjusting Room* where each piece is separately weighed, and those found too light, are condemned for remelting; and those which are too heavy are reduced, by filing, to the standard weight. All the planchets thus adjusted, are then re-taken to the Coiner's Office, and, with the filings and light planchets, are carefully weighed, and that weight must tally with the gross amount of the planchets delivered to the adjusters during the day.

The work of "adjusting" is performed by females of whom from ten to fifteen are employed, according to the amount of labor to be accomplished.

From the adjusting room the planchets are taken to the *Milling Room*, where they are dropped into a tube, belonging to the "milling machine," and by means of a revolving circular steel plate, with a groove in the edge, and a corresponding groove in a segment of a circle, the planchets are borne rapidly round, horizontally, by which process the edges are thickened, and the diameter of the planchet accurately adjusted to fit the collar of the "coining press." After "milling" they are returned to the Coiner's office and again weighed, to ascertain if the weight is correct.

They are then sent to the *Annealing Room*, where they are put into square cast-iron boxes, with double covers, carefully cemented with fire-clay, and placed in the annealing furnace, where they are subjected to a red heat for about an hour, when they are taken out and poured into a "pickle" containing diluted sulphuric acid. By this process they are softened and cleansed; and after they are rinsed with hot water they are well dried in saw-dust heated by steam, taken out and returned to the Coiner's office, where they are again weighed, and afterwards carried to the *Coining Room*, to be "stamped."

Adjusting Room where ladies weigh each blank planchet and use files to remove metal if overweight.

Milling machine imparting raised rims to blank planchets.

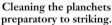

Cleaning the planchets
preparatory to striking.

Coining presses.

This process is performed by dropping the planchets into the tube in front of the machine, from whence they are carried by "feeders" to the "collar," into which they are dropped upon the lower die: the head die then descends, and by its immense power displaces every particle of gold in the planchet, and gives the impression upon both sides of the coin and the fluting on the edge, at the same moment. At every motion, the "feeders" not only take a planchet to the collar, but at the same time push the coin, previously struck, and now perfect, from the lower die, which rises and falls for the purpose at each revolution of the wheel, from whence the coin slides into a box underneath.

From the Coining Room they are again taken to the Coiner's office where they are weighed, counted and delivered to the Treasurer for payment to depositors.

There is one piece always taken out of about every sixty thousand dollars, coined into double-eagles, and a similar amount from smaller coins, which are sent to Philadelphia, and carefully preserved for examination at the "judgment day," as it is curiously and expressively called, which takes place annually at Philadelphia, under the superintendence of commissioners appointed by the U.S. government.

We are surprised at the aggregate amount of coin produced in so short a time, in such a small and very inconvenient building; for, it seemed to us that every man was more or less in the others' way; and wherever the fault may lie, we think of it very questionable economy, that requires a remedy without delay.

WHAT HAPPENED TO THE GOLD?

Without a doubt, the most enigmatic figure in the history of the San Francisco Mint is John Michael Eckfeldt, at least when viewed in retrospect. He was born in Philadelphia on February 6, 1831, son of George Eckfeldt of the Philadelphia Mint, another employee who traced his lineage back to Adam Eckfeldt, who had worked at the Mint since 1792. John Michael Eckfeldt was at the San Francisco Mint from day one, having worked earlier for Curtis, Perry & Ward in the United States Assay Office of Gold. At the Mint he was coiner, a key position paying $3,000 per year. He was admired by some employees, disliked by others. In the summer of 1857 anonymous articles against him and others were printed in the California *Chronicle* over such signatures as "Howard," "Anti-Humbug," and "Hawkeye."

In January 1859 his position was changed from coiner to assistant coiner. William Schmolz became chief coiner, also at $3,000. In the 1860s Eckfeldt helped supervise the construction of the Carson City Mint and in the early 1870s with the construction of the new Mint building.

Then the worst happened: Superintendent O.H. LaGrange reported on October 8, 1874, "Eckfeldt shot himself at 9 a.m. due to depression of spirits caused by overwork on new Mint machinery." For some time he had been in poor health with severe kidney problems and had been unable to work regularly. Funeral services were held at the Masonic Temple in Oakland at 2 o'clock in the afternoon on October 10.

In the meantime, in August 1857, it was found that a staggering $152,000 in gold bullion was missing. Melter and refiner Agoston Haraszthy was the prime suspect. Earlier with Wass, Molitor & Company, he had been with the San Francisco Mint since its opening in 1854. He protested his innocence and suggested that it might have been lost in waste up the chimney, as gold particles were found on nearby rooftops. Various Mint employees testified against him and he was discharged, soon to open an assay office in the city. Not simplifying matters, coiner Eckfeldt and assistant coiner Hiram T. Graves were indicted by a grand jury for perjury. After deliberation and reconsideration, another grand jury voted unanimously to dismiss the charges against the pair. In the meantime in the case against Haraszthy, after four years of court wrangling, the embezzlement charge was found by a jury to have no merit. However, a cloud remained over him for the rest of his life.

Surprise! In 1891 when a house was being demolished in Oakland across the bay, about $150,000 in gold was found. Among the earlier owners was John Michael Eckfeldt.[2]

THE SAN FRANCISCO MINT IN 1861

Bancroft's Handbook of Mining, 1861, included "How Gold is Coined," by John S. Hittell. After giving details of gold refining, assaying, making into bars, and rolling out strips he continued with:

> The rolled-out strip, not quite as thick as the coin, is taken thence to the cutting machine, which, by a punch, cuts out from the bar round pieces, a little larger than the coin.[3] These pieces are called blanks. They are carried to the Annealing Room and annealed and washed with soap and water. They are then taken to the Adjusting Room. Here each blank is weighted separately, and made the exact weight for the coin. If too heavy, the blank is filed down; if too light it is thrown into a box to be re-melted. The work in this room is done entirely by women.
>
> The adjusted blanks are run through the milling machine, which compresses the blank to the exact diameter of the coin and raises the edge. The purpose of making the edge thicker is to make the coin pile neatly, to protect the figures and to improve the general appearance. About two hundred and fifty blanks are milled in a minute.
>
> The milled blanks are carried back to the annealing room, placed in an air-tight cast iron box and placed in the furnace to be annealed, so that they may take the impression well. When they are at a cherry red they are taken out and poured immediately into water with a little sulphuric acid. This softens and cleanses the gold. The blanks are taken out, washed with cold water, put into hot water again, taken out, mixed in with saw-dust, which is then sifted off and the blanks are dried and perfectly clean.
>
> They are again taken to the coining and milling room, and stamped. The coining machine is elegant and massive. The blanks are placed in a tube or pipe, and from

this the machine takes them one by one, puts them between the dies, stamps them, throws them out of the die and carries them down into a box, and they are then delivered to the Treasurer and are ready for circulation.

About one-fourth of the gold yield of the coast goes to the San Francisco Mint, the only Mint on the coast. The treasure exported from San Francisco in 1860, was $42,325,916; the gold deposits of the Mint during the same period amounted to $11,219,209. About $31,000,000 were therefore exported in unrefined bars and in dust. A small portion of the export was silver. The number of deposits made at the Mint in 1860 was 4,841, showing that the average value of the deposits was $2,271. The gold coinage of the year amounted to $11,178,000, of which $10,899,000 was in double eagles, and $179,000 in smaller pieces. Of silver, $264,000 was coined in half dollars, quarter dollars and dimes; and $216,678 in refined silver bars. No three cent pieces, copper or nickels have ever been coined in California, nor are they used in the country; and half dimes are very rare.

THE CIVIL WAR YEARS

During the Civil War the San Francisco Mint kept up production. Coins circulated at par on the West Coast, while in the East and Midwest they were available only at a sharp premium in terms of Legal Tender Notes and National Bank notes (see chapter 4 for more details). In the West, such bills brought into the area traded at a deep discount in inverse ratio to the premium in the East. This changed in 1864 when the Supreme Court of California decreed that Legal Tender Notes were indeed lawful money, but it was not required that "every kind of lawful money could be tendered in the payment of every obligation."[4] The government decreed that the salaries of federal employees at the Custom House, Mint, and elsewhere would henceforth be paid in Legal Tender Notes. This meant that in local commerce what was $4 per day was now equal to $1.50 per day in spending power. Employees were threatened with arrest if they refused this arrangement. Samuel L. Clemens (later known as Mark Twain) was a writer for the *San Francisco Call*, whose offices were next to the San Francisco Mint. On September 25 he wrote:

> A report is abroad that the Branch Mint is about to close—that the employees, being no longer able to support themselves and families on the mere prospect of getting the salaries due them paid some day or other, have given notice that unless their accounts are previously squared, they will quit work in a body on the 30th instant. These reports were not without foundation.
>
> We are glad to be able to state, however, that the Mint is not going to stop, nor the men be allowed to suffer much longer for the moneys due them. Within two weeks, or at farthest, three, all cause of complaint will be removed, and the employees themselves have been satisfied of this fact. We get our information at headquarters.

In October, an adjustment was made by Superintendent Robert B. Swain whereby salaries in paper money were increased to give the employees amounts equal to what gold and silver coins could buy.

SILVER EXPORTED

The *Annual Report of the Director of the Mint* for the fiscal year ending June 30, 1864, told of why the vast production of silver in the Comstock Lode of Nevada had not resulted in an increase of domestic silver coinage. The San Francisco Mint was ignored:

The history of the development of any mining region is a romance full of surprising incidents; and none is more so than that of Nevada Territory. From the first discovery of silver, in June 1859, to the present moment, that country has been a scene of excited search, toil and speculation, of rapid fortunes, severe losses and disappointments . . .

In a time of peace, shall we have a share of the gold and silver of our own mines, for our currency and use, or must it be carried off to the plethoric vaults of European banks and capitalists? This latter is precisely the direction all the silver has taken thus far. . . . None of the Nevada silver is coined here [at the Philadelphia Mint where the director prepared his report], and but little at San Francisco, where it first goes. Our correspondents at that port inform us that it is all shipped to England, partly in rich ores that can more economically be smelted there, and partly in metallic bars . . .

THE MINT IN 1865

Samuel Bowles, a Massachusetts newspaper editor and writer, visited the West in the summer of 1865 and penned a report that included this:

Of all the government institutions in San Francisco, the Mint is the most interesting and important. Already it is the greatest manufactory of coin in the nation, and its comparative importance in this respect is destined to increase. It coins now about $20,000,000 in gold and silver a year, against $5,000,000 coined at all the other government mints in the country including the parent Mint at Philadelphia.

The coinage here for June and July [1865] was nearly $3,000,000 a month, and the aggregate for this year is likely to go up to $24,000,000 . . . The Mint here is in charge of one of the best merchants of the city, Mr. R.B. Swain, but it has no adequate accommodations. It is crowded into the back and upper rooms of an old and ordinary block in the principal business street. But, provision has been made by Congress for a distinct and appropriate building.

The metals are received at the Mint in all manner of half-worked forms, in dust, nuggets, rough bars, silver and gold mixed together, and more or less dross with all. Each parcel is kept distinct, first assayed to discover its exact value, and then worked over, the dross expelled, and the silver and gold separated. Fire, water, and chemicals are the means employed. The processes are simple enough and exquisitely entertaining as you follow them with eye and an intelligent explanation [given by an employee]. The results are returned to the owner either in solid bars, bearing an official stamp of their value, or in freshly made coin.[5]

Another commentary emphasized this:

There is no paper money in San Francisco. . . . In our banks there are great piles of double eagles, but no bank-notes are visible. Wherever you go, or whatever you do, or whatever you buy, you see only gold and silver; people do not think of paper. All large sums are paid in double eagles, and three fourths, if not nine-tenths, of our coin is of that size, which is far more convenient than the smaller coins common in other countries. A large proportion of our shipment of treasure abroad is in double eagles, and nine-tenths of the gold coined at the San Francisco Mint is in pieces of that size . . .[6]

The *Report of the Director of the Mint* for the fiscal year ended June 30, 1865 included this:

Under the efficient management of the superintendent of the branch mint at San Francisco, its operations have been well and successfully performed. The coinage of the past year has been very large. The monthly deposits of bullion are increasing, and it is confidently predicted that the yield of the mines for the current year will largely exceed that of any former period. The past has been a success; the future is full of encouragement.

It is gratifying to know that Congress, fully appreciating the magnitude and importance of the mineral wealth of the Pacific States, has made an appropriation for the erection of a new mint-building at San Francisco. The present building is not only unsafe, but wholly inadequate for the increasing business of that branch mint. The new structure should be, in architecture, capacity, machinery, and every particular, adapted to the present and future of California and the Pacific States.

Continuing Activity

The 1866 issue of the *Annual Report of the Director of the Mint* emphasized that conditions at the San Francisco Mint were very unsatisfactory, with the staff working in cramped, poorly ventilated quarters. Walter S. Denio, melter and refiner at the Mint, died on February 10 of lung congestion caused by noxious fumes. For a time, all refining operations were stopped. Director James Pollock stated:

> I cannot too earnestly urge upon the government the importance of erecting a new Mint building at San Francisco. The present building is not only wholly unfitted for the large and increasing business of that Branch Mint, but unsafe, and unworthy of the great mineral wealth of the Pacific States. The appropriation made by Congress should be applied at once to the erection of a building which in architecture, size, capacity, machinery, and every useful modern appliance should be equal to the present and future of California. The management of this Branch during the past year has been efficient, and its operations, under the direction of the superintendent, well and carefully performed.

In 1870 the Carson City Mint opened for business. Located about 15 miles from Virginia City and the Comstock Lode, many in the region felt it would be the ideal facility to coin silver and gold. Logically, it should have been. However, a combination of politics and the high cost of operation resulted in most Comstock Lode metal being sent to the San Francisco Mint for coinage.

Coinage of the First San Francisco Mint, 1854–1874

Coins of the first San Francisco Mint were produced under trying circumstances for the most part, in cramped quarters outlined in the narrative above. Production was from gold found during the California "Rush" and, after the late 1850s, gold and silver brought from the Comstock Lode in and near Virginia City, Nevada.

There was no numismatic interest in such coins at the time they were issued, with the result that today high-grade coins are found only as a matter of rare chance, some double eagles from recovered sunken treasure being exceptions (see page 278).

Half Dimes

Half dimes were latecomers to the San Francisco silver coinage scene. Made from 1863 to 1873, these circulated mostly on the West Coast. They were interchangeable at par with Spanish medios (6-1/4 cents), some of which remained in circulation there.

DESIGN TYPE AND DATES MINTED

Liberty Seated, Variety 4, Legend on Obverse:
1863-S, 1864-S, 1865-S, 1866-S, 1867-S, 1868-S, 1869-S, 1870-S, 1871-S, 1872-S, 1873-S.

Dimes

The first Liberty Seated dimes were struck in San Francisco in 1856, two years after gold coinage commenced and a year after silver coins were first made. These circulated regionally. Today, all range from scarce to rare in Mint State. Liberty Seated dimes of the 1860s through 1874 are similarly elusive in Mint State.

DESIGN TYPES AND DATES MINTED

Liberty Seated, Variety 2, Stars on Obverse:
No Arrows, Reduced Weight: 1856-S, 1858-S, 1859-S, 1860-S.

Liberty Seated, Variety 4, Legend on Obverse:
1861-S, 1862-S, 1863-S, 1864-S, 1865-S, 1866-S, 1867-S, 1868-S, 1869-S, 1870-S, 1871-S, 1872-S.
Variety 5, Arrows at Date, Increased Weight: 1873-S, 1874-S.

Quarter Dollars

Liberty Seated quarters from the second San Francisco Mint can generally be found in worn grades, Very Fine being about par. Mint State coins range from scarce to rare, and at the gem level are almost impossible to find for some years.

DESIGN TYPES AND DATES MINTED

Liberty Seated, Variety 3, Arrows at Date, No Rays, Reduced Weight: 1855-S. *Arrows Removed, Weight Remains:* 1856-S, 1857-S, 1858-S, 1859-S, 1860-S, 1861-S, 1862-S, 1864-S, 1865-S.

Liberty Seated, Variety 4, Motto Above Eagle: 1866-S, 1867-S, 1868-S, 1869-S, 1870-S, 1871-S, 1872-S. *Variety 5, Arrows at Date, Increased Weight:* 1873-S, 1874-S.

Half Dollars

The comments concerning quarters are applicable to half dollars as well. Most specialists, such as members of the Liberty Seated Collectors Club (online at www.lsccweb.org), are usually content with coins grading Very Fine or Extremely Fine.

DESIGN TYPES AND DATES MINTED

Liberty Seated, Variety 3, Arrows at Date, No Rays, Reduced Weight: 1855-S. *Arrows Removed, Weight Remains:* 1856-S, 1857-S, 1858-S, 1859-S, 1860-S, 1861-S, 1862-S, 1863-S, 1864-S, 1865-S, 1866-S.

Liberty Seated, Variety 4, Motto Above Eagle: 1866-S, 1867-S, 1868-S, 1869-S, 1870-S, 1871-S, 1872-S, 1873-S. *Variety 5, Arrows at Date, Reduced Weight:* 1873-S, 1874-S.

Silver Dollars

In 1870 a silver dollar of that year and mint was put in the cornerstone of the second San Francisco Mint. Subsequently there was a small production run that was not entered in the records. Today only about 10 are known. In early 1873 the San Francisco Mint reported the coinage of 700 Liberty Seated silver dollars, but no example has ever come to light.

The Mint coined Morgan-design dollars continuously from 1878 to 1904 (which silver supplies authorized by Congress ran out). *The Annual Report of the Director of the Mint* for the fiscal year dated 1885 told of the distribution of silver dollars as of June 30, 1885. The San Francisco Mint had 32,029,467 coins on hand and had distributed 3,516,033. These kept piling up in vaults.

DESIGN TYPES AND DATES MINTED

Liberty Seated, No Motto: 1859-S.

Liberty Seated, With Motto IN GOD WE TRUST: 1870-S, 1872-S, 1873-S.

Trade Dollars

Trade dollars were made in quantity for two years. Nearly all were used in the trade with China, although some circulated domestically as legal tender.

DESIGN TYPE AND DATES MINTED
Trade Dollar: 1873-S, 1874-S.

Gold Dollars

Gold dollars of the second San Francisco Mint are an interesting specialty. 1854-S coins are usually well struck, something of which cannot be said for 1856-S. Later issues are rare in Mint State. The 1870-S is seldom seen and is a rarity on an absolute basis, but those that do exist are mainly Mint State.

DESIGN TYPES AND DATES MINTED
Liberty Head (Type I): 1854-S.

Indian Princess Head, Small Head (Type II): 1856-S.

Indian Princess Head, Large Head (Type III): 1857-S, 1858-S, 1859-S, 1860-S, 1870-S.

$2.50 Quarter Eagles

The first quarter eagle, the 1854-S, is one of the most famous gold rarities in the Liberty Head series across all denominations. Later issues are usually available in Very Fine grade, seemingly the default listing as these circulated widely in commerce. Conservatively graded Mint State coins are seldom seen, even when great collections come to market.

DESIGN TYPE AND DATES MINTED

Liberty Head: 1854-S, 1856-S, 1857-S, 1859-S, 1860-S, 1861-S, 1862-S, 1863-S, 1865-S, 1866-S, 1867-S, 1868-S, 1869-S, 1870-S, 1871-S, 1872-S, 1873-S.

$3 Gold Coins

The $3 coins of the first four listings are collectible in worn grades but are elusive in Mint State. Just two examples were struck of the 1870-S, of which only one survives—on loan today at the Harry W. Bass Jr. Gallery at American Numismatic Association Headquarters in Colorado Springs.

DESIGN TYPE AND DATES MINTED

Indian Princess Head: 1855-S, 1856-S, 1857-S, 1860-S, 1870-S.

$5 Half Eagles

The first half eagle, the 1854-S, is a famous rarity with just two (probably) or three known to exist, one of which is in the National Numismatic Collection at the Smithsonian Institution. The others through 1866 are collectible, but are usually seen with significant wear. Those of 1866 and later are nearly always seen in circulated grades, but of a slightly higher average level. High-level Mint State coins are rarities across the board.

DESIGN TYPES AND DATES MINTED

Liberty Head, Variety 1, No Motto Above Eagle: 1854-S, 1855-S, 1856-S, 1857-S, 1858-S, 1859-S, 1860-S, 1861-S, 1862-S, 1863-S, 1864-S, 1865-S, 1866-S.

Liberty Head, Variety 2, Motto Above Eagle: 1866-S, 1867-S, 1868-S, 1869-S, 1870-S, 1871-S, 1872-S, 1873-S, 1874-S.

$10 Eagles

All Liberty Head eagles are collectible in worn grades with Very Fine being about par. Conservatively graded Mint State coins are rarities.

DESIGN TYPES AND DATES MINTED

Liberty Head, No Motto Above Eagle:
1854-S, 1855-S, 1856-S, 1857-S, 1858-S, 1859-S, 1860-S, 1861-S, 1862-S, 1863-S, 1864-S, 1865-S, 1866-S.

Liberty Head, Motto Above Eagle:
1866-S, 1867-S, 1868-S, 1869-S, 1870-S, 1871-S, 1872-S, 1873-S, 1874-S.

$20 Double Eagles

All of the double eagles listed here are collectible in circulated grades, with the 1861-S Paquet and 1866-S, No Motto, being scarce.

Today, Liberty Head double eagles are very popular—a scenario inspired in part by coins, many in Mint State, recovered from three treasure-laden ships: SS *Central America* (lost in 1857), SS *Brother Jonathan* (1865), and SS *Republic* (1865). Most of the *Central America* and *Brother Jonathan* coins have S mintmarks. A description of each of these ships can be found in the front pages of *A Guide Book of United States Coins*.

DESIGN TYPES AND DATES MINTED

Liberty Head, Without Motto on Reverse: 1854-S; 1855-S; 1856-S; 1857-S; 1858-S; 1859-S; 1860-S; 1861-S; 1861-S, Paquet Reverse; 1862-S; 1863-S; 1864-S; 1865-S; 1866-S.

Liberty Head, Motto Above Eagle, Value TWENTY D.: 1866-S, 1867-S, 1868-S, 1869-S, 1870-S, 1871-S, 1872-S, 1873-S, 1874-S.

SECOND SAN FRANCISCO MINT (1874–1937)

THE NEW MINT

In 1870 the cornerstone was laid for the second San Francisco Mint. In it were put examples of coins of that year, including two 1870-S $3 pieces, a variety not coined for general circulation and thus a rarity when it was recognized by numismatists years later.

In 1878 it became known that gold was disappearing. Suspicion pointed to a night watchman, James Henry Smith, who had been observed selling bullion to dealers. Taking the law into their own hands they broke into Smith's house and tore up some of the flooring, without success. Then they dug around in the yard, and beneath a bed of pansies found $6,000 in gold that had been stolen from the refining process at the Mint. It turned out that he had stolen about $20,000 worth of gold in all. Smith was convicted and sent to jail.[7]

An audit in 1885 revealed that the melter and refiner, by reducing gold coins and ingots very slightly, but still keeping within legal tolerances, was able to secure an unknown extra amount of gold, some of which seemed to have been to his personal benefit. The report was filed on July 1, 1885, and on August 1, Colonel E.F. Burton, superintendent of the San Francisco Mint, was suspended, and Judge Israel Lawton was appointed as his replacement.

In 1891 the average cost of minting coins at the San Francisco Mint was $0.0239— slightly more than two cents. Interest in collecting mintmark coins arose in the 1890s when Augustus G. Heaton published a treatise on the subject. It turned out that 1894-S dimes were nowhere to be found. In that year and in 1895 Mint officers fielded a number of inquiries. It developed that just 24 were coined.

The second San Francisco Mint as it appeared in the early 1900s. This building was in use from 1874 to 1937.

Construction during the early 1870s.

Another view of the Mint in progress.

The Mint as seen from the south.

The Mint in the 1880s.

The Receiving Room for
deposits of silver and gold bullion.

The Melting Room for silver and gold bullion.

Vats of acid in the Refinery Department.

Refinery furnaces.

The Assaying Department.

The Weighing Room in
the Assaying Department.

The Engine Room, the source of power used to
drive heavy equipment via a system of belts.

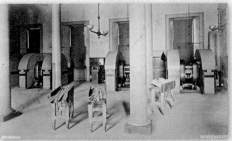

The Rolling Room where silver and
gold ingots were reduced into thin strips.

The Cutting Room where planchets were punched from thin metal strips.

The Adjusting Room where ladies sit in front of balance scales to weigh blank planchets. Those that are overweight are adjusted by filing metal off. Those that are underweight are set aside to be melted.

The Coining Room with four knuckle-action presses in view.

The Weighing Room where quantities of silver and gold coins were weighed prior to being put into cloth bags to be shipped.

APRIL 18, 1906

At 5:15 in the morning on April 18, 1906, a severe earthquake struck the city of San Francisco. The shock broke the mains of the Spring Valley Water Company. For a short time, citizens milled about assessing the situation. There had only been some structural damage so far, but within a short time fire broke out in several areas of the city. Firemen and others were helpless as it spread. Crews used dynamite to level some buildings in the path of the flames, but the effort was of no use. Three days later when the conflagration ended, the central part of the city was a scene of burned-out buildings, fallen and partially standing brick walls, and rubble everywhere.

Flames surrounded the Mint. Using water from its own well, workmen hosed down the roof. The Mint was the only building left standing in its district. Plate glass windows were lost, some window frames were damaged, and some stone work was flaked, causing damage estimated at $53,400. The *Annual Report of the Director of the Mint* gave details:

> The building was saved through the recent establishment of a fire-protection plant within the building. About 50 of the officers and employees of the Mint succeeded in reaching the institution and, with the exception of two or three, they remained in the building fighting the fire until all danger was passed, and in their efforts to protect the building there was a constant battle from early morning until quite late in the afternoon. An abundance of water was supplied from the artesian well in the Mint and forced to the various parts of the building by a steam pump.
>
> As the Sub-Treasury had been destroyed and every bank in the city lay in ruins, the Mint was the only financial institution left intact, consequently it immediately

became the financial center and nucleus for the resumption of business, and the point of distribution of financial relief. The assistant treasurer was given offices and vault room in the building and supplied with money with which to resume business. The banks organized a Union Bank, embracing all the leading banks of the city, and they were afforded space and vault room, by which they were able to transact a regular banking business.

Besides this, at the request of the president, we became the depository and treasury for the relief funds until the banks were able to return to business in their own quarters. In addition to all this the Mint officials handled, in round numbers, $40,000,000—money that was transferred by telegraph through a system of transferring funds from various parts of the East to individuals and banks and corporations in this city, made possible by an order of Secretary of the Treasury Shaw.

This was one of the greatest relief measures instituted, and was received with expressions of gratitude by all. This work was the most arduous of all our labors, employing nearly all of the clerical force from early morning until late in the evening. Of course we had no system or method for the transaction of that kind of business, it being entirely foreign to our usual line of work, but all the money was transferred and distributed without loss, error, or unnecessary delay.

It was not until May 23 that bankers collectively agreed that business could be resumed in various temporary offices set up in the city.

In 1908 Indian Head cents were struck for the first time at the San Francisco Mint, and in 1912 the facility made its first nickel five-cent pieces. The demand for minor coinage was much less on the West Coast than in the Midwest and East.

In 1915 a hydraulic press used to coin medals at the Philadelphia was shipped to the San Francisco Mint on loan. Weighing 14 tons and having a striking pressure capacity of 450 tons, it was used to coin 1,506 examples of each of the two large-diameter $50 gold coins, round and octagonal, struck for the Panama-Pacific International Exposition. Later commemorative coins were struck at the second San Francisco Mint through and including 1937 when coinage at that facility ceased. Included were 1925 Fort Vancouver half dollars for which the S mintmark had been inadvertently omitted from the obverse die.

The San Francisco Mint after the earthquake.

Another view of the destruction.

Following the provisions of the 1918 Pittman Act, many silver dollars in storage at the San Francisco Mint were melted into bullion and the metal sent to England. Large quantities of silver dollars were shipped to China in 1919 from quantities on hand, the sale price being $1.35 per pure ounce of silver (a dollar was 90-percent pure). In 1921 the coinage of silver dollars resumed, after a lapse since 1904.

Double eagles were struck intermittently at the San Francisco Mint from onward, most of them going into storage and not released. The same thing happened with 1920-S and 1930-S eagles. Commemoratives were made there on many occasions through the end of the use of the facility. When the new building was ready in 1937 equipment was removed from the premises.

LATER USE OF THE BUILDING

After 1937 the building, which locals nicknamed the "Granite Lady," was home to several government agencies. In 1968 it was declared surplus, no longer needed. Its fate was unknown. Repairs needed to be made, and no viable ideas for its maintenance were given.

In 1972 President Richard Nixon announced that it had been transferred to the care of the General Services Administrations, and plans were announced to restore it and make it available for public use.

In April 1973 the people in the current San Francisco Mint who processed mail orders for Proof sets and other collector items and also those working on computer systems were transferred to rooms in the back of the old Mint. On July 16 Mint Director Mary Brooks officially opened the premises as a museum, with some rooms restored to the appearance they had in the 19th century. The cost was about $4.5 million, financed by the sale of coins and medals to collectors. A film titled *The Granite Lady* was made for the 1976 American Bicentennial and featured some of the Mint employees as actors.

On January 28, 1977, the old Mint was tagged as California Registered Historical Landmark No. 875, seemingly a long overdue recognition. In October 1979 tremors from a distant earthquake in Nevada caused some shifting at the Mint. By that time the rare coin displays in the museum had been removed due to a dispute about security and insurance. On January 3, 1994, the Old San Francisco Mint Museum was to have been closed due to the expense and unprofitability of the Mint Sales Center, the cost of security (rare coins were on display), and less than expected public interest. Secretary

The old San Francisco Mint in August 1958.

of the Treasury Lloyd Bentsen gave the museum a 90-day extension. Further extensions took place, and Secretary Bentsen allowed the building to stay open to the end of the year. It was closed on December 30, 1994. Safety repairs were made later, and other proposals were considered. The building stands today with no plans to destroy it.

COINAGE OF THE SECOND SAN FRANCISCO MINT, 1874–1937

When it opened for business in 1874 the second San Francisco Mint was state of the art in equipment. From that time onward it produced a steady stream of silver and gold coins, augmented by the first coinage of copper cents in 1908 and the first coinage of nickels in 1912.

In the 1870s it was the prime producer of trade dollars for shipment to Eastern Asia, followed by a steady output of Morgan-design silver dollars from 1878 to 1904 and again in 1921, after which Peace dollars were made intermittently from 1922 to 1935. During the Klondike Gold Rush era of the late 1890s, it was the prime facility for converting gold to coins, mainly double eagles.

Small Cents

The first minor coins struck at any branch mint were the 1908-S Indian Head cents. Lincoln cents have been made intermittently from 1909 to date, with the low-mintage 1909-S, V.D.B., gaining great numismatic fame.

DESIGN TYPES AND DATES MINTED

Indian Head, Oak Wreath With Shield: *Variety 3,* *Bronze:* 1908-S, 1909-S.

Lincoln, Wheat Ears Reverse: *Variety 1, Bronze:* 1909-S, (with and without V.D.B.), 1910-S, 1911-S, 1912-S, 1913-S, 1914-S, 1915-S, 1916-S, 1917-S, 1918-S, 1919-S, 1920-S, 1921-S, 1923-S, 1924-S, 1925-S, 1926-S, 1927-S, 1928-S, 1929-S, 1930-S, 1931-S, 1935-S, 1936-S, 1937-S.

Nickel Five-Cent Pieces

After a brief coinage of 1912-S Liberty Head nickels in December of that year, Indian Head or Buffalo nickels were made from 1913 to 1837 except for 1922 and 1932 to 1934.

DESIGN TYPES AND DATES MINTED

Liberty Head: 1912-S.

Indian Head or Buffalo, Variety 1, FIVE CENTS on Raised Ground: 1913-S.

Indian Head or Buffalo, Variety 2, FIVE CENTS in Recess: 1913-S, 1914-S, 1915-S, 1916-S, 1917-S, 1918-S, 1919-S, 1920-S, 1921-S, 1923-S, 1924-S, 1925-S, 1926-S, 1927-S, 1928-S, 1929-S, 1930-S, 1931-S, 1935-S, 1936-S, 1937-S.

Dimes

The second San Francisco Mint struck dimes of the Liberty Seated, Barber, and Mercury types for most of the years of its existence.

DESIGN TYPES AND DATES MINTED

Liberty Seated, Variety 5, Legend on Obverse, Arrows at Date, Increased Weight: 1874-S. *Arrows Removed, Weight Remains:* 1875-S, 1876-S, 1877-S, 1884-S, 1885-S, 1886-S, 1887-S, 1888-S, 1889-S, 1890-S, 1891-S.

Barber or Liberty Head: 1892-S, 1893-S, 1894-S, 1895-S, 1896-S, 1897-S, 1898-S, 1899-S, 1900-S, 1901-S, 1902-S, 1903-S, 1904-S, 1905-S, 1906-S, 1907-S, 1908-S, 1909-S, 1910-S, 1911-S, 1912-S, 1913-S, 1914-S, 1915-S, 1916-S.

Winged Liberty Head or "Mercury": 1916-S, 1917-S, 1918-S, 1919-S, 1920-S, 1923-S, 1924-S, 1925-S, 1926-S, 1927-S, 1928-S, 1929-S, 1930-S, 1931-S, 1935-S, 1936-S, 1937-S.

Twenty-Cent Pieces

The 1875-S twenty-cent piece was the largest mintage issue by far. Most if not all were distributed on the West Coast. The public confused the coins with quarters, and the mintage in San Francisco was discontinued. Small mintages took place in Philadelphia and Carson City (most of which were melted). Today the 1875-S is the most available issue of this short-lived denomination.

DESIGN TYPE AND DATE MINTED

Liberty Seated: 1875-S.

Quarter Dollars

Quarters of the Liberty Seated, Barber, Standing Liberty, and Washington types were struck at the San Francisco Mint intermittently. Certain of these were later recognized as numismatic rarities, the 1878-S, 1901-S, and 1918-S, 8 Over 7, being leading examples.

DESIGN TYPES AND DATES MINTED

Liberty Seated, Variety 5, Motto Above Eagle, Arrows at Date, Increased Weight: 1874-S. *Arrows Removed, Weight Remains:* 1875-S, 1876-S, 1877-S, 1878-S, 1891-S.

Barber or Liberty Head: 1892-S, 1893-S, 1894-S, 1895-S, 1896-S, 1897-S, 1898-S, 1899-S, 1900-S, 1901-S, 1902-S, 1903-S, 1905-S, 1907-S, 1908-S, 1909-S, 1911-S, 1912-S, 1913-S, 1914-S, 1915-S.

Standing Liberty, Variety 1, No Stars Below Eagle: 1917-S.

Standing Liberty, Variety 2, Stars Below Eagle, Pedestal Date: 1917-S; 1918-S, 8 Over 7; 1918-S; 1919-S; 1920-S; 1923-S; 1924-S.

Standing Liberty, Variety 2, Stars Below Eagle, Recessed Date: 1926-S, 1927-S, 1928-S, 1929-S, 1930-S.

Washington: *Silver:* 1932-S, 1935-S, 1936-S 1937-S.

Half Dollars

Liberty Seated half dollars were made in San Francisco from 1874 to 1878, after which there were enough in commerce that no others were needed (although the Philadelphia Mint had reduced production in later years). The 1878-S is a numismatic rarity. Barber half dollars were made every year from 1892 to 1915 in sufficient numbers that none are notable rarities, although some are elusive in high levels of Mint State. Walking Liberty Half dollars were produced from 1916 to 1921, then intermittently to 1933, after which time they were made every year through 1937. Among these the 1921-S, although the highest production that year of the three mints, is recognized as the rarest in Mint State.

DESIGN TYPES AND DATES MINTED

Liberty Seated, Variety 5, Motto Above Eagle, Arrows at Date, Reduced Weight: 1874-S. *Arrows Removed, Weight Remains:* 1875-S, 1876-S, 1877-S, 1878-S.

Barber or Liberty Head: 1892-S, 1893-S, 1894-S, 1895-S, 1896-S, 1897-S, 1898-S, 1899-S, 1900-S, 1901-S, 1902-S, 1903-S, 1904-S, 1905-S, 1906-S, 1907-S, 1908-S, 1909-S, 1910-S, 1911-S, 1912-S, 1913-S, 1914-S, 1915-S.

Liberty Walking: 1916-S, 1917-S, 1918-S, 1919-S, 1920-S, 1921-S, 1923-S, 1927-S, 1928-S, 1929-S, 1933-S, 1934-S, 1935-S, 1936-S, 1937-S.

Silver Dollars

Morgan dollars were struck every year from 1878 to 1904 and again in 1921. The 1893-S is recognized as the key variety among San Francisco dates, especially in high grades. Peace silver dollars were made from 1922 to 1928 and again in 1934 and 1935. In Mint State the 1934-S is the most elusive variety.

DESIGN TYPES AND DATES MINTED

Morgan: 1878-S; 1879-S; 1880-S,
80 Over 79; 1880-S, 8 Over 7;
1880-S; 1881-S; 1882-S; 1883-S;
1884-S; 1885-S; 1886-S; 1887-S;
1888-S; 1889-S; 1890-S; 1891-S;
1892-S; 1893-S; 1894-S; 1895-S;
1896-S; 1897-S; 1898-S; 1899-S;
1900-S; 1901-S; 1902-S; 1903-S;
1904-S; 1921-S.

Peace: 1922-S, 1923-S, 1924-S,
1925-S, 1926-S, 1927-S,
1928-S, 1934-S, 1935-S.

Trade Dollars

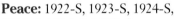

Trade dollars were made in quantity each year for export to China. All are readily collectible today.

DESIGN TYPE AND DATES MINTED

Trade Dollar: 1874-S, 1875-S,
1876-S, 1877-S, 1878-S.

$2.50 Quarter Eagles

Liberty Head quarter eagles were struck from 1875 to 1879.

DESIGN TYPE AND DATES MINTED

Liberty Head: 1875-S, 1876-S, 1877-S, 1878-S, 1879-S.

$5 Half Eagles

Half eagles of the Liberty Head design with IN GOD WE TRUST on the reverse were made from 1874 to 1906. Indian Head half eagles were struck each year from 1908 to 1916.

DESIGN TYPES AND DATES MINTED

Liberty Head, Variety 2, Motto Above Eagle:
1874-S; 1875-S; 1876-S; 1877-S; 1878-S; 1879-S;
1880-S; 1881-S; 1882-S; 1883-S; 1884-S; 1885-S;
1886-S; 1887-S; 1888-S; 1892-S; 1893-S; 1894-S;
1895-S; 1896-S; 1897-S; 1898-S; 1899-S; 1900-S;
1901-S, Final 1 Over 0; 1901-S; 1902-S; 1903-S;
1904-S; 1905-S; 1906-S.

Indian Head: 1908-S, 1909-S, 1910-S, 1911-S, 1912-S, 1913-S, 1914-S, 1915-S, 1916-S.

$10 Eagles

Liberty Head eagles with IN GOD WE TRUST were struck each year from 1874 to 1907. Indian Head eagles of the Saint-Gaudens design were made each year from 1908 to 1916 and again in 1920 and 1930. The last two dates are rare today.

DESIGN TYPES AND DATES MINTED

Liberty Head, Motto Above Eagle: 1874-S,
1876-S, 1877-S, 1878-S, 1879-S, 1880-S,
1881-S, 1882-S, 1883-S, 1884-S, 1885-S,
1886-S, 1887-S, 1888-S, 1889-S, 1892-S,
1893-S, 1894-S, 1895-S, 1896-S, 1897-S,
1898-S, 1899-S, 1900-S, 1901-S, 1902-S,
1903-S, 1905-S, 1906-S, 1907-S.

**Indian Head, Variety 2, Motto on
Reverse:** 1908-S, 1909-S, 1910-S, 1911-S,
1912-S, 1913-S, 1914-S, 1915-S, 1916-S,
1920-S, 1930-S.

$20 Double Eagles

Liberty Head double eagles of the with-motto design and with the denomination as TWENTY D. were struck from 1874 to 1876, followed by coins with TWENTY DOLLARS from 1877 to 1907. Each of these dates was made in quantity, some more than others.

Saint-Gaudens double eagles were made each year from 1908 to 1916, the 1908-S being numismatically scarce. Coinage resumed with the rare 1920-S, then 1924-S through 1927-S made in quantity, but then they were mostly exported or melted. The 1930-S was the last mintage, a rarity today.

DESIGN TYPES AND DATES MINTED

Liberty Head, Motto Above Eagle, Value TWENTY D.: 1874-S, 1875-S, 1876-S.

Liberty Head, Motto Above Eagle, Value TWENTY DOLLARS: 1877-S, 1878-S, 1879-S, 1880-S, 1881-S, 1882-S, 1883-S, 1884-S, 1885-S, 1887-S, 1888-S, 1889-S, 1890-S, 1891-S, 1892-S, 1893-S, 1894-S, 1895-S, 1896-S, 1897-S, 1898-S, 1899-S, 1900-S, 1901-S, 1902-S, 1903-S, 1904-S, 1905-S, 1906-S, 1907-S.

Saint-Gaudens, With Motto IN GOD WE TRUST: 1908-S, 1909-S, 1910-S, 1911-S, 1913-S, 1914-S, 1915-S, 1916-S, 1920-S, 1922-S, 1924-S, 1925-S, 1926-S, 1927-S, 1930-S.

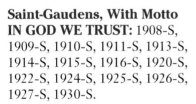

THIRD SAN FRANCISCO MINT (1937 TO DATE)

THE NEW MINT

The need and utility for a new Mint building with modern facilities was realized in the 1930s, the Depression era in which the government encouraged municipal and federal construction projects. A site on a rocky hillside at Hermann and Buchanan streets was chosen. After about two years of work, the imposing granite structure with 33,000 square feet of floor space was officially dedicated on the afternoon of May 15, 1937. Mint Director Nellie Tayloe Ross and San Francisco Mint Superintendent Peter J. Haggerty welcomed several thousand guests, many of whom toured the new facilities which resembled a fortress when viewed from a distance.

On the first floor of the million-dollar structure were the two largest vaults of reinforced concrete two feet thick and with doors weighing 22 tons, wired with microphones to transmit any sounds to a security center. Prior to May 15 about $595 in silver and gold had been transferred from the old Mint at Fifth and Mission Streets. The coining presses and related equipment were waiting to be moved.

In 1949 there were six machines available to roll ingots into planchet strips. Twenty-three presses were on line, the oldest of which dated back to 1868, but was now operated by electricity. The typical press could stamp about 5,000 coins per hour. At the time that a contingent of American Numismatic Association members visited there were 18 presses in operation turning out about 1,500,000 coins in an eight-hour shift. The pressure on each press was adjustable. The smallest-diameter coin, the dime, required 40 tons, the largest made at the time, the half dollar, 90 tons. Silver dollars, which had not been struck since 1935, required 110 tons.[8]

The new Mint was in constant production of coins, including occasional commemorative issues, through 1955. In that year only cents and dimes were made. The Treasury Department felt that the Denver Mint could handle coinage requirements for the West Coast, and the presses were stilled. Coinage equipment and related machinery were removed, and space was remodeled for the use of other government agency. A small space was reserved for assaying silver and gold deposits.

The third San Francisco Mint of 1937, the facility in use today.

THE SAN FRANCISCO ASSAY OFFICE

Beginning on August 1, 1958, the name of the facility was changed to the San Francisco Assay Office. The title of the person administrating the production was changed from superintendent to officer in charge. John F. Brekle was the first in this post and served from that date until October 1969.

In 1964 with an impending coin shortage the Assay Office was fitted with equipment to produce planchets for precious-metal denominations (changed to clad in 1965). Planchets were also made for cents and nickels. In 1965 this activity began under the Coinage Act of July 23, 1965. Planchets for cents and nickels were shipped to the Denver Mint to be coined. On September 1, 1965, the Assay Office began striking Lincoln cents. Mint Director Eva Adams was on hand and pushed a button on one of four presses to begin the coinage. Adams felt that numismatists were responsible as a coin shortage developed and decided to punish them by not allowing mintmarks on San Francisco or Denver coins, a policy continued through 1967 (accordingly, branch-mint coins of these years are listed under Philadelphia Mint totals in *A Guide Book of United States Coins*). The Assay Office struck more than 15 million 1964 mintmarkless silver quarters in 1965 and another 4,640,865 early in 1966.

In 1965 the staff of the Assay Office was increased, eventually rising to about 300 from just 35 on hand before coinage was resumed. From the Denver Mint ten coin presses were brought to San Francisco and installed. In that year cents, nickels, dimes, and quarters were struck there. Special Mint Sets, as they were called, with special satiny finish were made there from 1965 to 1967, still without mintmarks. In addition, millions of coins were made for foreign countries under contract.

The congressional Act of June 24, 1967, provided for the restoration of mintmarks beginning in 1968. In that year the coinage also included 1968-S Proof sets. From then through today, Proof sets have been made there instead of in Philadelphia.

The S mintmark was used on circulating cents and nickels through 1973. These were made in relatively low quantities, causing many to be hoarded. Mary Brooks, Mint director at the time, mandated that mintmarks be removed to prevent this.

SAN FRANCISCO MINT NAME RESTORED

On March 31, 1988, the facility regained its traditional name, the San Francisco Mint. In 1992 a number of Proof 1992-S half dollars turned up in general circulation, including in Las Vegas casinos. An investigation found that a San Francisco Assay Office police officer had been stealing commemorative and Proof coins from at least December 1990. He was arrested on April 3 and pleaded guilty.

In recent decades the production at the San Francisco Mint has emphasized coins for collectors. These have included Proofs, quarters of the various series from State (1999) onward, and commemoratives. None have been made for general circulation.

THE SAN FRANCISCO MINT TODAY

The City of San Francisco bought the structure of the Old Mint from the federal government in 2003 for the sum of $1 (paid for with an 1879 Morgan dollar struck at the mint). The plan was to reopen the facility as a historical center to be called the San Francisco Museum at the Mint. Work began in the fall of 2005, with ground broken for the renovation of the mint's central court into a galleria enclosed by glass.

Two commemorative coins (a silver dollar and a gold $5 piece) minted in 2006 raised money for the "San Francisco Museum and Historical Society for rehabilitating the Historic Old Mint as a city museum and an American Coin and Gold Rush Museum." A surcharge of $10 for each silver dollar sold, and $35 for each gold coin, went to this cause. Some 67,100 Uncirculated silver dollars and 160,870 Proofs were distributed, along with 17,500 Uncirculated and 44,174 Proof gold pieces. Representative Nancy Pelosi, who along with Mike Castle of Delaware authored the House legislation to issue the coins, said, "The Old Mint is a key part of San Francisco's history and along with history comes a sacred responsibility to preserve and protect it for future generations. Just as the Old Mint's vaults protected millions of dollars in government gold, these commemorative coins will help preserve our Granite Lady. . . . Today [the first day of striking] begins a bright new future for the Old Mint."

Erik Christoffersen, executive director of the San Francisco Museum and Historical Society, expressed the group's ambitions. "The Granite Lady will be restored to its original 1874 state of glory, and the entire space will be open to the public for the first time in the building's history. It will be the home to the first dedicated museum where San Franciscans, Bay Area residents, and visitors from around the world can come to learn how San Francisco and the surrounding Bay Area came to be what it is today."

In 2011 the first phase of renovations was completed and the Historical Society started raising money for the second phase, intended to include permanent exhibitions. Mustering the estimated $60 million needed for the renovation proved difficult. In 2014 the Historical Society was evicted by the City of San Francisco for failure to make progress on the rehabilitation project. The city closed the Old Mint in March 2015 and later in the year selected a new tenant: local production company Activate San Francisco Events. "The city will waive the $25,000 a month rent in exchange for the company spending money on the Old Mint's upkeep and maintenance," reported Emily Green in *SFGate*, "including fixing its broken elevator, installation of new lighting, and redoing the landscaping." In March 2016 the Old Mint was reopened with "San Francisco History Days," an event that attracted 6,500 visitors who came to learn from community historians, archivists, genealogists, and other researchers and enthusiasts from 60 historic organizations. The Old Mint will continue to be used for special events, some private and some public, until a new permanent tenant is found.

Uncirculated 2015 quarters struck at the San Francisco Mint. These are circulation strikes (not Proofs). The San Francisco coins are scarce compared to those of Philadelphia and Denver.

Two 2015 quarters struck at the San Francisco Mint: an Uncirculated circulation strike, and a Proof. Note their S mintmarks.

Meanwhile, the "new" San Francisco Mint at 155 Hermann Street continues as an active production facility. It does not accommodate visitors with public tours, as all its space is needed for personnel and machinery. "The United States Mint at San Francisco plays an important role in our nation's coinage," the Mint's literature states. "Although it does not currently produce circulating coins, it is the exclusive manufacturer of regular Proof and Silver Proof coin sets that set the standard for numismatic excellence with their brilliant artistry, fine craftsmanship, and enduring quality."

In 2012 the San Francisco Mint started production of Uncirculated S-mintmark quarters in relatively small quantities. This coinage has continued every year since. The coins, of each year's normal America the Beautiful ("National Park") designs, are made for collectors and not for regular circulation; they can be purchased directly from the Mint for a premium above face value. Unlike the Uncirculated S-mintmark Bicentennial quarters of 1976, which contained silver and were sold only in packaged sets, S-Mint America the Beautiful quarters are of normal copper-nickel composition and are sold in bags and rolls. Thus, the 2012-S coins are considered the first circulation-strike quarters made at San Francisco since 1954. As an example of their small mintages, approximately 745,000 circulation-strike 2015 Saratoga National Historical Park quarters were made in San Francisco, compared to 223,000,000 in Philadelphia and 215,800,000 in Denver.

April 2015: Q. David Bowers with Larry Eckerman, at that time the facility's plant manager (since retired), in front of a Grabener press making Uncirculated S-Mint Saratoga National Historical Park quarters. A bit of San Francisco Mint humor in the background—a hand-cranked telephone with a sign: "Stamping Press Repair Mechanics Direct Bell Line."

Q. David Bowers with Michael Levin, inventory manager and historian of the San Francisco Mint.

Mint workers wiping down blanks for Proof Native American dollars, in the early stages of planchet preparation. Their work is quick, skillful, and efficient. The mint has its own laundry facilities to clean the massive number of towels used in this operation.

Native American dollar planchets waiting to be struck into coins.

A San Francisco Mint technician working with Proof 2015 Native American dollars.

Proof Kennedy half dollars struck at the San Francisco Mint.

A blister pack for a San Francisco Proof set, before its final assembly.

COINAGE OF THE THIRD SAN FRANCISCO MINT, 1937 TO DATE

This coinage includes some mintmarkless issues produced under the title of the San Francisco Assay Office in the early 1960s as described previously. Of the coinage from 1937 to date all issues are readily collectible, although a few are elusive in higher levels of Mint State. Coinage of circulating pieces ended in 1955. When production of mint-marked coins began again in 1968, mintages were mostly of coins for collectors, with a few exceptions.

Cents

All dates from 1937 onward are readily collectible today.

DESIGN TYPES AND DATES MINTED

Lincoln, Wheat Ears Reverse: *Variety 1, Bronze:*
1937-S, 1938-S, 1939-S, 1940-S, 1941-S, 1942-S, 1944-S,
1945-S, 1946-S, 1947-S, 1948-S, 1949-S, 1950-S, 1951-S,
1952-S, 1953-S, 1954-S, 1955-S. *Variety 2, Zinc-Coated Steel:* 1943-S.

Lincoln, Memorial Reverse: *Copper Alloy:* 1968-S,
1969-S, 1970-S, 1971-S, 1972-S, 1973-S, 1974-S, 1975-S,
1976-S, 1977-S, 1978-S, 1979-S, 1980-S, 1981-S, 1982-S.
Copper-Plated Zinc: 1983-S, 1984-S, 1985-S, 1986-S,
1987-S, 1988-S, 1989-S, 1990-S, 1991-S, 1992-S, 1993-S,
1994-S, 1995-S, 1996-S, 1997-S, 1998-S, 1999-S, 2000-S, 2001-S, 2002-S, 2003-S,
2004-S, 2005-S, 2006-S, 2007-S, 2008-S.

Lincoln, Bicentennial: *Birth and Early Childhood,
Formative Years, Professional Life, Presidency:* 2009-S.

Lincoln, Shield Reverse: 2010-S, 2011-S, 2012-S, 2013-S,
2014-S, 2015-S, 2016-S.

Nickel Five-Cent Pieces

All dates from 1937 onward are readily collectible today.

DESIGN TYPES AND DATES MINTED

Jefferson: *Nickel:* 1938-S, 1939-S, 1940-S, 1941-S, 1946-S, 1947-S, 1948-S, 1949-S, 1951-S, 1952-S, 1953-S, 1954-S. *Designer's initials (FS) added:* 1968-S, 1969-S, 1970-S, 1971-S, 1972-S, 1973-S, 1974-S, 1975-S, 1976-S, 1977-S, 1978-S, 1979-S, 1980-S, 1981-S, 1982-S, 1983-S, 1984-S, 1985-S, 1986-S, 1987-S, 1988-S, 1989-S, 1990-S, 1991-S, 1992-S, 1993-S, 1994-S, 1995-S, 1996-S, 1997-S, 1998-S, 1999-S, 2000-S, 2001-S, 2002-S, 2003-S. *Wartime Silver Alloy, Large Mintmark:* 1942-S, 1943-S, 1944-S, 1945-S.

Westward Journey: *Peace Medal, Keelboat:* 2004-S.

Westward Journey, Modified Portrait: *American Bison, Ocean in View:* 2005-S.

Jefferson Modified: 2006-S, 2007-S, 2008-S, 2009-S, 2010-S, 2011-S, 2012-S, 2013-S, 2014-S, 2015-S, 2016-S.

Dimes

All dates from 1937 onward are readily collectible today. Silver coins struck from 1992 to date are Proof strikes.

DESIGN TYPES AND DATES MINTED

Winged Liberty Head or "Mercury": 1937-S, 1938-S, 1939-S, 1940-S, 1941-S, 1942-S, 1943-S, 1944-S, 1945-S

Roosevelt: *Silver:* 1946-S, 1947-S, 1948-S, 1949-S, 1950-S, 1951-S, 1952-S, 1953-S, 1954-S, 1955-S, 1992-S, 1993-S, 1994-S, 1995-S, 1996-S, 1997-S, 1998-S, 1999-S, 2000-S, 2001-S, 2002-S, 2003-S, 2004-S, 2005-S, 2006-S, 2007-S, 2008-S, 2009-S, 2010-S, 2011-S, 2012-S, 2013-S, 2014-S, 2015-S, 2016-S. *Clad:* 1968-S, 1969-S, 1970-S, 1971-S, 1972-S, 1973-S, 1974-S, 1975-S, 1976-S, 1977-S, 1978-S, 1979-S, 1980-S, 1981-S, 1982-S, 1983-S, 1984-S, 1985-S, 1986-S, 1987-S, 1988-S, 1989-S, 1990-S, 1991-S, 1992-S, 1993-S, 1994-S, 1995-S, 1996-S, 1997-S, 1998-S, 1999-S, 2000-S, 2001-S, 2002-S, 2003-S, 2004-S, 2005-S, 2006-S, 2007-S, 2008-S, 2009-S, 2010-S, 2011-S, 2012-S, 2013-S, 2014-S, 2015-S, 2016-S.

Quarter Dollars

All dates from 1937 onward are readily collectible today. The State and later issues are particularly interesting and are inexpensive. Silver coins struck from 1992 to date are Proof strikes.

DESIGN TYPES AND DATES MINTED

Washington: *Silver:* 1937-S, 1938-S, 1939-S, 1940-S, 1941-S, 1942-S, 1943-S, 1944-S, 1945-S, 1946-S, 1947-S, 1948-S, 1950-S, 1951-S, 1952-S, 1953-S, 1954-S, 1992-S, 1993-S, 1994-S, 1995-S, 1996-S, 1997-S, 1998-S. *Clad:* 1968-S, 1969-S, 1970-S, 1971-S, 1972-S, 1973-S, 1974-S, 1977-S, 1978-S, 1979-S, 1980-S, 1981-S, 1982-S, 1983-S, 1984-S, 1985-S, 1986-S, 1987-S, 1988-S, 1989-S, 1990-S, 1991-S, 1992-S, 1993-S, 1994-S, 1995-S, 1996-S, 1997-S, 1998-S.

Washington, Bicentennial: *Copper-Nickel Clad:* 1776–1976-S [1975, 1976]. *Silver Clad:* 1776–1976-S [1976].

State Quarters (5 designs per year): *Clad:* 1999-S, 2000-S, 2001-S, 2002-S, 2003-S, 2004-S, 2005-S, 2006-S, 2007-S, 2008-S. *Silver:* 1999-S, 2000-S, 2001-S, 2002-S, 2003-S, 2004-S, 2005-S, 2006-S, 2007-S, 2008-S.

District of Columbia and Territories Quarters Series (6 designs): *Clad:* 2009-S. *Silver:* 2009-S.

America the Beautiful (5 designs per year): *Clad:* 2010-S, 2011-S, 2012-S, 2013-S, 2014-S, 2015-S, 2016-S. *Silver:* 2010-S, 2011-S, 2012-S, 2013-S, 2014-S, 2015-S, 2016-S.

Half Dollars

All dates from 1937 onward are readily collectible today. Kennedy half dollars were mostly hoarded, then later made only for numismatic sale. Silver coins struck from 1992 to date are Proof strikes. In 2014 Kennedy dollars were struck in both standard Proof finish and an Enhanced Proof finish.

Design Types and Dates Minted

Liberty Walking: 1937-S, 1939-S, 1940-S, 1941-S, 1942-S, 1943-S, 1944-S, 1945-S, 1946-S.

Franklin: 1949-S, 1951-S, 1952-S, 1953-S, 1954-S.

Kennedy: *Silver Clad:* 1968-S, 1969-S, 1970-S. *Copper-Nickel Clad:* 1971-S, 1972-S, 1973-S, 1974-S, 1977-S, 1978-S, 1979-S, 1980-S, 1981-S, 1982-S, 1983-S, 1984-S, 1985-S, 1986-S, 1987-S, 1988-S, 1989-S, 1990-S, 1991-S, 1992-S, 1993-S, 1994-S, 1995-S, 1996-S, 1997-S, 1998-S, 1999-S, 2000-S, 2001-S, 2002-S, 2003-S, 2004-S, 2005-S, 2006-S, 2007-S, 2008-S, 2009-S, 2010-S, 2011-S, 2012-S, 2013-S, 2014-S, 2015-S, 2016-S. *Silver:* 1992-S, 1993-S, 1994-S, 1995-S, 1996-S, 1997-S, 1998-S, 1999-S, 2000-S, 2001-S, 2002-S, 2003-S, 2004-S, 2005-S, 2006-S, 2007-S, 2008-S, 2009-S, 2010-S, 2011-S, 2012-S, 2013-S, 2014-S, 2015-S, 2016-S.

Kennedy, Bicentennial: *Copper-Nickel Clad:* 1776–1976-S [1975, 1976]. *Silver Clad:* 1776–1976-S [1976].

Dollars

All dates from 1971 intermittently onward are readily collectible. The later Native American and Presidential dollars have the S mintmark "hidden" on the edge and not easily viewable, if at all, when in an album or holder. Some Presidential dollars were struck in both standard Proof finish and Reverse Proof finish.

Design Types and Dates Minted

Eisenhower, Eagle Reverse:
Copper-Nickel Clad: 1971-S, 1972-S, 1973-S, 1974-S, 1977-S, 1978-S. *Silver Clad:* 1971-S, 1972-S, 1973-S, 1974-S.

Eisenhower, Bicentennial:
Copper-Nickel Clad: 1776–1976-S [1975, 1976]. *Silver Clad:* 1776–1976-S [1976].

Susan B. Anthony: 1979-S, 1980-S, 1981-S.

Sacagawea: 2000-S, 2001-S, 2002-S, 2003-S, 2004-S, 2005-S, 2006-S, 2007-S, 2008-S.

Native American (new reverse yearly):
2009-S, 2010-S, 2011-S, 2012-S, 2013-S, 2014-S, 2015-S, 2016-S.

Presidential (4 designs per year): 2008-S, 2009-S, 2010-S, 2011-S, 2012-S, 2013-S, 2014-S, 2015-S, 2016-S.

CARSON CITY MINT, 1870-1893

SILVER IN THE MOUNTAINS

GOLD AS THE REASON FOR BRANCH MINTS

In America in 1859 there was a federal mint operating in Philadelphia, where all dies were made and business matters were coordinated, and in branches in Charlotte, Dahlonega, New Orleans, and San Francisco. Of the branches, all but New Orleans were located near sources of silver or gold.

The monetary system was on a *de facto* gold standard, where such coins being of full weight and value in bullion content are equal to the denomination in dollars. Silver coins had been worth their melt-down value, more or less, from the first coinage in the

The Carson City Mint. (*Frank Leslie's Illustrated Newspaper*, February 23, 1878)

early 1790s until vast discoveries of gold in California from 1848 onward made gold more plentiful in world markets. Silver, which had been on the approximate ratio that 1 ounce in gold was equal in value to 15.5 ounces of silver, began to rise in value in 1849. By 1850 it took more than a dollar's worth of metal to make a silver dollar, and as such, silver coins in circulation were taken out of commerce by speculators and exchange houses to melt. This was remedied by the Act of February 21, 1853, in which the silver content of coins from the half dime to the half dollar was reduced. After that time it was not profitable to melt such coins.

Silver dollars remained at their earlier weight and were used as bullion coins, mostly in the export trade. A depositor of silver desiring dollars in return would deliver, say, $1.03 worth of metal to one of the mints. He would then receive a dollar in return that was valued at $1.03 or more when exported.

The discoveries of silver and gold ore were always a boon to commerce, as earlier related for the gold "rushes" to North Carolina, Georgia, and California. By 1859 there had been no silver ore discoveries to the extent that commerce would be affected.

SILVER DISCOVERIES

By 1859 the western part of the Utah Territory had been extensively traveled, starting in a large way when "Forty Niners" crossed the alkali plains, and then headed into the Sierra-Nevada on their way to California. In the early days the route was dangerous, and many wagons were abandoned and people perished. By 1859 travel was more routine, with improved routes and trails and sellers of goods and provisions along the way. In the course of these journeys several finds of gold had been found, including in Gold Cañon on the Walker River near the border with California, which was worked for several years. However, no large-scale production ensued.

As to the discoverers of the silver ore deposits on the western slope of Mount Davidson in the western Utah Territory, what became known in history as the Comstock Lode, there is no clear record. Ethan Allen Grosh and his brother, Hosea Ballou Grosh, trained mineralogists who had experience in the California Gold Rush, are candidates. After their find in 1857, Hosea died. Allen, accompanied by an associate, Richard Maurice Bucke, went westward to California with ore samples hoping to find investors to develop the claim. Another associate, Henry T.P. Comstock, remained in a cabin on site to guard the area against intruders. Allen died of hypothermia on December 19, 1857. Bucke suffered from frostbite, went through surgery by an amateur physician, and continued to California, never to return to the site.

Upon learning of the death of the Grosh brothers, Comstock claimed the find as his own, although he was not an expert in mineralogy and others were digging in the area. Comstock was the first to file a formal claim on the Grosh land as well as on nearby land. On January 28, 1859, four men filed a claim on an outcropping on Gold Hill that had traces of gold and silver, while other miners also made discoveries. Comstock made threats of prior claims and was able to arrange a worthwhile interest in several properties. On June 12, a major strike of manganese sand with gold and bluish-gray quartz was made. Upon assay on June 27 it was found to be rich with silver sulfides. The ore was about 75-percent silver and 25-percent gold.

Without knowing the full extent of the deposit, most of the investors including Comstock sold their interests. Patrick McLaughlin had a 1/6 stake that he sold to George Hearst, who made a fortune. His son, William Randolph Hearst, became the most prominent newspaper publisher in America. Comstock went on to operate two retail stores that became losing propositions.

After the Comstock Lode was widely publicized it launched a great rush of prospectors seeking to establish claims—some coming from California where by that time the gold-mining industry was mainly in the hands of corporations, and others coming from the East.

The district developed rapidly. Virginia City became the epicenter of activity. Mining supplies, consumer goods, and other necessities were brought from California on the backs of mules and by wagons. Mining was deep-rock by shafts and tunnels. Problems included flooding of lower levels. In 1861 Nevada Territory was carved out of the western reaches of Utah Territory.

THE CALL FOR A NEVADA MINT

The Committee of Ways and Means in Congress on January 26, 1863, ordered this to be printed:

> *The Committee of Ways and Means, to whom was referred that part of the report of the Secretary of the Treasury commending "to the consideration of Congress the expediency of establishing an assay office or branch mint at some convenient point in Nevada Territory;" also House bill No. 663, entitled "An act to establish a branch mint of the United States in Nevada Territory," respectfully report:*
>
> That, in the opinion of the committee, there is an urgent necessity for a branch mint in said Territory. From the report of the Secretary of the Interior, in answer to a resolution of Congress calling for information upon the mineral resources of that Territory, it appears that the gold and silver mines there are found stretching from the Washoe, in the southwestern part of Nevada, across the Territory from west to east, and radiating in northerly and southerly directions from the great central discoveries of Washoe. The principal clusters of gold and silver discoveries are in the counties of Washoe, Storey, Lynn, and Ormsby, surrounding Carson City, the capital of the Territory. Extending south some thirty or forty miles from Virginia City and Gold Hill, between the forks of Carson River, has been discovered what is known as the "Silver Mountain." About ten miles have been staked off, in which it is said that ore has been discovered, running to a great depth, and rivalling in richness that found in any other district on the eastern slope. Still further south, on the western side of the Territory, the Esmeralda and Van Horn mining districts are found. Here the discoveries of ore have warranted expensive processes of working. Returning to the central discoveries around Carson, and travelling due east across the Territory, at a distance of about 177 miles from Carson, near the extreme eastern part of Nevada, the Reese river mining district is found, and still further is the Simpson Park district, both abounding in rich discoveries of gold and silver ore. Near the geographical centre of the Territory, and surrounding it, are the Humboldt, Eldorado, Echo, Sacramento, and several other mining districts. The Humboldt, especially, is believed to be of great richness in mineral product.
>
> In August last it was estimated that there were one hundred and forty mills in operation in the Territory, and it is fair to presume that at the present time the number has been increased to two hundred. The discoveries of the precious metals in Nevada warrant the belief that in a few years it will in that respect surpass even the "golden State." The first discovery of rich quartz leads in this Territory was in June, 1859. The fact was not generally known until September of the same year. A population of some thirty or forty thousand now people the Territory, and millions of dollars have been expended in prospecting and working the mines, erecting

quartz mills, with massive reducing machinery and ingenious separating processes. Several thriving towns and numerous villages have sprung up. The great business of the Territory is and must be mining.

A very small portion of the soil is adapted to agriculture. All the supplies for the western part of the Territory are drawn from California. The most valuable mines are veins or loads of quartz, generally dipping into the earth at an angle of about forty-five degrees, from four to eight feet in width, frequently containing from one to three veins of rich ore, varying in width from one to twenty feet. Such are found in the Comstock mine, at Virginia City, and at Gold Hill. These mines are found by the practical and energetic "prospectors," and they are generally well paid for their labor. Shafts sunk or tunnels run, to strike deeper in such places, develop richer products. The solid ledges generally yield richer qualities of ore, but require great labor and expense to work them. The machinery used in working the solid ledges is generally made of iron, of great strength, and is manufactured at San Francisco, and freighted thence to Sacramento in vessels, from whence it is carried overland across the mountains.

The expense of transportation is, of course, very great. It is estimated that the cost of these mills averages about $30,000. There are two exceptions, however: The mills of "Gould, Curry & Co.," and of the "Ophir Company," which it is said cost $1,000,000 each. The entire cost of the mills in use in the Territory in August last, was $6,200,000.

The mines are of a permanent character, and the richness of the veins increase as the bowels of the earth are penetrated. The rich yields will not be likely to cease so long as drainage can be effected by pumping or tunneling.

The estimated yield of gold and silver in the Territory at the present time, is $2,000,000 per month, with a constant prospect of an increase, as new machinery is put in operation.

There seems to be a necessity for the establishment of a mint in a region so central and so rich in the precious metals. An additional reason may be urged from the peculiar situation of the Territory, and its dependence on distant markets, from whence all its supplies are drawn. All the supplies of food, clothing, machinery, &c., are imported into the Territory. But little, save the precious metals, is produced in the Territory. Payment for these supplies, and the necessary transportation, is made in the Territory.

The cost of transporting the bullion from the mines to California is from five to six percent. The returns are received in about thirty days, with an additional cost of two percent in carrying back the coin. The transportation of silver is still more expensive. The bullion used in commerce can be shipped abroad in that shape as well, if not better, than in coin. The coinage of the amount that would find its way out of the Territory, in payment of supplies and transportation, it is believed, will reach at least half a million per annum. Notwithstanding this yield of precious metals, the people of the Territory are barely paying their way. A large proportion of the population is yet engaged in prospecting and opening the mines, and getting up the necessary machinery to work them. The whole country will be greatly benefited by a speedy development of these mines. A market will at once be opened for all kinds of manufactured goods. Very many of the miners are compelled to have their stock worked in mills doing custom work. They sell much of their bullion to dealers, at a discount, to enable them to raise means to procure their own machinery. They cannot wait until it is carried to San Francisco and returned.

The business taken to San Francisco is beyond the capacity of the Mint there, by reason of the new discoveries.

This is more especially the case in the article of silver. The mines of Oregon and Washington Territory have largely increased the work of the Mint at San Francisco, and it is believed that a large appropriation must be made to increase the capacity of the Mint at San Francisco, or a new mint established to meet the events of the Pacific coast and the interior country. The latter is recommended.

The same report noted:

Allowing 18 days as the average time of a trip, and the number of teams and teamsters employed, amounts to 2,772, and of animals, 14,652. At the present date, October 22, 1862, the price of freight is seven to eight cents per pound.

Estimating the yearly average of freight over the Placerville road to be 120 *tons per day*, at an average price of *six cents per pound, and the total amount paid for freight alone, amounts to* $5,256,000 upon this one road. A four horse, or mule team, which makes the trip in about sixteen days, pays for tolls $22 75; a six horse or mule team pays $30 toll. Averaging the time at eighteen days, the tolls at $25 per trip, and we find that the enormous sum of $693,000 per year is paid for tolls by freight teams. The returns show that the stages average 37 passengers per day, which, at $30 per passenger, amounts to $405,150. It is believed, however, that the total receipts of the stage line exceeds this sum. It will be observed that 68 additional travelers per day, or nearly double the number carried by stage, pass over this road, at least one-half of whom would probably take the cars were a railroad completed.

From an entirely reliable source I have ascertained that the total amount of silver bullion brought down by Wells, Fargo's Express, for the ten months of 1862, is over 150,000 pounds, and may be safely stated at 200,000 pounds for the entire year. Its value is not, of course, known, gold being mixed with it, but it is safe to estimate it at $30 per pound, or a total value of $6,000,000. This is only what comes by express, and does not indicate the amount actually taken out, and retained there, or sent down by private conveyance. It is estimated by Wells, Fargo & Co. that this amount will be doubled for the year 1863, and in 1866 reach twenty-five millions of dollars.

Carson City, from the fact that it is the central point of a vast area of mineral wealth, and that it is peopled by an industrious and loyal people, would indicate it as being the proper point for the establishment of a branch mint. In view of the facts above stated, of the recommendation of the Secretary of the Treasury, and of the united opinion of the Pacific delegation in its favor, the committee report back the bill with an amendment, and a recommendation that it do pass.

On October 21, 1863, the director of the Mint issued this statement:

Measures were taken early in the month of April last, to organize and put into operation the Branch Mint authorized by law to be established at Denver, Colorado Territory. The time required to prepare the building purchased for Mint purposes, and to have the necessary machinery, apparatus, &c., constructed in the East and transported to so distant a point, prevented the opening of the Branch Mint for business until the latter part of September, (ult.), when operations were commenced, and are now being successfully carried on. The institution is confined for the present to the melting, refining, assaying and stamping of bullion, the same being returned to the depositor in the form of unparted bars bearing the government stamp of weight and fineness. The institution will no doubt prove of great advantage to the mining and other interest of that region of our country.

Owners of the leading mines became known as "Silver Barons" and typically lived in mansions in San Francisco. In 1864 Nevada achieved statehood and became known as the Battle Born State as it was in the time of the Civil War, a name later superseded by the Silver State. In the late 1860s the Central Pacific Railroad connected Reno with points east and west, and goods arrived by wagon from Virginia City.[1] Later, the Virginia & Truckee Railroad connected Reno with Virginia City.

THE CARSON CITY MINT IN OPERATION

THE MINT AT CARSON

What was nearly always officially referred to as the Branch Mint at Carson (or less often, Carson City), later the Carson City Mint, was established by the Act of March 3, 1863, which provided also for the appointment of a superintendent at $2,000 a year and an assayer, a melter-refiner, and a coiner for $1,800 each annually. Carson was the capital of the territory, then the state.

The Comstock Lode was located approximately 15 miles away and by that time was America's richest silver bonanza. In addition, large quantities of gold were extracted from the earth in the district.

On July 17, 1866, plans for the Mint drawn by Alfred Bult Mullett, supervising architect for the Treasury Department (from 1866 to 1874), arrived in Carson City as did various other documents. Construction of the Mint commenced on September 18 with the laying of the cornerstone. Events would prove that this was far in advance of the facility being put in operation.

In the *Annual Report of the Director of the Mint*, for the fiscal year ending June 30, 1869, James Pollock noted that the branch mint at Carson was approaching completion—the machinery was nearly all in place and operations would soon begin. He stated that the peculiar characteristics of the bullion were such that the workers would have to be "practical, experienced, and scientific men" in order to refine it properly. Much of the bullion was expected to contain a mixture of silver and gold: "the bars of mixed bullion being officially stamped with both gold and silver proportions will be as salable in that form as if they were parted." Further:

> The operations of this mint will, in all probability, culminate in commercial bars, as *coin* already abounds in that region so extensively that their papers express alarm as to the prospects of redundancy. Practically it will be much more an assay office than a mint, and as such, fully meet the wants of the district. The power to make coin may be of occasional benefit: perhaps, in the future, at much advantage.

The coins in circulation in Nevada at that time were mostly from the San Francisco Mint.

An early account told of the building's specifications and use:

> Granite from the prison stone quarry. Pict style of architecture. Portico, Ionic. Hall, 12 feet in width; main hall 12x40; on the right of the entrance.
> Paying teller's office, 13x16 feet. Coining room, 19x19. Spiral staircase conducts above. Whitening room 10x14.5, with a vault in solid masonry 5x6. Annealing furnace and rolling room, 17x24. Gold and silver melting room 10x24. Melters' and refiners' office, 12x19 feet. Deposit melting room 14.5x19. Deposit weighing room,

19x19, with a strong vault 6.5x10.5 feet. Treasurer's office, 13x16, with a vault five feet square. Engine room, 16.5x53 feet. Beside which there is a cabinet, adjusting room, ladies' dressing room, humid assay room, assayer's office, assayer's room, watchman's room, two store rooms, attic, and basement. As a preventive against fire the floors are double, with an inch of mortar between. The foundations are seven feet below the basement floor and laid in concrete. Building two and a half stories high.

The machinery for the Mint arrived November 22, 1868. The Mint has a front of 90 feet on Carson Street . . .[2]

COINAGE BEGINS

The Carson City Mint was ready to do business in December 1869. Dies dated 1869 were sent by the Philadelphia Mint and were received at Carson City by October 21, 1869. How many were sent is not known—no inventory listing of 1869 and 1870 dies has been found. The 1869-dated dies were not used, and the reverses were probably held for 1870 and later use. Abram Curry, a principal in the Gould & Curry mill in Virginia City, was chosen as the first superintendent. The presses and other equipment were tested on November 1 and were found to be in good order. Difficulty occurred in filling Mint offices as well as staff positions, as salaries allowed by the Treasury Department were often less than men were already making in the region.

The first Carson City Mint coins were silver dollars minted from 1870-dated dies on February 10, 1870, a quantity of 3,747 pieces. All were struck using a press made in Philadelphia by Morgan & Orr. Each 1870-CC dollar bore the distinctive CC mintmark on the reverse. On February 11, Andrew Wright received the first delivery of CC dollars, a quantity of 2,300 coins. Wright, a watchman or guard at the mint, was undoubtedly entrusted with their safekeeping and/or paying them out. An additional three coins were saved for the Assay Commission.

The *Annual Report of the Director of the Mint,* for the fiscal year ending June 30, 1870 noted the following:

> The branch mint at Carson City, Nevada, is now in operation. In May 1869 the fitting up of this branch for business as a mint was commenced, and completed in December of that year. On the 8th of January 1870, it was opened for the reception of bullion. The superintendent [Abram Curry] in his report, says, "Since that time the business has been steadily increasing, and, with the facilities afforded other institutions of its kind, will do a large business in refining and coining. This branch mint, to make it efficient and successful, requires a bullion fund equal to the legitimate demands of business, and the just expectations of its depositors. No such fund has, as yet, been provided for the institution.

In reference to the future of this branch and its influence in developing the mineral resources of the country, the report says:

> The mining interests of the country from whence the larger portion of bullion is received, are improving rapidly: new mines are being developed and larger quantities of bullion produced as the cost of working the ore is becoming reduced. With proper arrangement and facilities afforded this branch, it will increase its business materially during the next fiscal year and give much aid in developing the mining interests of this and adjoining states.

The deposits at this branch during the year, were, gold, $124,154.44: gold coined, $110,576.05: silver deposits and purchases, $28,262.16: silver coined, $19,793.00. Total deposits and purchases, $152,416.60: total coinage, $132,369.05: total number of pieces, 38,566. The report is very encouraging, and it is earnestly desired that the present anticipations of its officers may be fully realized in the future prosperity of this branch.[3] I cannot forbear repeating the declaration made in my last annual report, that the policy of the government in relation to the development of the mineral wealth of our country should be liberal and generous.

CONTINUING COINAGE

Coins minted until the Virginia and Truckee Railroad was opened were shipped by horse-drawn wagon 30 miles over very rough roads to the railhead in Reno. Stored in cloth bags, the coins were extensively marked by the time they arrived at Reno, more so at their final destinations.

Among silver denominations dimes, quarters, and half dollars were made until 1878. Twenty-cent pieces were struck with CC mintmarks in 1875 and 1876. Liberty Seated silver dollars were made until 1873, after which there was a lapse for this denomination until Morgan dollars were coined in 1878. In the meantime trade dollars were struck continuously from 1873 to 1878.

The Carson City Mint in the 1870s. (Stereograph view No. 3477 by John Calvin Scripture)

The Carson City business district with the mint to the right.
(*Frank Leslie's Illustrated Newspaper*, February 23, 1878)

In 1870 the minting of gold commenced, limited to the higher $5, $10, and $20 denominations. For the first several years most gold coins were circulated locally and regionally, which remained true for later $5 and $10 (but not $20) coins. Many double eagles were shipped overseas.

The Carson City Mint struck silver coins from dimes to trade dollars from 1870 to 1878. From 1879 onward the only coins made of silver in Carson City were Morgan dollars. Gold $5, $10, and $20 coins were made from 1870 to 1885, after which time the presses became still. Most if not all of the coinage operations at the Carson City Mint could be performed more efficiently and at lower cost at the new San Francisco Mint, which opened in 1874. Coinage ceased, and the Carson City facility was used only for refining, assaying, and storage.

In 1880 there were three presses at the Mint. The largest was the Ajax, rated at 152 tons of pressure and used to strike silver dollars, trade dollars (lastly in 1878), and double eagles at the rate of 80 to 98 per minute. The second press was smaller and could be used for striking the larger-diameter coins as well, and in earlier times dimes to half dollars (which were last made in 1878). The smallest press could strike 140 coins per minute, but was not in use after 1878.

From 1878 to 1881 an extension was built on the rear of the Mint, but this was demolished after the facility ceased operations

THE LATER YEARS

The *Annual Report of the Director of the Mint*, 1889, commented that the business of the Carson City Mint during the *fiscal year*, which ended June 30, 1889, was confined to that of an assay office. Further:

> Since the commencement of the present fiscal year the mint at Carson has been reopened for coinage and is now in full operation. Samuel C. Wright was appointed superintendent and took charge July 1, 1889.
>
> P.B. Ellis was appointed by the President assayer, vice Joseph R. Ryan, July 1, 1889. E.B. Zabriskie was appointed melter and refiner, July 12, 1889, vice [melter and refiner] John H. Dennis. Charles H. Colburn was appointed coiner, July 1, 1889.

More was told in the *Annual Report of the Director of the Mint*, for the fiscal year that ended on June 30, 1890, an uncertain beginning for a new era of Carson City coinage:

> The mint at Carson was reopened for coinage on July 1, 1889, but, owing to the dilapidated condition in which the building and machinery was found, after four years of idleness, repairs and betterment of the building and overhauling and repairing the machinery were necessary, and consequently the coinage of gold and silver was not commenced until October 1, 1889. . . . The melter and refiner received, during the year, bullion containing 183,635.672 standard ounces of gold. He made 83 melts of gold ingots, of which 6 were condemned.
>
> He returned to the superintendent in settlement, at the close of the fiscal year, an excess of 3.322 standard ounces of gold. The same officer received, during the year, bullion containing 1,812,222.15 standard ounces of silver. He made 1,358 melts of silver ingots, of which 39 were condemned. He returned to the superintendent in settlement, at the close of the year, an excess of 921.80 standard ounces of silver.
>
> The coiner received from the superintendent 192,722.350 standard ounces of gold. There were coined in his department and delivered to the superintendent 92,460 double eagles of the value of $1,849,200, being 51.5% of good coin produced from ingots operated on. He had a gold wastage of 6.689 standard ounces.

The same officer operated upon 2,331,896 standard ounces of silver and delivered to the superintendent 1,438,000 standard silver dollars, being 54% of good coin produced from ingots operated upon. He had a silver wastage of 378.98 standard ounces.

The "new era" did not last very long, although mintage quantities of Morgan dollars and double eagles were generous, less so for half eagles and eagles.

At the Carson City Mint in 1890 three men conspired to steal silver ingots. One miscreant was James T. Jones, assistant minter and refiner. Another was James Heney, who worked with processing metal. They made their own ingots, gold-plated silver bars with a generous amount of worthless lead added, and stamped them with false information as to weight and fineness, then switched them for real gold bars. When the Mint's stock of gold ingots was tapped to process into strips for planchets the deception was discovered. Two men went to prison for eight years at hard labor and the third, name not given, had died.[4]

REQUIEM

In 1891 the average cost of minting coins at the Carson City Mint was $$0.0564+, which was more than twice the cost at the other three mints. This helps explain why the operation of the facility met with continuing criticism by the Treasury Department.

The *Annual Report of the Director of the Mint*, 1893, told of the end of coinage at the Carson City Mint:

> By direction of the secretary of the Treasury coinage operations at the mint at Carson City were suspended on June 1, 1893, and the force employed in the coiners department dispensed with. A corresponding reduction was also made in other departments of the mint. The business of the Carson mint is now conducted on the same basis as that of the Assay Office at New York: depositors of gold receiving payment either in coin or fine bars, as preferred, and of silver, in unparted or fine bars. . . .
>
> The Mint at Carson City being of limited capacity, and the amount of gold deposited and silver purchased there being small as compared with the amount of gold deposited and silver purchased at the San Francisco Mint, which possesses a large coinage capacity, the expenses for coinage were much greater at Carson than at San Francisco. . . .
>
> Upon the suspension of coinage operations at the Carson City Mint the presses and other machinery used in the coinage department were painted and leaded under the supervision of Mr. Charles H. Colburn, the retiring coiner, to prevent corrosion [reflective of humid conditions within the Mint building]. At the close of the fiscal year 1893 the bullion, coin and other moneys with which the superintendent was charged was weighed and counted by Messrs. W.E. Morgan and A.A. Hassan, of the Bureau of the Mint, and the amount found to be correct."

EPILOGUE FOR THE CARSON CITY MINT

There was a close alliance on several levels between the silver-mining interests in Virginia City and the banking and social salons of San Francisco, and the Carson City Mint was simply a nuisance. Of the various mints, it was probably the least necessary, as if it had not been constructed, the facilities of the San Francisco Mint would have served to process Comstock Lode metal, which it did in any event. Of course, today numismatists realize that this worked to their benefit, for in general Carson City issues of all silver and gold denominations have lower mintages than those struck at Philadelphia, New Orleans, and San Francisco, have a special romantic appeal, and are very interesting to collect.

From the standpoint of care and minting, a special nod must be given to the coiners at Carson City, for their silver dollars in particular, but certain other issues as well, are usually better struck, with finer detail, than are those from other minting institutions.

There was thought of additional coinage later, but that did not happen. In 1900, certain that the days of coining silver and gold had ended, dies and certain equipment were shipped to the Philadelphia Mint.

The Annual Report of the Director of the Mint, for the fiscal year ended June 30, 1907, noted that the Carson City Mint, then conducted only as an assay office, in fiscal year 1907 employed seven men and had deposits of just $12,112.28 worth of silver (coining value) and $811,415.95 worth of gold. No coins had been struck there since 1893.

In 1911 quantities of bagged silver dollars were sent to the Treasury Building in Washington for storage. In 1942 the building became the Nevada State Museum which today is a prime attraction.

COINS OF THE CARSON CITY MINT

NUMISMATIC INTEREST

In contrast to today, there was a time generations ago when mintmarked coins, including Carson City issues, attracted virtually no numismatic interest at all. The beginning coinage of Liberty Seated dollars in 1870 was not newsworthy in collecting circles. That changed in 1893 when Augustus G. Heaton, an artist, poet, writer, and one-time president of the American Numismatic Association, published *A Treatise on Mint Marks*, available for $1 by mail. This gave 17 "Points of Attractiveness" for mintmarked coins.

Heaton had no way of knowing that in 1893 the Carson City Mint would cease coinage forever. So far as is known, only one numismatist—John M. Clapp, a Pennsylvania oilman—ordered one each of the silver and gold coins minted in Carson City in the last year.[5]

It was not until well into the 1900s that there was widespread numismatic interest. By that time the opportunity to acquire high-grade coins with CC mintmarks was largely lost. Exceptions were certain Morgan silver dollars that had been stored for many years and occasionally came to light.

COINAGE OF THE CARSON CITY MINT, 1870–1893

Silver coins of Carson City range from collectible, such as certain dimes, quarters, and half dollars of the 1875 to 1877 years and Morgan silver dollars except 1879-CC and 1889-CC, to scarce or rare. In particular the issues of the first several years are rare.

Among gold coins the half eagles and eagles are mostly scarce and rare, and in high grades some dates are unknown. Double eagles are collectible, with the 1870-CC being the key issue with only a few dozen known. In mid- to high-level Mint State nearly all are rarities. The Carson City Coin Collectors of America group publishes an informative journal, *Curry's Chronicle*, named after the first superintendent. More information can be found on the Internet at www.carsoncitycoinclub.com.

Dimes

Carson City dimes of 1871 and 1872 are very rare in high grades.[6] There are no records of any having been saved for numismatic purposes at the time. Nearly all Carson City dimes show significant wear. Only a handful exist in lower ranges of Mint

State. The 1873-CC, Without Arrows, is unique and traces its modern pedigree to the Louis E. Eliasberg Collection auctioned in 1996. The 1873-CC and 1874-CC with arrows at the date are extremely rare in Mint State. The same reverse die was used to strike all Carson City dimes from 1871 through 1874.

From 1875 to the end of the run in 1878 the various issues are readily available in Mint State. The 1876 was made in two varieties: with the mintmark above the bow on the reverse and the mintmark below the bow.

DESIGN TYPES AND DATES MINTED

Liberty Seated, Variety 4, Legend on Obverse: 1871-CC, 1872-CC, 1873-CC. *Variety 5, Arrows at Date, Increased Weight:* 1873-CC, 1874-CC. *Arrows Removed, Weight Remains:* 1875-CC, 1876-CC, 1877-CC, 1878-CC.

Twenty-Cent Pieces

Produced only from 1875 to 1878, and for circulation only in one truly significant year, 1875, the twenty-cent piece is the most short-lived of all regular American coinages.

The 1875-CC coins were released into circulation locally and regionally, however, no record exists of any numismatic interest in them. In 1876 at the Philadelphia Mint 14,640 were coined and at Carson City 10,000. Many if not most of the Philadelphia coins were later melted. It is thought that the 10,000 1876-CC coins went to the melting pot, but that perhaps 20 or so were saved, possibly including pieces sent for the Assay Commission ceremony held early in 1877. It was not until later that any particular notice was given, with the New York Coin and Stamp Company auction of the Davis Collection on January 20–24, 1890, being perhaps the earliest auction appearance of note. Today only about 20 or so are known, mostly in Mint State.

DESIGN TYPE AND DATES MINTED

Liberty Seated: 1875-CC, 1876-CC.

Quarter Dollars

Quarters were made at the Carson City Mint continuously from the first year of coinage operation in 1870 to the last in 1893, except for 1874 and, later, 1886 to 1888 when no coins of any denomination were made. The total of 8,340 1870-CC quarters, the first year, was very small, indicating that no great demand was perceived for them. Most or nearly all were placed into circulation regionally. Today the 1870-CC is the second rarest Liberty Seated quarter and is virtually impossible to find in high grades. Only one Mint State coin, the Eliasberg specimen, is known of the estimated 40 to 50 of this issue known today. The 1871-CC and 1872-CC quarters are key issues and in Mint State are especially rare with just three known of the first and two of the second. The 1873-CC, No Arrows, is the rarity of rarities not only among Carson City quarters, but for the entire denomination from 1796 to date. Only six are known today from the mintage of 4,000. Quarters from 1875 through 1878 are easily enough collected in high grades, although in gem Mint State the 1875-CC is scarcer than the others.

DESIGN TYPES AND DATES MINTED

Liberty Seated, Variety 4, Motto Above Eagle: 1870-CC, 1871-CC, 1872-CC, 1873-CC. **Variety 5, Arrows at Date, Increased Weight:** 1873-CC. *Arrows Removed, Weight Remains:* 1875-CC, 1876-CC, 1877-CC, 1878-CC.

Half Dollars

Half dollars were struck at the Carson City Mint from 1870 to 1878. In 1873 No Arrows and With Arrows varieties were made. The 1870-CC half dollar, the first of that denomination and mint, was struck to the extent of 54,617 pieces, the lowest for any silver denomination. Half dollars of the early years are scarce today.

Availability changes dramatically with 1875-CC through 1877-CC, and in 1878 the mintage of Carson City half dollars fell to just 62,000.

DESIGN TYPES AND DATES MINTED

Liberty Seated, Variety 4, Motto Above Eagle: 1870-CC, 1871-CC, 1872-CC, 1873-CC. *Variety 5, Arrows at Date, Increased Weight:* 1873-CC, 1874-CC. *Arrows Removed, Weight Remains:* 1875-CC, 1876-CC, 1877-CC, 1878-CC.

Silver Dollars

Silver dollars struck in Carson City are of two types: Liberty Seated from 1870 to 1873 and Morgan from 1878 to 1889. Among the early issues the 1870-CC is seen most often, but in terms of overall American numismatics can be called scarce. The rarest is 1873-CC. Morgan design dollars were made from 1878 to 1885 and again from 1889 to 1893. The rarest issues are 1889-CC followed at a distance by 1879-CC. The 1893-CC is slightly scarce. The others are readily available in Mint State due to the dispersal of coins in the 1900s held by the Treasury.

DESIGN TYPES AND DATES MINTED

Liberty Seated, With Motto IN GOD WE TRUST: 1870-CC, 1871-CC, 1872-CC, 1873-CC.

Morgan: 1878-CC; 1879-CC;
1880-CC, 80 Over 79;
1880-CC, 8 Over 7; 1880-CC;
1881-CC; 1882-CC; 1883-CC;
1884-CC; 1885-CC; 1889-CC;
1890-CC; 1891-CC; 1892-CC;
1893-CC.

Trade Dollars

Trade dollars were struck at the Carson City Mint continuously from 1873 until 1878, when the denomination was terminated on February 28 by the Bland-Allison Act which provided for the Morgan silver dollar. All of the Carson City trade dollars are readily collectible today, although the 1878-CC is viewed as rare in any grade and very rare in Mint State.

Design Type and Dates Minted

Trade Dollar: 1873-CC,
1874-CC, 1875-CC, 1876-CC,
1877-CC, 1878-CC.

$5 Half Eagles

In Carson City the series of gold coins commenced with the half eagle in 1870. From that time until the Carson City Mint ceased production in 1893, 709,617 $5 coins were struck with a total face value of $3,548,085. These were made for local and regional use. Most gold bullion that was received at the mint was converted to double eagles as it was more economical to make these than larger numbers of $5 and $10 coins totaling the same face value.

Design Type and Dates Minted

Liberty Head, Variety 2, Motto Above Eagle:
1870-CC, 1871-CC, 1872-CC, 1873-CC, 1874-CC,
1875-CC, 1876-CC, 1877-CC, 1878-CC, 1879-CC,
1880-CC, 1881-CC, 1882-CC, 1883-CC, 1884-CC,
1890-CC, 1891-CC, 1892-CC, 1893-CC.

$10 Eagles

Similar to Carson City half eagles, eagles from this mint include many rare issues, especially among the earlier dates, and notable landmark rarities at the Mint State level. The rarest is the first, 1870-CC. For the rest of the eagle series Mint State coins range from unknown to incredibly rare to available. The 1891-CC is an exception and is actually *common* in the context of the series, with 1,500 to 1,700 Mint State coins estimated.

DESIGN TYPE AND DATES MINTED

Liberty Head, Motto Above Eagle: 1870-CC, 1871-CC, 1872-CC, 1873-CC, 1874-CC, 1875-CC, 1876-CC, 1877-CC, 1878-CC, 1879-CC, 1880-CC, 1881-CC, 1882-CC, 1883-CC, 1884-CC, 1890-CC, 1891-CC, 1892-CC, 1893-CC.

$20 Double Eagles

Among Carson City gold coins double eagles are the only denomination that can be completed in even low-level Mint State. In all grades the $20 coins are much more plentiful as vast amounts were exported to foreign countries, especially to Latin America and Europe. Beginning in the late 1940s, millions were repatriated. Most were later issues, but enough Carson City twenties were brought back that assembling a set with the early years in VF to AU grades and the later years in Mint State is easy to do. The key issue is 1870-CC with only 45 to 50 in existence, with only a single Mint State coin reported, the whereabouts of which is unknown today.

DESIGN TYPES AND DATES MINTED

Liberty Head, Motto Above Eagle, Value TWENTY D.: 1870-CC, 1871-CC, 1872-CC, 1873-CC, 1874-CC, 1875-CC, 1876-CC.

Liberty Head, Motto Above Eagle, Value TWENTY DOLLARS: 1877-CC, 1878-CC, 1879-CC, 1882-CC, 1883-CC, 1884-CC, 1885-CC, 1889-CC, 1890-CC, 1891-CC, 1892-CC, 1893-CC.

10

DENVER MINT, 1862 TO DATE

CLARK, GRUBER & CO. PURCHASED

On February 28, 1861, Congress created the Colorado Territory from the western section of Kansas. On July 2, 1861, a Republican convention, the Republicans having been the winning party in the 1860 presidential election, was held in Denver. An appeal was made to Congress to establish a branch mint in Denver to process the gold in the districts. Agreeing with the proposal was Milton Edward Clark, who managed the Clark, Gruber & Co. bank and mint in Denver (see chapter 12). To further this he obtained letters of endorsement from prominent businessmen. Clark, Gruber & Co. partner Austin M. Clark went to Washington and met with Secretary of the Treasury Salmon P. Chase. The secretary sent financial statements and examples of coins from Clark, Gruber & Co. to James Pollock, director of the Mint in Philadelphia. The response was favorable, and a bill to put a branch mint in Denver was introduced in Congress and approved on April 21, 1862.

The Denver Mint in the 19th century was housed in facilities that replaced the Clark, Gruber & Co. bank and private mint purchased in 1862. The Mint is the building at the far upper right with a tower.

On June 11 Secretary of the Treasury Salmon P. Chase appointed a committee to investigate the purchase of Clark, Gruber & Company for the branch mint: Samuel H. Ebert, secretary of Colorado Territory; district attorney Samuel E. Brown; and George W. Browne. Dr. Oscar D. Munson, earlier connected with the San Francisco Mint, conferred with the committee regarding technical matters. The group found the existing mint to be worth $25,000. This amount was made as a formal offer to Clark, Gruber & Co. on November 25, and they accepted. Clear title to the land could not be obtained as there was no land office or registry in Denver. This problem was resolved in a simple manner: on March 3, 1863, Congress resolved that in the absence of records, title to the land become the property of the government. The purchase became final in April, after which steps were taken to outfit the building, to obtain machinery in the East, and to take some other measures. Plans did not yet include the striking of coins. Major construction took place, and the building was vastly remodeled to the point that the original structure was unrecognizable.

THE DENVER MINT OPENS FOR BUSINESS

By the end of September 1863 all was in place, and receipt of gold deposits, assaying, and refining commenced. The Mint was in new hands and reorganized before the melting of gold and issuance of bars began. The *Annual Report of the Director of the Mint* for the fiscal year ending June 30, 1864, included this:

> The Branch Mint at Denver, Colorado Territory, during the past year has been successfully engaged in melting, refining, assaying, and stamping gold bullion, returning the same to the depositor in the form of unparted bars, bearing the government stamp of weight and fineness. The number of bars stamped was 532 amounting in value to $486,329.97.

Another view of the early Denver Mint.

The efficiency and usefulness of this branch would be greatly increased if a safe and expeditious mode of transportation could be secured. An overland route of 600 miles is a formidable obstacle in the way of commercial intercourse with our eastern cities and markets. In addition, the hostility of the Indian tribes along the route, doubtless instigated by rebel [Confederate] emissaries and bad white men, has increased the difficulty and dangers of intercommunication and the transportation of bullion to the Atlantic markets. These difficulties will probably be obviated in due time, and that institution will then assume her proper position as a branch Mint.

The *Report of the Director of the Mint* for the fiscal year ended June 30, 1865 included this:

Branch mint at Denver, gold deposits; $541,559 04; silver, $7,050 81; total deposits, $548,609 85. Number of stamped bars, 469; value, $545,363. The report of the superintendent of this branch represents its operations during the year as successful and encouraging. It is engaged in melting, refining, assaying and stamping gold bullion, returning the same to the depositor in the form of unparted bars, bearing the government stamp of weight and fineness.

In my last annual report in reference to this branch mint I remarked that "the efficiency and usefulness of this branch would be greatly increased if a safe and expeditious mode of transportation could be secured. . . .

The superintendent at Denver constantly urges the necessity for a prompt introduction of the system of purchase and exchange, as contemplated in the act of Congress, to which reference has been made; and, concurring in the necessity for such action, I most respectfully ask the early and favorable consideration of this subject by your department.

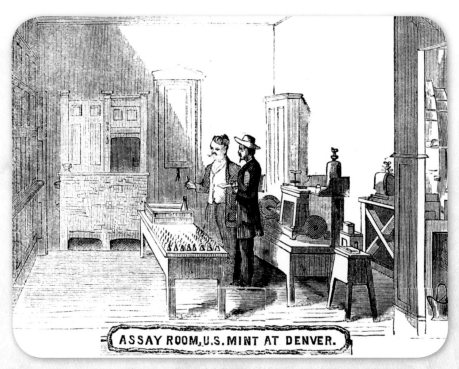

The Assay Room of the Denver Mint in 1866.

In 1869 Congress passed a resolution that the facility formally be known as the Assay Office, more or less continuing to conduct operations as they had been. Gold continued to be returned to depositors in the form of unparted (not refined) bars of the fineness as received. Information concerning the fineness and weight were stamped on the face.

Tourists and other visitors were welcome at the Denver Mint. Displays included coins (the nature of which are not known today, as the Mint was not yet striking coins), mineral specimens, mining implements, and fossils. Annual reports for the era reflect that the Mint was a place busy with the refining, assaying, and storage of gold.

NEW MINT FOR COINAGE PROPOSED

In its issue of October 1887, *Mining Reporter* called for a coinage facility:

> The time that is required to accommodate any needed change in the existing order of things in the United States is well illustrated by the fact that Denver has no coinage mint. Gold bullion is still sent from all the West to New York to be refined, and from there sent to Philadelphia and New Orleans to be coined. The most of that which is coined at New Orleans is then returned to Philadelphia, and from there distributed to the regions where it is required. That the proceeding is necessarily expensive is manifest. It is [a procedure] which was adopted when the country was new, and when conditions were very different from those which prevail at present, and these changed conditions call for a change in the methods of conducting the coinage business.
>
> Denver is the metropolitan city of the mining region of the United States. It is central in location and possesses a much needed facility as a distributing point for the mint products. Everything in the nature of chemicals used in refining gold and silver is produced here as cheaply as it is produced in any portion of the Union. The work of coinage can be done as cheaply here as it can in New Orleans or Philadelphia, and to do it here would save large sums of money each year to the government. It would, too, be a great saving to the miners, and as they form no inconsiderable portion of the people of this nation, they have some right to be considered. It is certainly not right to compel miners to pay for shipping bullion to New York to be refined, when it could be refined at Denver, especially when it would be economy for the government to treat it here.

On February 20, 1895, Congress passed an act providing for a new Mint to be established in Denver for the coinage of silver and gold. By that time silver mining was a very important industry in Colorado, and gold mining in the Cripple Creek District had been underway for several years. Untold wealth was being extracted from the ground, mainly by hard-rock mining. The gold industry was prosperous. The silver industry was having hard times due to oversupply of the metal in regard to the demand for it. The "Silver Question," what to do about the falling price of silver, was the preemptive political topic of the time.

THE NEW MINT BUILDING

On March 2, 1895, Congress passed its budget for fiscal-year 1896, which included $100,000 for the purchase of a site and the beginning of construction for the Denver Mint coining facility. The secretary of the Treasury was authorized to contract for the completion of a building whose cost, including site, vaults, heating, and other essentials, was not to exceed $500,000. On April 22, 1896, a fine piece of real estate was purchased for $60,261.71.

A handsome four-story building was erected of gray granite quarried in Colorado. The structure measured 175 feet wide by 100 feet deep and 90 feet high. Below ground was an area a story and one-half high. The total floor space comprised 70,000 square feet.

The new Mint was outfitted in elegant style, with the reception area and main halls having polished Vermont marble on the floors and some of the walls. Six vaults for bullion were wired with the latest alarm system. A separate vault to store coins was 20 feet wide, 60 feet deep, and 10 feet high, with 22 compartments, and fitted with a steel door weighing 75 tons. Five coin presses were brought from Philadelphia, the largest of which was capable of turning out 100 silver dollars or double eagles per minute. Power was provided by electric motors, though a gas plant in the lower level was coal-fired and served to provide fuel for heat. A very large bin for storing coal was constructed to insure a supply should miners go on strike. The turn of the 20th century was a period of labor unrest in Colorado in particular, but also in other locations.

Additional appropriations were made, finally totaling $800,228.01 by the time that the new building was essentially completed and partially occupied in 1904.[1]

The new Mint as pictured in *The Annual Report* for fiscal year 1906.

The Denver Mint at its opening in 1906.

OPERATIONS BEGIN AT THE MINT

Beginning on January 1, 1905, ingots, raw gold, and other bullion from various Colorado sites, the most important being the Cripple Creek District on the west side of Pikes Peak, were shipped to the Denver Mint to be stored, rather than being sent to Philadelphia. From the current *Annual Report:*

> The receipts of bullion at the Denver Mint for the whole of 1904 were $14,968,135.26, and for the ten months, 1st January to 31st October, inclusive, were 812,127,749.53. The receipts for the ten months, 1st January to 31st October, 1905, showed an increase over the corresponding period of 1904 of $10,815,561 26, being 822,943,310, or about £4,723,000 sterling. On the basis that a ton of gold is worth $602,874-86, the gold bullion in the Denver Mint on the 31st October weighed over thirty-seven tons. I was shown by one of the chief officials through the mint a few days ago, and, among other things, saw this interesting accumulation of bullion.
>
> The establishment has at present about seventy-five employees, which number, after the commencement of coinage operations, it is expected will be increased to about 250. This new coinage mint will have a marked and very beneficial effect on Denver as a financial, milling and mining machinery, and supply centre.

THE MINT OPENS FOR COINAGE

Appointed for the new Mint were these officials who were already in place in the old facility: Frank M. Downer, superintendent; Joseph W. Milson, melter and refiner; and A.R. Hodgson, assayer. On November 1, 1905, a press was put into operation and tokens the diameter of the double eagle were struck in a ceremony sponsored by Mint officials and attended by the governor of the state and other people of interest at the time.

At 10:59 a.m., Thursday, February 1, 1906, Superintendent Frank M. Downer gave a signal, and as part of a public ceremony the first official Denver Mint coins were struck. On hand was a crowd who heard commentaries and watched the machinery in motion.[2] On February 9, 1906, George F. Heath, editor of *The Numismatist*, was sent the message from Downer: "Replying to your inquiry I would say that this institution has coined a few half and quarter dollars. No Proof coins have been struck and there probably will be none." In the first year silver dimes, quarters, and half dollars were struck as were gold half eagles, eagles, and double eagles.

The Annual Report of the Director of the Mint told this about the Denver Mint:

> Coinage operations on February 1, 1906, under most auspicious circumstances, having a thoroughly modern equipment in every part. Experts of the Mint service from Philadelphia, New Orleans, Washington, and San Francisco had been detailed at Denver for months previous, engaged in arranging all the necessary apparatus required for successful operation.
>
> The building is practically a five-story building, three of the stories being above the street level and two below. In the lower basement are installed the boilers, pumps, fuel economizer, induced-draft apparatus, air compressors for the artesian-well air lift, house-water pumps and tanks for water and air under compression, motor-driven telescopic ash hoist, and supply tanks for the storage of 20,000 gallons of crude oil for furnace use. The boilers are of the water-tube type and are fitted with Hawley down-draft furnaces. An engine-driven draft fan induces draft through a Green economizer to a steel stack, which extends through a stack and

pipe shaft to about 15 feet above the roof of the building. High and low pressure steam mains are run in the boiler room, the low-pressure mains operating the auxiliaries. The Webster system of vacuum pumps is used on the heating system. Two compound air compressors are employed to force the water to a storage tank from an artesian well that is 700 feet deep. The storage tank has a capacity of 30,000 gallons and can be pumped full in about four hours' time. The artesian well is the only source of water supply for the entire building.

In the Coining Room are installed three coining presses, three upsetting or edge-rounding machines for blanks, and a complete scale equipment for weighing heavy drafts of coins, together with counting boards and bagging equipment. The size of the Coining Room is such that twice the number of machines that are now in place can be installed in the future. This fact is also true of the rolling and Cutting Room.

ACTIVITY AT THE MINT

In 1906 the Denver Mint began striking coins under contract with various foreign countries, a specialty that became very important in coming years. The first were 4.8 million five-peso gold coins for Mexico.

In 1909 after the New Orleans Mint closed, certain equipment was moved to Denver and mintage operations were ramped up. In that fiscal year carvings on marble slabs over the main doorway and 13 windows were completed under the aegis of John Gibson. Three murals painted by Vincent Adriente depicted allegories of Commerce, Mining, and Manufacturing. The Great Seal of the United States was positioned on the floor of the vestibule.[3] By actual count by the guards and attendants, in calendar year 1910 over 70,000 people visited the Mint, the heaviest attendance being in August. Due to the installation of modern equipment in recent years the work force had been cut from 185 to just 87 employees.[4]

In 1911 the Mint received bullion from many sources including the federal assay offices at Boise, Deadwood, Helena, and Salt Lake City. In 1911 the first cents were struck at Denver, and in 1912 the first nickels. In the spring of 1913 an improved refining apparatus was installed. In fiscal year 1914 there were 67,027 visitors. During the World War the demand for coins was intense, and "the coin shipments were very heavy and made to most of the important cities from Boston to San Francisco," this despite other mints on the East and West coasts.[5]

In 1918 Enos P. Schell, age 68, a counter of coins at the Denver Mint, was indicted for theft. Eighty-eight "defective" half dollars had been found in his desk and in his work trousers. He had been employed by the Treasury Department, including at the San Francisco Mint, since 1875. The jury found him guilty. In the fiscal year ended June 20, 1918, the Mint turned out 115,000,000 coins, or three times the amount of any previous year. Most demand was for cents needed in change when wartime taxes created odd retail pricing. For a time starting in April 1919 the Mint produced cents exclusively. The *Denver Post* commented wryly that any would-be robber would have to have a "fleet of moving vans." The guard force at the Mint had been reduced from six to three men.[6] A few months later, coinage of other denominations was resumed. In November 1919 700,000 cents, nickels, and dimes were turned out each day.

Morgan silver dollars were struck at all three mints in 1921, Denver, Philadelphia, and San Francisco, to build up Treasury stocks as backing for Silver Certificates. In Denver the first such coin was struck on May 3. Production in the following months created a storage problem, and a pit in the basement was converted to a vault to store 30,000,000 coins. For a time these were minted at the rate of 75,000 per day in three shifts. On May 26, 1921, recently inaugurated President Warren Harding appointed Robert J. Grant to replace Thomas Annear, who had served in the post since August 1913.

ORVILLE HARRINGTON

Orville Harrington, a graduate of the Colorado School of Mines with the class of 1898 and considered to be a mining engineer of high talent, went to work in the Butterfly-Terrible and Happy Thought mines. Seeking a change he joined the staff of the Denver Mint in February 1906. He was troubled with his right leg from when he was eleven years and shot by accident in his right hip. From that time on, he endured pain. Harrington worked in the Refining Department of the Mint for three and one-half years, then requested a furlough. He went to Cuba to take the position of superintendant of the Minas de Matahamdre copper mines. By 1918 he was partly paralyzed. At St. Luke's Hospital his right leg was amputated four inches below the knee, after which he was fitted with a wooden (or cork) prosthesis. Later, he went back to Denver to take the position of secretary at the School of Mines, a position "with many titles and little pay," he told Frances Wayne, a *Denver Post* reporter.

When Harrington went back to the Denver Mint in September 1919 he worked in the refining room. "His liberal education and stern application to his duties made him one of the most trusted employees," a Mint official said. Meanwhile, he and his wife

One-legged gold thief Orville Harrington haunts the public exhibits at the Denver Mint.

Lydia, from a prominent family in Texas, were mentioned now and again in society columns and related news. By the time of the described scenario the couple had two children: a girl, age four, and a boy, eleven months.

In early 1920 Mint officials felt something was amiss. Gold ingots were missing so Secret Service agents were called in. After some scouting they confronted Orville Harrington, who calmly confessed to Rolland R. Goddard that he had pilfered gold over a period of six weeks. The agents accompanied him to his home, where Lydia opened the door and saw her husband in handcuffs, to her utter amazement. Harrington led the men down to the basement, where they spent several hours battering down a brick wall that had been put up to hide most of the gold. From there they went to the large orchard behind the house where more gold was buried under a cement sidewalk. The last was his undoing, as earlier the agents, watching from a distance, saw him digging a hole and burying something.

By four in the afternoon the recovery of the gold ingots was complete and Harrington was led off to jail. He told the *Post* reporter that he stole in order to support his family in a fine life style not possible with his $4 per day Mint salary. As to how much gold he had sold, he did not say.[7]

He was convicted and sentenced to 10 years in federal prison and was sent to Leavenworth, Kansas, to serve time. Interestingly, Harrington became a cult figure of sorts—in the best tradition of outlaws of the old West. Visitors to the Denver Mint today can view a special display devoted to his caper.

The news played nationwide in the papers, causing a chill at the Treasury Department. Security at the mints had always been a challenge, but most thefts were small and handled without publicity.

On August 24, 1921, the Denver Mint received instructions from the Treasury Department that henceforth the building was to be closed to visitors. There was more to come.

A DARING ROBBERY

At 10:40 in the morning of Monday, December 18, 1922, at the front entrance to the Mint employees were finishing loading 50 sacks each containing $4,000 in paper money onto a truck to be taken to the Federal Reserve Bank and from there to be distributed to banks in Denver and Wyoming. The bills had been placed by the Federal Reserve in a Mint vault for security, believed to be safer than at the Federal Reserve building. The Mint people were attacked by armed men, one was killed, and after grabbing the money the bandits fled in an automobile with Colorado license 31001 and disappeared. This was sensational news—the first such event in the history of the Mint system. For the next several years there were many attempts to find the crooks and the money, but to no avail. The incident prompted all mints to be closed to visitors, and at the Philadelphia Mint the coin collection there to be transferred in due course on loan to the Smithsonian Institution.

The identity of the robbers and the disposition of the money remained a mystery until years later. On December 1, 1934, following a lengthy investigation, it was announced that five men and two women—all now dead or in prison—had been involved in the robbery. Harvey Bailey, who was convicted of another crime, the kidnapping of Oklahoma City millionaire Charles P. Urschel, drove the escape car in 1922 and in 1934 was serving a life sentence in Alcatraz Prison in San Francisco Bay. Robert Leon Knapp, alias Robert Burns, was dead, as was Frank McFarland, "the Memphis Kid." It was revealed that the bullet-riddled body of Nicholas Trainor, alias Nick Sloan, had been found frozen in a residential garage in Denver on January 14, 1923, a month after the robbery. The two women, Florence Sloan, who had been considered the "queen" of the gang, and "Margaret Burns," who posed as the wife of "Robert Burns," really Knapp, were found shot to death and burned in an automobile near Red Wing, Minnesota, in 1933. The money had been spent long since.[8]

CONTINUING BUSINESS

In the recession year of 1922 the Denver Mint was the only facility that struck one-cent pieces. On November 12, 1923, Denver Mint Superintendent Grant was appointed by President Calvin Coolidge to be director of the Mint at the office in Washington. On December 10 Frank E. Shepard was named as his successor.

In 1924 the Mint produced quarter eagles, the first of this denomination coined by any mint since 1915. No more of this denomination would be made in Denver from then until the quarter eagle was officially discontinued in 1929. In 1927 the Mint turned out 180,000 double eagles. There was hardly any call for them in commerce, and nearly all went into storage. Peace-design silver dollars were also minted, the last year until coinage was resumed in 1934 for one year only.

Although mintage quantities at the Denver Mint were reduced during the early years of the Depression, 1932 was reported as the busiest year since 1928 in terms of silver and gold bullion deposits.

In 1933 the Denver Mint struck its first commemorative coins—1933-D Oregon Trail half dollars for distribution at the Century of Progress exposition held in Chicago, although the design of the coins, first minted in Philadelphia and San Francisco in 1926, had nothing to do with the 1933 event. The last gold coins minted at Denver had been 1931-D double eagles. After the Treasury called in gold coins from the public, starting in April 1933, the Denver Mint continued conducting a large business in receiving silver and gold bullion. The gold was sold to the federal government for $35 per pure ounce, the new level set in January 1934. Nearly $27,000,000 in gold was deposited in the first seven months of 1934.

In 1937 an addition was made to the Mint, adding about 6,000 square feet and comprising a basement and two upper floors. The main part of the building was remodeled, and ten new coining presses were installed.

THE 1940s AND 1950s

During the World War II years in the early 1940s, mintage quantities increased at the three mints. Silver and gold deposits continued to be an important business. In October 1945 there was a ground-breaking ceremony for an addition of three stories measuring 161 by 96 feet to house melting and rolling equipment to process copper ingots. Such ingots weighing up to 420 pounds could be reduced by rollers into thin strips suitable for making planchets. Galleries were installed on which visitors could stand and watch various processing and coining operations below.

Construction of the Denver Mint addition in the mid-1930s.

In February 1956 the Denver Mint observed its 50th anniversary with Superintendent Alma K. Schneider present. She gave a short presentation on the history of the establishment and complimented Dr. Nolie Mumey, a Colorado numismatist and historian, for his historical display. Dr. P.W. Whiteley (whose collection of Lesher "dollars" struck in Colorado in 1900 and 1901 was the finest ever), local coin dealer Dan Brown, photographic print collector Fred Mazzula, and the First National Bank of Denver also mounted displays.

CHANGES OF THE 1960s

In 1963 the American Numismatic Association held its summer convention in Denver. Mint Superintendent Fern Miller arranged tours for up to 560 people to tour the facility for each of four days.

In 1965 the international price of silver bullion rose to the point that silver coins cost more than face value to produce. Sensing an opportunity for profit, citizens began hoarding coins in circulation. Soon there was a nationwide shortage that by extension included cents and nickels, never mind that those denominations were not involved in the rise of metal prices. Mint Director Eva Adams decided to place the blame squarely on numismatists, who in actuality had little to do with it. As punishment she declared that henceforth coins made at branch mints (meaning Denver, as San Francisco was not minting at the time) would not have mintmarks. To all appearances Denver coins of all denominations appeared to have been made in Philadelphia. Cent and nickel planchets processed at the San Francisco Assay Office were shipped to Denver for coining.

To handle the increased demand for coins to alleviate the shortage the Denver Mint ordered 12 new presses—five made in Oakland, California, by the Columbia Machine Company, and seven from the Farracute Machine Company in New Jersey, a long-time maker of presses for American mints and those in foreign countries.[9] Bliss-manufactured presses on hand were modified to strike two, then four cents or nickels at a single stroke. On hand were 16 presses that the Department of Defense had used to stamp ammunition cases, which had been modified for coinage.[10]

In the meantime, on June 10, 1967, Matt Rothert, president of the American Numismatic Association, staff, and various dignitaries and onlookers attended the dedication of the ANA Home and Headquarters in Colorado Springs. The Denver Mint was represented by Assistant Superintendent Charles Miller.

Director Adams "got religion," spent time interfacing with numismatists, and learned that collectors were not at fault. She actually became interested in the hobby and won a seat on the Board of Governors of the American Numismatic Association. On January 4, 1968, with Superintendent Marian A. Rossmiller she held a special ceremony at the Denver Mint for the striking of 1968-D coins with mintmarks.

In 1969 the machine shop at the Denver Mint made six Janvier transfer lathes to reduce electrotypes to steel hubs to be used for coinage. These were shipped to the Philadelphia Mint to replace equipment installed more than 60 years earlier. After the opening of the fourth Philadelphia Mint that year there was some talk that the Denver Mint with its older equipment might be redundant, and that Philadelphia could take care of coinage demand. Should the Denver Mint be closed? Nothing happened with such discussions.

On July 3, 1971, Congress voted to appropriate $1.5 million to purchase land for the construction of a new Denver Mint. Bureau of the Mint officials inspected various sites and picked two, one of which was eliminated when the Burlington Northern Railroad stated it was going to use the area to lay more tracks. The other site was in the Park Hill district. On July 16, 1974, Mint Director Mary Brooks announced that construction was to begin in late 1975 or early 1976 and was to be completed around 1980.

LATER YEARS

In 1971, Columbia presses of 350-ton capacity were used for the first time to strike Eisenhower dollars. From the 1970s onward the Denver Mint was a featured day trip each year for students of the American Numismatic Association Summer Seminar. The superintendents of the Mint welcomed collectors.

In the summer of 1979, in connection with a renovation project, the Denver Mint expanded its popular shop for visitors to buy current coins and medals, the latter made at the Philadelphia Mint. On July 1, 1982, at a special ceremony Mint Director Donna Pope pushed a button to strike the first 1982-D George Washington commemorative half dollar and Denver Mint Superintendent Nora Hussey struck the second. These were the first commemorative coins to be struck since 1954.

After much discussion within Congress and by various agencies the project of building a new Mint was canceled in favor of upgrading existing facilities. On January 19, 1984, the Park Hill land was sold for $3,351,356.05 to an Alaskan company. In that year began another expansion of the present Mint, including docks, facilities for increased coinage, and other improvements.

At an ANA convention in early 1989 Superintendent Cynthia Grassby Baker spoke of the pride her 350-person staff takes in minting more than seven billion coins each year. She encouraged collectors to visit the Mint and also expressed her willingness to arrange special tours of the impressive facility.

On September 9, 1991, an expanded visitors' center was dedicated in a special ceremony. Exhibits included scarce and rare coins, issues of Clark, Gruber & Co., and mementoes of Colorado's mining history. Many of these were on loan from the United Bank of Denver.

On September 22, 1994, ground was broken for another addition to the Mint, this to provide space for a die shop. Opened on May 13, 1996, the new facility began producing coinage dies for both the Denver and San Francisco mints, an activity that had taken place at the Philadelphia Mint before that time. This increased the Bureau of the Mint's die production capacity to 150,000 per year.

In the 1990s high-speed Schuler presses were installed to strike 700 to 750 coins per minute—or faster than the eye could see. These replaced older press that held four dies and could turn out 200 coins per minute. The Schuler presses also became standard at the other mints.

THE DENVER MINT TODAY

The Denver Mint offers a free 45-minute guided tour that teaches visitors about the history of the U.S. Mint and how Denver produces coins for the American public. The tour, which starts at the Cherokee Street entrance, runs Monday through Thursday (except for federal holidays) from 8:00 a.m. to 3:30 p.m., and is recommended for visitors seven years of age and older. The Mint also coordinates a free speakers bureau for local schools and community groups, covering a wide range of topics that can be tailored to specific needs.

Naturally, security is very tight at the Denver Mint (as it is in Philadelphia, which also offers public tours). Visitors go through an x-ray screening for metal detection. And to answer a frequent question: No, there are no free samples!

The Denver Mint gift shop, with its entrance located on Cherokee Street, is open Monday through Thursday. It provides an expansive selection of modern and classic coins, Mint memorabilia, coin-collecting books, and other souvenirs.

As a production facility, today's Denver Mint manufactures all denominations of circulating coins, makes its own coinage dies from hubs produced in Philadelphia, packages the Denver "D-Mint" portion of the Mint's annual Uncirculated coin sets,[11] and mints commemorative coins as authorized by the U.S. Congress. It also stores silver bullion.

In 2013, in response to lower demand from the Federal Reserve, the U.S. Mint cut production at its Philadelphia and Denver facilities from three shifts to two. In that year the mints combined produced 10.7 billion coins. In fiscal year 2015 they made 16.2 billion coins, and both Philadelphia and Denver converted back to three shift operations to accommodate increased production.

The Denver Mint reduced water usage in fiscal year 2015 and was awarded its fifth environmental Gold Award in April by the Metro Wastewater Reclamation District, the wastewater treatment authority for much of metropolitan Denver and parts of northern Colorado. The Gold Award recognized the mint for being in full compliance with strict pretreatment requirements and having a demonstrated commitment to environmental excellence.

Throughout the federal government, the workforce is aging and beginning to retire in increasing numbers. President Barack Obama's requirements for hiring reform gave the Mint a priority of hiring the right people for the right jobs, and in recent years many new hires have been military veterans, including those who are disabled. The Treasury Department's goal is for 6 percent of all new hires to be disabled veterans, and 27.2 percent of the Mint's employees met this classification in 2015. "When we added a third shift to production facilities in Denver and Philadelphia, we needed several more employees to fill the additional shifts," the Mint's annual report noted. "Working with the Center for Women Veterans, the Colorado Department of Labor, and Employment Veterans Outreach, the Mint held career fairs in both locations, focusing on veterans and in particular, female veterans. From these events, the Mint hired enough staff to make the third shift possible, so we could fulfill the Federal Reserve's orders for circulating coinage."

The grand lobby hallways of the Denver Mint.

Randall Johnson, acting plant manager of the Denver Mint, in front of posters showing proper coinage die alignments.

The security room overlooking the Denver Mint's old entryway. A guard kept watch, armed with a Thompson machine gun, a Winchester rifle, pistols, and tear-gas guns at the ready. Just in case, he could step on a plate that would activate a flood of tear-gas to fill the entire lobby. This precaution never proved necessary.

Steel rods waiting to be cut and shaped into coinage dies.

Quality control is very strict at the Denver Mint. These dies have been condemned for imperfections.

A Denver Mint worker observing a huge roll of metal strip from which planchets will be punched for coinage.

Blank cent planchets waiting to be sucked up and filtered into the production line then stamped into coins.

A Schuler press set up to strike Jefferson nickels.

The press setup for Eisenhower Presidential dollars destined to be packaged in Mint Uncirculated sets. These Mint set coins are struck under pressure higher than that used for normal circulation-strike coins. Another distinction: they are made from planchets that are "spalecked"—specially cleaned and finished in machines made by Spaleck of Germany.

Packaged coins from Philadelphia and from Denver, ready to be combined into a single Uncirculated set.

A bulk bag holding $50,000 face value of freshly struck quarters. Hundreds of these bags are stored inside the Denver Mint's basement at any given moment, waiting for delivery to Federal Reserve banks.

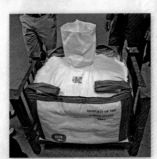

Denver's "Numismatic Automated Packaging Machine," an assembly line where coins are packaged together without being touched by human hands.

Coinage of the Denver Mint, 1906 to Date

The Denver Mint has been a steady producer of coins from 1906 to date. Certain coins of the early 1960s are without mintmarks and appear as Philadelphia coins. Today the Denver Mint welcomes interested visitors to see coins being produced.

Cents

Lincoln cents, the first minor coins to be struck at Denver, were made in 1911, after which production of the design continued non-stop into the late 20th century, the only one of the three mints for which this was true, as only Denver struck cents in 1922. Mintmarks were removed from 1965 to 1967.

Design Types and Dates Minted

Lincoln, Wheat Ears Reverse: *Variety 1, Bronze:*
1911-D, 1912-D, 1913-D, 1914-D, 1915-D, 1916-D,
1917-D, 1918-D, 1919-D, 1920-D, 1922-D, 1924-D,
1925-D, 1926-D, 1927-D, 1928-D, 1929-D, 1930-D,
1931-D, 1932-D, 1933-D, 1934-D, 1935-D, 1936-D,

1937-D, 1938-D, 1939-D, 1940-D, 1941-D, 1942-D, 1944-D, 1945-D, 1946-D,
1947-D, 1948-D, 1949-D, 1950-D, 1951-D, 1952-D, 1953-D, 1954-D, 1955-D,
1956-D, 1957-D, 1958-D. *Variety 2, Zinc-Coated Steel:* 1943-D.

Lincoln, Memorial Reverse: *Copper Alloy:* 1959-D,
1960-D, 1961-D, 1962-D, 1963-D, 1964-D, 1968-D,
1969-D, 1970-D, 1971-D, 1972-D, 1973-D, 1974-D,
1975-D, 1976-D, 1977-D, 1978-D, 1979-D, 1980-D,
1981-D, 1982-D. *Copper-Plated Zinc:* 1982-D, 1983-D,

1984-D, 1985-D, 1986-D, 1987-D, 1988-D, 1989-D, 1990-D, 1991-D, 1992-D,
1993-D, 1994-D, 1995-D, 1996-D, 1997-D, 1998-D, 1999-D, 2000-D, 2001-D,
2002-D, 2003-D, 2004-D, 2005-D, 2006-D, 2007-D, 2008-D.

Lincoln, Bicentennial: *Birth and Early Childhood,*
Formative Years, Professional Life, Presidency: 2009-D.

Lincoln, Shield Reverse: 2010-D, 2011-D, 2012-D,
2013-D, 2014-D, 2015-D, 2016-D.

Nickel Five-Cent Pieces

The first nickels were made at Denver in 1912, the twilight of the Liberty Head design. After then they became a mainstay of production and were made for most but not all years from that time to the present. Those struck from 1965 to 1967 were without mintmarks.

DESIGN TYPES AND DATES MINTED

Liberty Head: 1912-D.

Indian Head or Buffalo, Variety 1, FIVE CENTS on Raised Ground: 1913-D.

Indian Head or Buffalo, Variety 2, FIVE CENTS in Recess: 1913-D; 1914-D; 1915-D; 1916-D; 1917-D; 1918-D, 8 Over 7; 1918-D; 1919-D; 1920-D; 1922-D; 1924-D; 1925-D; 1926-D; 1927-D; 1928-D; 1929-D; 1934-D; 1935-D; 1936-D; 1937-D; 1938-D.

Jefferson: *Nickel:* 1938-D, 1939-D, 1940-D, 1941-D, 1942-D, 1946-D, 1947-D, 1948-D, 1949-D, 1950-D, 1951-D, 1952-D, 1953-D, 1954-D, 1955-D, 1956-D, 1957-D, 1958-D, 1959-D, 1960-D, 1961-D, 1962-D, 1963-D, 1964-D. *Designer's initials (FS) added:* 1968-D, 1969-D, 1970-D, 1971-D, 1972-D, 1973-D, 1974-D, 1975-D, 1976-D, 1977-D, 1978-D, 1979-D, 1980-D, 1981-D, 1982-D, 1983-D, 1984-D, 1985-D, 1986-D, 1987-D, 1988-D, 1989-D, 1990-D, 1991-D, 1992-D, 1993-D, 1994-D, 1995-D, 1996-D, 1997-D, 1998-D, 1999-D, 2000-D, 2001-D, 2002-D, 2003-D. *Wartime Silver Alloy, Large Mintmark:* 1943-D, 1944-D, 1945-D.

Westward Journey: *Peace Medal, Keelboat:* 2004-D.

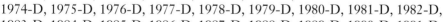

Westward Journey, Modified Portrait: *American Bison, Ocean in View:* 2005-D.

Jefferson Modified: 2006-D, 2007-D, 2008-D, 2009-D, 2010-D, 2011-D, 2012-D, 2013-D, 2014-D, 2015-D, 2016-D.

Dimes

Barber dimes were minted in Denver from 1906 to 1916, followed by Mercury dimes from 1916 to 1945, except for 1922, 1930, 1932, and 1933. The 1916-D is considered to be the key issue in that series. Roosevelt dimes have been made since 1946.

DESIGN TYPES AND DATES MINTED

Barber or Liberty Head: 1906-D, 1907-D, 1908-D, 1909-D, 1910-D, 1911-D, 1912-D, 1914-D.

Winged Liberty Head or "Mercury": 1916-D; 1917-D; 1918-D; 1919-D; 1920-D; 1921-D; 1924-D; 1925-D; 1926-D; 1927-D; 1928-D; 1929-D; 1931-D; 1934-D; 1935-D; 1936-D; 1937-D; 1938-D; 1939-D; 1940-D; 1941-D; 1942-D, 42 Over 41; 1942-D; 1943-D; 1944-D; 1945-D.

Roosevelt: *Silver:* 1946-D, 1947-D, 1948-D, 1949-D, 1950-D, 1951-D, 1952-D, 1953-D, 1954-D, 1955-D, 1956-D, 1957-D, 1958-D, 1959-D, 1960-D, 1961-D, 1962-D, 1963-D, 1964-D. *Clad:* 1968-D, 1969-D, 1970-D, 1971-D, 1972-D, 1973-D, 1974-D, 1975-D, 1976-D, 1977-D, 1978-D, 1979-D, 1980-D, 1981-D, 1982-D, 1983-D, 1984-D, 1985-D, 1986-D, 1987-D, 1988-D, 1989-D, 1990-D, 1991-D, 1992-D, 1993-D, 1994-D, 1995-D, 1996-D, 1997-D, 1998-D, 1999-D, 2000-D, 2001-D, 2002-D, 2003-D, 2004-D, 2005-D, 2006-D, 2007-D, 2008-D, 2009-D, 2010-D, 2011-D, 2012-D, 2013-D, 2014-D, 2015-D, 2016-D.

Quarter Dollars

Quarters have been struck at the Denver Mint since 1906, with the exception of just a few years. Among Washington quarters the 1932-D is the key issue. The State and later issues 1999 to date are wonderfully diverse, popular, and inexpensive.

DESIGN TYPES AND DATES MINTED

Barber or Liberty Head: 1906-D, 1907-D, 1908-D, 1909-D, 1910-D, 1911-D, 1913-D, 1914-D, 1915-D, 1916-D.

Standing Liberty, Variety 1, No Stars Below Eagle: 1917-D.

Standing Liberty, Variety 2, Stars Below Eagle, Pedestal Date: 1917-D, 1918-D, 1919-D, 1920-D, 1924-D.

Standing Liberty, Variety 2, Stars Below Eagle, Recessed Date: 1926-D, 1927-D, 1928-D, 1929-D.

Washington: *Silver:* 1932-D, 1934-D, 1935-D, 1936-D, 1937-D, 1939-D, 1940-D, 1941-D, 1942-D, 1943-D, 1944-D, 1945-D, 1946-D, 1947-D, 1948-D, 1949-D, 1950-D, 1951-D, 1952-D, 1953-D, 1954-D, 1955-D, 1956-D, 1957-D, 1958-D, 1959-D, 1960-D, 1961-D, 1962-D, 1963-D, 1964-D. *Clad:* 1968-D, 1969-D, 1970-D, 1971-D, 1972-D, 1973-D, 1974-D, 1977-D, 1978-D, 1979-D, 1980-D, 1981-D, 1982-D, 1983-D, 1984-D, 1985-D, 1986-D, 1987-D, 1988-D, 1989-D, 1990-D, 1991-D, 1992-D, 1993-D, 1994-D, 1995-D, 1996-D, 1997-D, 1998-D.

Washington, Bicentennial: 1776–1976-D [1975, 1976].

State Quarters (5 designs per year): 1999-D, 2000-D, 2001-D, 2002-D, 2003-D, 2004-D, 2005-D, 2006-D, 2007-D, 2008-D.

District of Columbia and Territories Quarters Series (6 designs): 2009-D.

America the Beautiful (5 designs per year): 2010-D, 2011-D, 2012-D, 2013-D, 2014-D, 2015-D, 2016-D.

Half Dollars

Half dollars have been struck at the Denver Mint since 1906, with the exception of just a few years. Commemoratives have been made intermittently since 1933. Among Liberty Walking halves 1921-D is the key issue.

DESIGN TYPES AND DATES MINTED

Barber or Liberty Head: 1906-D, 1907-D, 1908-D, 1911-D, 1912-D, 1913-D, 1915-D.

Liberty Walking: 1916-D, 1917-D, 1918-D, 1919-D, 1920-D, 1921-D, 1929-D, 1934-D, 1935-D, 1936-D, 1937-D, 1938-D, 1939-D, 1941-D, 1942-D, 1943-D, 1944-D, 1945-D, 1946-D, 1947-D.

Franklin: 1948-D, 1949-D, 1950-D, 1951-D, 1952-D, 1953-D, 1954-D, 1957-D, 1958-D, 1959-D, 1960-D, 1961-D, 1962-D, 1963-D.

Kennedy: *Silver:* 1964-D. *Silver Clad:* 1968-D, 1969-D, 1970-D. *Clad:* 1971-D, 1972-D, 1973-D, 1974-D, 1977-D, 1978-D, 1979-D, 1980-D, 1981-D, 1982-D, 1983-D, 1984-D, 1985-D, 1986-D, 1987-D, 1988-D, 1989-D, 1990-D, 1991-D, 1992-D, 1993-D, 1994-D, 1995-D, 1996-D, 1997-D, 1998-D, 1999-D, 2000-D, 2001-D, 2002-D, 2003-D, 2004-D, 2005-D, 2006-D, 2007-D, 2008-D, 2009-D, 2010-D, 2011-D, 2012-D, 2013-D, 2014-D, 2015-D, 2016-D.

Kennedy, Bicentennial: 1776–1976-D
[1975, 1976].

Dollars

Morgan dollars were made in quantity in 1921, followed by Peace dollars intermittently 1922 and later (including 316,076 struck but not released in 1964), and in the modern era Eisenhower, Susan B. Anthony, Sacagawea, Native American, and Presidential dollars. The last two series have the D mintmarks on the edges where they are nearly impossible to see if in a holder or album.

DESIGN TYPES AND DATES MINTED

Morgan: 1921-D.

Peace: 1922-D, 1923-D, 1926-D,
1927-D, 1934-D, 1964-D.

Eisenhower, Eagle Reverse:
Copper-Nickel Clad: 1971-D,
1972-D, 1973-D, 1974-D,
1977-D, 1978-D.

Eisenhower, Bicentennial:
1776–1976-D [1975, 1976].

Susan B. Anthony: 1979-D, 1980-D, 1981-D, 1999-D.

Sacagawea: 2000-D, 2001-D, 2002-D, 2003-D, 2004-D, 2005-D, 2006-D, 2007-D, 2008-D.

Native American (new reverse yearly):
2009-D, 2010-D, 2011-D, 2012-D, 2013-D, 2014-D, 2015-D, 2016-D.

Presidential (4 designs per year): 2008-D, 2009-D, 2010-D, 2011-D, 2012-D, 2013-D, 2014-D, 2015-D, 2016-D.

$2.50 Quarter Eagles

$2.50 gold coins were made for just three years. The 1911-D is considered to be the key issue.

DESIGN TYPE AND DATES MINTED
Indian Head: 1911-D, 1914-D, 1925-D.

$5 Half Eagles

Half eagles of two designs were made intermittently from the opening of the Mint in 1906 until 1914.

DESIGN TYPES AND DATES MINTED

Liberty Head, Variety 2, Motto Above Eagle: 1906-D, 1907-D.

Indian Head: 1908-D, 1909-D, 1910-D, 1911-D, 1914-D.

$10 Eagles

Eagles were made from 1906 to 1911 and then again in 1914.

DESIGN TYPES AND DATES MINTED

Liberty Head, Motto Above Eagle: 1906-D, 1907-D.

Indian Head, Variety 1, No Motto on Reverse: 1908-D.

Indian Head, Variety 2, Motto on Reverse: 1908-D, 1909-D, 1910-D, 1911-D, 1914-D.

$20 Double Eagles

Double eagles were minted intermittently from 1906 to 1931. Those after 1923 are considered to be key issues by numismatists, with the 1927-D being a major rarity and the 1931-D being very elusive.

DESIGN TYPES AND DATES MINTED

Liberty Head, Motto Above Eagle, Value TWENTY DOLLARS: 1906-D, 1907-D.

Saint-Gaudens, Arabic Numerals, No Motto: 1908-D.

Saint-Gaudens, With Motto IN GOD WE TRUST: 1908-D, 1909-D, 1910-D, 1911-D, 1913-D, 1914-D, 1923-D, 1924-D, 1925-D, 1926-D, 1927-D, 1931-D.

WEST POINT MINT, 1984 TO DATE

The United States Silver Bullion Depository, often referred to as the West Point Bullion Depository and today designated as the West Point Mint, officially opened on June 13, 1938. Its purpose was to store silver bullion.

Meanwhile in Fort Knox, Kentucky, a recently opened building kept most of the nation's gold reserve. This included large amounts of bullion melted down from coins. Some of these were coins that the public was forced to surrender beginning in 1933 by Executive Order 6102, which forbid the hoarding of gold coin, bullion, and certificates by private citizens. Others were Treasury coins that had never been released. Double eagles ($20 gold pieces) from 1850 to 1933 were melted by the millions.

Later, part of the nation's gold reserve was stored at West Point as well. Since a huge deposit of 142 truckloads of gold (57 million troy ounces) made in the early 1980s for the production of the American Arts gold medallions, West Point has been second only to Fort Knox in storage of the precious metal. Today it safeguards one quarter of the nation's reserves.

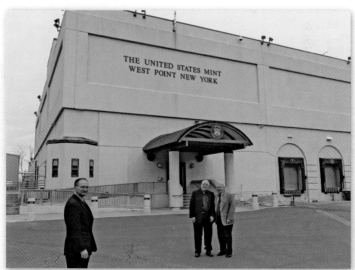

Whitman publisher Dennis Tucker, author Q. David Bowers, and Mint Director of Corporate Communications Tom Jurkowsky in front of the West Point Mint, April 2014.

The original building had 44,000 square feet of floor space. The *Annual Report of the Director of the Mint* for fiscal year 1938 noted:

> The Silver Bullion Depository, constructed on a site formerly included in the military reservation at West Point, was completed during the fiscal year 1938. It will be operated as an auxiliary of the New York Assay Office to be occupied at once for the storage of silver bullion previously temporarily placed in the assay office at New York, the old Assay Office building at New York, the Sub-Treasury building, and at rented vaults.

In 1971 2.9 million Carson City silver dollars were transferred from the Treasury Building in Washington to the West Point Silver Bullion Depository in seven guarded semi-trailer trucks for storage prior to their sale to the public by the General Services Administration. On December 6 a special ceremony was held there as GSA officials accepted the coins from Treasury officials. The coins were packaged at West Point for sale, a process expected to take about a year.

From 1973 to 1986 Lincoln cents were made at the Depository—about seven million coins per day during those years. Lacking mintmarks identifying their origin, they appear similar to Philadelphia Mint coins. In 1974 West Point struck Bicentennial quarters to help meet anticipated demand for coinage. In 1980 the facility began striking gold pieces for the American Arts gold medallion program, which continued through 1984, to be replaced in 1986 with the American Eagle bullion coin program.

A five-year program of American Arts gold medallions was minted at West Point.

On Tuesday, September 13, 1983, the Depository had a special ceremony in which 1984-W $10 gold coins were struck. Not since the opening of the Denver Mint in 1906 had there been a new mintmark for numismatists to contemplate. On hand for the ceremonies were Secretary of the Treasury Donald Regan, former U.S. Treasurer Angela Buchanan, several outstanding Olympic athletes, a number of officials from the Treasury Department and West Point, and, gratifyingly, dozens of coin collectors, coin dealers, and numismatic writers. Mint Director Donna Pope graciously allowed several guests to operate the coining press, and Q. David Bowers, president of the American Numismatic Association at the time, struck the ninth coin.

On October 18, 1985, more than 200 guests were on hand at the Depository for the ceremonial striking of the Statue of Liberty $5 gold commemorative. As the years went by, more commemoratives and bullion coins were struck there, but no regular circulating coins with mintmarks.

By presidential decree, on March 31, 1988, the Bullion Depository name was changed to the West Point Mint. In 2013 the minting and related production staff comprised about 155 people, plus an undisclosed number of security and police officers. On a visit there in June 2013, *Coin World* journalist Paul Gilkes reported that minting was done by three shifts working around the clock seven days a week, turning out American Silver Eagle bullion coins and certain of the gold issues.[1] Plant Manager Ellen McCullom greeted him and arranged for him to view the different aspects of the operations.

From October 1999 to October 2002 the Mint was vastly expanded by the addition of a second floor and other modifications, more than doubling the facility's original size to 94,000 square feet. A celebratory ribbon-cutting ceremony was held April 11, 2003, with Mint Director Henrietta Holsman Fore in attendance.

Today the West Point Mint is a very busy facility. At present there are no accommodations for public tours, for security reasons.

Plant Manager Ellen McCullom at the West Point Mint.

THE WEST POINT MINT TODAY

"The red letter date for West Point was March 31, 1988, when it gained official status as a United States Mint," the Mint notes in its official description of the facility. "Today, it is still a storage facility, but also manufactures, packages, and ships gold and silver commemorative coins, and American Eagle bullion coins, in Proof and Uncirculated condition. Its Platinum Eagles have been very popular since their first issuance in 1997. In 2000, it struck the first ever gold and platinum bi-metallic coin."

The West Point Mint as seen from Route US-9W, December 2008.

Whitman publisher Dennis Tucker examines blank silver planchets that will be used to make American Silver Eagles, July 2014. Each tray holds 462 planchets.

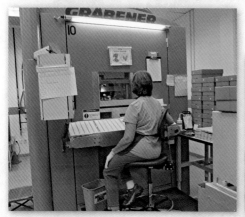

A mint worker monitoring the striking of 2014 American Silver Eagles on a Gräbener press.

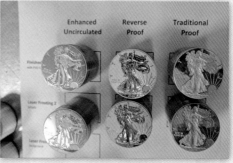

Dies with different finishes create collectible coins.

Proof American Silver Eagles in protective capsules, waiting to be packaged for sale to collectors.

A tray of freshly struck 2014 Baseball Hall of Fame gold commemorative coins.

.9999 fine American Gold Buffalo coins lining up as they leave the minting press.

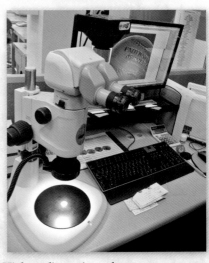

High-quality optics and computer programs used for quality control at the West Point Mint.

A Spaleck machine that tumbles and polishes silver planchets for the Burnished American Silver Eagles, and a close-up of the media used in it.

A vintage bullion crate in the main gold vault of the West Point Mint, with stacks of gold bars visible in the background. Each bar weighs 400 ounces. In May 2014 the vault held a total fine weight of just over 1,420,600 ounces of gold, valued at $1.85 billion. The silver in the vault at the time was worth $55.2 million.

COINAGE OF THE WEST POINT MINT, 1984 TO DATE

The coinage at the West Point Mint has consisted of mintmarkless circulation strikes turned out in the 1960s and commemoratives and other numismatic-trade coins with W mintmarks, as well as various medals and also bullion coins in silver, gold, and platinum. The production of designs intended for circulation has been limited to just a few issues as described below, all of which were sold at a premium to collectors. If and when the U.S. Mint begins production of Proof palladium coins, it will take place at the West Point Mint.

Today West Point creates national medals—like this silver one that memorializes the 10th anniversary of the 9/11 terrorist attacks—as well as bullion and commemorative coins.

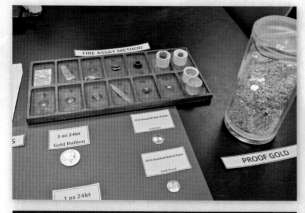

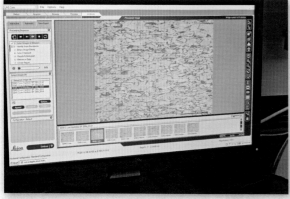

The West Point Mint is fully equipped to assay (test the quality of) gold, with equipment ranging from the old-fashioned to high-tech.

Dimes

In 1996 the West Point Mint struck its first circulation-design coins since the 1960s: 1996-W dimes in special packaging that observed the 50th anniversary of the Roosevelt dime. The dies were hastily buffed or polished, including on the portrait—producing pieces that did not come close to the quality that other mints were turning out for numismatic sale. The mintage totaled 1,457,000 pieces. Striking at the West Point Mint has improved exponentially since the 1996-W dime. The 2015-W silver Proof dime was made for a commemorative set featuring the 2015 March of Dimes commemorative silver dollar.

DESIGN TYPE AND DATES MINTED

Roosevelt: 1996-W, 2015-W (Proof, silver).

Half Dollars

The West Point Mint struck half dollars for the numismatic trade. The 2014-W struck in gold set off a flurry of excitement when launched at the American Numismatic Association convention in Chicago in August of that year. Later, the market for the pieces settled.

DESIGN TYPES AND DATES MINTED

Kennedy: 2014-W (Reverse Proof, silver).

Kennedy, 50th Anniversary: 2014-W (Proof, gold).

Dollars

22-karat gold numismatic specimens of the Sacagawea dollar dated 2000-W were struck at the Philadelphia Mint in 1999 using a prototype reverse design with boldly detailed tail feathers on the eagle. These coins were never released and are still stored by the Treasury Department. The 2015-W Native American dollar of the Mohawk

Ironworkers design was a surprise to many in the hobby community. Included in the Mint's 2015 Coin and Currency Set, it features an enhanced finish. With only 90,000 minted and given its unique status as the only West Point dollar of a circulating design, it will likely draw more collector attention in the future. Mint publicity at the time of sale did not emphasize its West Point origin—an important and appealing distinction for coin collectors.

DESIGN TYPES AND DATES MINTED

Sacagawea: 2000-W (Proof, gold).

Native American: 2015-W.

OTHER COINAGE MINTS

PRIVATE AND TERRITORIAL MINTS

Templeton Reid
Milledgeville and Gainesville, Georgia (1830)

By the latter half of the 1820s news spread of gold discoveries in Georgia. Many fortune seekers came to the district. Milledgeville, then the state capital, was one of the centers of activity. Templeton Reid, a gunsmith and clockmaker, sensed an opportunity to fill a commercial need by converting gold dust, then traded by weight in the area, into coins. On July 24, 1830, an article appeared in the *Southern Recorder* which told of Reid's new enterprise:

Templeton Reid $2.50 gold coin coined in the summer of 1830, one of the first private gold coins from the Southern mining district.

> We have examined, during the past week, with great pleasure, an apparatus constructed by our very ingenious fellow citizen, Mr. Templeton Reid for the purpose of putting gold into a shape more convenient than that in which it is originally found. He makes with great facility and great neatness, pieces worth ten, five, and two and a half dollars. No alloy is mixed with it, and it is so stamped that it cannot be easily imitated. He sets out soon for the mines, and intends putting his apparatus into operation, as soon as he reaches them.
>
> About $1500 worth of Georgia Gold has been stamped by our ingenious townsman, Mr. Templeton Reid, with handsome dies, showing the actual value of each piece of metal, in parcels of $2.50, $5, and $10. . . . Mr. Reid informs us that the gold dust stamped by him will be taken at the Mint and at most of the banks for the value it purports on its face to bear. This will give it a pretty general currency, and make it answer the purposes of money. . . .

Shortly thereafter Templeton Reid moved to Gainesville, which was situated closer to the center of actual mining activity. Coins were made and put into circulation, including one specimen which, unfortunately for Reid, found its way to a disgruntled citizen who styled himself anonymously as "No Assayer" in a letter to the *Georgia*

Courier, August 16, 1830. The disgruntled letter-writer complained that Reid was making nearly a 7% profit on his coinage scheme, an amount considered to be exploitative. Apparently, Reid produced his coins from native metal and did not pay close attention to the metal's fineness, while this was an era in which federal gold coins were worth slightly more than face value in terms of gold content. A Reid $10 contained just $9.38 in gold, according to "No Assayer." This sealed

Milledgeville, the capital of Georgia in 1830.

the fate of the operation. Templeton Reid's private Georgia mint only operated for about three months. Coinage occurred during part of July, all of August and September, and part of October, 1830. Many of his coins were subsequently melted by the United States Mint, accounting in part for their extreme rarity today. Dexter C. Seymour, who studied the series intensively, suggested that only about 1,600 coins were produced in total, including approximately 1,000 $2.50 gold pieces, 300 $5 gold pieces, and 250 $10 gold pieces. Templeton Reid may have gone to California in 1849, for some dies bearing his name were made with that location as an imprint, but if he did travel there, facts concerning his activities in the West are not known today.

Christopher Bechtler and Family
Rutherfordton, North Carolina (1831–1852)

Christopher Bechtler, the elder, and several members of his family came from Germany to Rutherfordton in 1830. Accompanying him were his two sons, Charles and Augustus, and his nephew, who is known as Christopher Bechtler, the younger. The elder Bechtler, trained in the art of the gunsmith and goldsmith, established a jewelry store shortly after his arrival. There was gold-bearing earth on his property, and he sunk several shafts to

Gold dollar issued by Augustus Bechtler, who continued the family private coinage until 1852.

exploit it. His son Augustus shared his interests and abilities and was a capable assistant.

From 1831 to 1852 the Bechtler family coined $2,241,850.50 in gold coins. Additional coins were minted under a separate venture by Augustus after 1840. Bearing the imprint CAROLINA GOLD or GEORGIA GOLD, depending upon the state in which the coinage metal originated, the pieces were produced in the denominations of $1, $2.50, and $5 in minting facilities in the vicinity of Rutherfordton, North Carolina.

Realizing the service that the Bechtler minting operation was providing to miners and tradesmen of the area, the government made no effort to stop them. A Treasury investigation into the Bechtler coinage did provide the data which led Congress in 1835 to provide for the establishment of a branch mint at Charlotte. In 1838 the Charlotte Mint issued its first coins for circulation.

The travelogue of George W. Featherstonehaugh includes a visit to the Bechtler mint on September 20, 1837:

The morning was beautiful, but cool enough to make a nice wood fire agreeable in my bedroom, which was not too well protected against the wind. After breakfast, I walked a few miles to visit a German, of the name of [Christopher] Bechtler, who issued a gold coinage of which I had seen several pieces. He received me very civilly, and I passed a great part of the day with him at his cottage in the woods. Bechtler emigrated with a very clever young man, his son, from the Grand Duchy of Baden, where he had been a gun maker and goldsmith of some reputation, and had acquired a considerable knowledge in the management of metals. He had resided seven years in this country, and had established for himself a character for integrity, as well as skill in his profession. I found him rather mystical and imaginative, as many Germans are; and certainly if he had lived when alchemy flourished, he would have been a conspicuous operator in that inviting art. It was probably this bias that induced him to settle in the Gold Region of North Carolina, where his career had been a rather singular one, but hitherto distinguished for much good sense.

The greater part of the small streams in this part of the Gold Region, have more or less gold in them, so that all the settlers upon the streams were engaged, more or less, in washing for gold. Each of them possessing but a small quantity, and there being no general purchaser, it was an article not easily disposed of without taking the trouble to go great distances. Bechtler had also obtained some in the usual manner, and having made a die, coined his gold into five-dollar pieces, of the same intrinsic value as the half-eagles of the United States, which are worth five dollars each. He also coined pieces of the value of two dollars and a half, and stamped the value, as well as his own name, upon every piece that he coined. These, after a while, found their way to the Mint of the United States, were assayed, and found to be correct. This becoming known, all the gold-finders in his vicinity—and, indeed, from greater distances—began to bring their gold to his mint to be coined. At the period of my visit, his gold coinage circulated more freely than that of the United States, which was very scarce. He told me that his books showed that he had coined about two millions of dollars from the gold found by the settlers; putting his name, with its weight and quality to every piece.

On receiving the gold from the country people—which in this part of the Gold Region is alloyed with silver—he first reduced it to a common standard, then made the five-dollar pieces equal to those of the United States in value, and when coined, delivered it to the respective proprietors, deducting two per cent, for the seigniorage. It would be in his power to take improper advantage of the confidence placed in him; but I heard of no instance of his having attempted this. Some of the gold of this region is alloyed with platina, the specific gravity of which, compared with that of gold, is as 21 to 19. He might, therefore, have made the difference in weight up with platina, which would have put fourteen per cent, into his pocket. As a metallurgist, he had all the skill necessary to do this; but when I mentioned the possibility of this, as an argument against its being received into general circulation, he answered that it was what an honest man would not do, and that if any man were to do it, he would soon be found out, for the gold did not remain long in circulation, since it found its way very soon to the United States Mint, where it was necessary for him to keep a good character.

Bechtler's maxim was, that honesty is the best policy; and that maxim appeared to govern his conduct. I never was so pleased with observing transactions of business as those I saw at his house during the time I was there. Several country-people came with rough gold to be left for coinage: he weighed it before them and entered it in his book, where there was marginal room for noting the subsequent assay. To others he delivered the coin he had struck.[1]

Following the death of the elder Christopher Bechtler in 1842, the coinage business was conducted by his son Augustus. Augustus apparently continued production of coins for a year or so and then was succeeded by the younger Christopher Bechtler, the nephew of the original coiner. Apparently standards of honesty and quality declined, for Director of the Mint Robert M. Patterson made a report which stated:

> Assays repeatedly made at this mint showed that the coins thus fabricated [by Bechtler] are below the nominal value marked upon: yet they circulate freely at this value, and therefore it must be more advantageous to the miner to carry his bullion to the private rather than the public mint.

By this time the Charlotte and Dahlonega Mints had been in service for several years (since 1838) and had reduced the demand for Bechtler coins.

Augustus Bechtler died sometime prior to 1847. The younger Christopher Bechtler moved to Spartanburg in the early 1850s, at which time the Bechtler coinage was discontinued.

Norris, Gregg & Norris
Benicia City and Stockton, California (1849–1850)

The newspaper *Alta California* noted on May 31, 1849, the existence of "a five-dollar gold coin struck at Benicia City, though the imprint is San Francisco. In general appearance it resembles the United States coin of the same value, but it bears the private stamp of Norris, Gregg & Norris and is in other particulars widely different."

The firm was earlier located in New York City where the principals engaged in plumbing, steamfitting, and civil engineering. The new El Dorado beckoned, and the partners headed west. Gold coins of the $5 denomination were subsequently made in several varieties by the Norris, Gregg & Norris firm in California. Three of the pieces were assayed at the Philadelphia Mint and showed finenesses of .870, .880, and .892, and respective intrinsic gold values of $4.83, $4.89, and $4.955, not including the silver alloy (which if added to the computations would have given them each about $0.025 extra value).

Examples of the coinage with the imprint of San Francisco were made in large quantities and circulated extensively; these were probably the first such private issues to achieve popular distribution in the region. Varieties were made with plain or reeded edges. A variety dated 1850 and imprinted STOCKTON is unique.

It is not known by whom the dies were cut, but a strong possibility is that they were produced in New York before the partners sailed for California.

1849 Norris, Gregg & Norris $5 gold piece struck in Benicia City on San Francisco Bay.

Moffat & Company
San Francisco, California (1849–1853)

Moffat & Co., while not the first coiner of gold in California, became the most important private mint in San Francisco. At a time when the coinage of other assayers, bankers, and minters was being seriously questioned, the issues of Moffat were readily accepted by merchants. Later, the facilities of the firm were incorporated into the United States Assay

1849 Moffat & Co. $5 gold coin.

Office of Gold and, still later, the San Francisco Mint. The firm's name is from one of the partners, John Little Moffat (1788–1865), of New York City, who came to California in 1849 to recoup his fortune, which had dwindled in the recent past.

John Little Moffat, respected as an assayer in New York City with the firm of Wilmarth, Moffat & Curtis, departed that city on the bark *Guilford* February 14, 1849, taking with him assaying, refining, and coining equipment. He began business in San Francisco in the summer of 1849. Associated with him were Joseph R. Curtis, P.H.W. Perry, and Samuel H. Ward. Their office at Clay and Dupont streets was busy with the activity of trading in gold dust, refining it, and converting the metal to bars and ingots to ship to the East for sale. Moffat produced small, rectangular gold ingots ranging in value from $9.43 to $264. Most were of the value of $16. These are believed to have been first issued in June or July 1849. In time the firm issued $5, $10, and $20 coins.

United States Assay Office
Moffat & Company
San Francisco, California (1851)

Augustus Humbert $50 gold coin of the United States Assay Office.

In September 1850 Congress authorized the secretary of the Treasury to contract with a well-established assaying business in California to affix the stamp of the United States to bars and ingots, to assay gold, and to assign value to it. Moffat & Company, the most respected of the San Francisco coiners, received the commission. Augustus Humbert was appointed to the position of United States assayer. Humbert was a New York City maker of watch cases prior to his appointment. In preparation for the new franchise, in late 1850, Moffat & Co. curtailed most of their private business and prepared to issue coins under the government contract. New premises were secured on Montgomery Street between Clay and Commercial streets.

Humbert arrived in San Francisco in late January 1851, after which operations began immediately. At the same time the first octagonal $50 gold piece bearing his stamp was shown to the press, probably in the form of a trial piece brought from New York.

The *Daily Alta California* commented on the new $50 pieces on February 21, 1851:

> The new 50-dollar gold piece . . . was issued by Moffat & Co. yesterday. About three hundred of these pieces have already been struck off. . . . The coin is peculiar, containing only one face, and the eagle in the center, around which are the words 'UNITED STATES OF AMERICA.' Just over the eagle is stamped "887 THOUS." signifying the fineness of the gold. At the bottom is stamped '50 DOLLS.' The other face is ornamented with a kind of work technically called engine-turning, being a number of radii extending from the common centre, in which is stamped, in small figures, '50.' Around the edge is stamped the name of the United States Assayer. . . .

Early $50 coins had lettering hand-stamped on each of the eight edge sides and 50 hand-stamped on the obverse. Soon, this information was incorporated into the dies, and reeded edges were used.

While the federal standard for gold coinage was .900 fine, in San Francisco in 1851 this was difficult to attain with the refining processes then in use, and the Humbert coinage was of two finenesses, .880 and .887, the latter coins being slightly lighter in overall weight due to the smaller proportion of alloy. By 1852 coins of .900 fineness were being made, but other finenesses (.884 and .887) were employed as well. The remaining alloy was native silver (whereas under the government standard, copper was used).

Although the Humbert $50 pieces were clearly produced under government auspices, and although they were receivable for U.S. customs payment in San Francisco, in Philadelphia on April 23, 1851, Mint Director George N. Eckert perversely (it would seem) stated that while Augustus Humbert was the United States Assayer in California, his stamping of bars for owners of bullion did not make them legal tender.

United States Assay Office of Gold
Moffat & Company
Curtis, Perry & Ward
San Francisco, California (1852–1853)

On December 24, 1851, the *Daily Alta California* carried this notice:

> The firm heretofore known and existing under the name and style of Moffat & Co. is this day dissolved by mutual consent, the entire interest of the special partner, John L. Moffat, having been purchased by the remaining partners, who have the right to use the name of Moffat & Co.

The declaration was signed by John L. Moffat, Joseph R. Curtis, Samuel Ward, and Philo H. Perry. It was further stated that "the firm will hereafter consist of the undersigned remaining partners, and its business until further notice will be conducted under the name and style of Moffat & Co." The addendum was signed by Curtis, Perry, and Ward.

Coins were made with the imprint of Moffat & Company and also the United States Assay Office of Gold. Denominations of $10 and $20 were struck and circulated, the $20 gold piece with the United States Assay Office of Gold being made in large quantities. Some $20 coins were made in 1853 with the imprint of Moffat & Co.

United States Assay Office of Gold.

The *Daily Alta California* on March 22, 1853, contained a description of the business:

> The machinery made use of by Messrs. Curtis, Perry & Ward is of the same description, made by the same mechanics, and is as perfect in all respects, as that of the United States Mint at Philadelphia. The capacity of their press is such to enable them to coin $360,000 in $10 pieces and $720,000 in $20 per day, and it keeps up with their facility for drawing, cutting, and adjusting by being worked only a few hours per day.
>
> The mechanical execution of the coin itself is fully equal to that of the United States Mint, as will be seen by a comparison of the coins. Too much credit cannot be awarded to Messrs. Curtis, Perry & Ward for the radical change in the facilities for coinage offered by them to the people of this state while at the same time it is advantageous to them personally.

On December 4, 1853, the United States Assay Office of Gold ceased operations. Machinery and equipment were transferred to the new San Francisco Mint. (See chapter 8 for the San Francisco Mint.)

Miners' Bank
Wright & Co.
San Francisco, California (1849)

The firm of Wright & Co., exchange brokers, was located at the corner of Washington and Kearny streets, Portsmouth Square, San Francisco, beginning early in September 1849. On August 7 the firm had requested permission from local authorities to coin $5 and $10 gold coins, declaring they would be worth as much as federal issues. Authorization was not

1849 Miners' Bank $10 gold piece.

granted. In November, Wright & Co. reorganized. Composing the new company were Stephen A. Wright, John Thompson, Samuel W. Haight, and J.C.L. Wadsworth. Known as the Miners' (or Miners or Miner's—punctuation varied) Bank, the outfit was housed in a wooden frame structure for which the incredible sum of $75,000 rent per year was paid.

It is believed that the $10 coins were produced in the autumn, apparently before the November 1849 reorganization was completed, for William P. Hoit, assayer of the New Orleans Mint, reported on December 13, 1849, that he had assayed a Miners' Bank $10 piece nearly two months earlier, and that he had found it to be worth only $9.65. In 1849, prior to the adoption of the California Constitution that prohibited such, the company issued paper money on which Miner's had an apostrophe.

Miners' Bank $1 note signed and issued,
March 1, 1849. (Smithsonian Institution)

Alta California reported this on April 11, 1850: "The issue of the Miners' Bank is a drug on the market. Brokers refuse to touch it at less than 20 percent discount. . . ." On December 14, 1850, the Miners' Bank dissolved. As the Miners' Bank $10 pieces no longer circulated at par, the pieces in the hands of the public went to bullion dealers at a discount and were melted. Within a few years they were rare.

J.S. Ormsby
Sacramento, California (1849)

In April 1849 Dr. J.S. Ormsby was a member of a group of adventurers from Pennsylvania who headed westward from St. Joseph, Missouri, where thousands of gold seekers had camped to await favorable spring conditions before traveling across the prairie and desert lands to California. A newspaper account noted that as of April 14 the party included L.P. Ormsby; Major William M. Ormsby

J.S. Ormsby $5 gold
piece, Sacramento.

(said to be from Peru; other information not given); J.K. Trumbull (of Kentucky); and A. McLain, J. Moats, J. Shutt, M.L. Detter, J. McManus, and Samuel Stauffer (all of Westmoreland, Pennsylvania). The Ormsby group was outfitted with four wagons, each drawn by six mules and filled with supplies for the long trek.[2] This contingent, along with many others, was "ready to move" and had been awaiting the arrival of warm-enough weather to do so.

Presumably, the entourage experienced many of the rigors recorded by others on the California Trail, although no diary of the journey has been seen. After arriving in California, the Ormsbys settled in Sacramento and opened an assaying, refining, and coining business, J.S. Ormsby & Co., on K Street. Coins were struck of the $2.50 and $5 denominations.

Pacific Company
Broderick & Kohler
San Francisco, California (1849)

David C. Broderick and F.H. Kohler operated the Broderick & Kohler assaying office in San Francisco. Although facts are scarce, it has been suggested by some numismatic historians, including Edgar H. Adams, that they made the low-value Pacific Company coins. Most are crudely struck, probably by a sledgehammer rather than in a coining press. Whatever the situation, by

1849 Pacific Company $5 gold piece.

early 1850 Broderick had left the firm to pursue other interests (he soon became prominent in state politics), and under the name of F.D. Kohler & Co., the other partner was highly regarded as a private assayer. In 1850 he was appointed California State Assay, after which he issued gold ingots, but not coins.

Dubosq & Co.
San Francisco, California (1850)

The *Philadelphia Evening Bulletin*, January 18, 1849, noted that the *Grey Eagle* had sailed and gave a passenger list, including Theodore Dubosq Sr., Theodore Dubosq Jr., and Henry A. Dubosq, and went on to state:

> Mr. Theodore Dubosq, jeweler, North Second Street, we understand takes out with him machinery for melting and coining gold, and stamping it with a private mark, so as to establish a currency which will afford the greater convenience and facility for dealing in the raw material.

The ship arrived in San Francisco on May 18, following a stop in Valparaiso. Although Dubosq arrived at a time in which coins were rare in circulation and other private minters met with success, it is not known whether gold coins were actually struck by him in the year 1849, although trial pieces in copper of that date are known to have been made. It is likely

1850 Dubosq & Co. $10 gold piece.

that Dubosq brought with him 1850-dated dies made in Philadelphia, possibly by James B. Longacre. The denominations were $5 and $10. It is known that Longacre was involved with Duosq's venture in some way by 1850, per entries in Longacre's diary. On April 16, 1850, he went to New Brunswick, New Jersey, to "meet Mr. [Peter F.] Cross." On the 17 he wrote that he: "Gave Mr. Cross the dies (1 pair) with the necessary directions to be made for Dubosq & Co."[3]

By 1850 Dubosq had formed the partnership of Dubosq & Goodwin, using the trade style Dubosq & Co. Today the only Dubosq & Co. gold coins known bear the date 1850, although it seems reasonable that at least some of these were struck in 1849 and perhaps additional pieces in 1851.[4] Coinage continued apace in early 1851, as evidenced by data published in San Francisco, noting that from January 1 to March 31 of that year Dubosq & Co. had produced $150,000 worth of gold. As no specimens bearing the date 1851 are known, these pieces may have been dated 1850. In 1852 1850-dated dies for a Dubosq $10 piece were employed to strike pieces for Wass, Molitor & Co. The obverse die was changed to read W.M. & Co., and the date was changed to read 1852—a remarkable situation. The Dubosq and most other privately minted gold coins, except for those of Moffat & Co. and Humbert, fell out of favor after coins were sent to Augustus Humbert at the Moffat office to be assayed. Although most were near full gold value, they were unfairly discredited in the newspapers, and minting ceased.

Baldwin & Co.
San Francisco, California (1850–1851)

The firm of Baldwin & Co. was founded on March 15, 1850, as the successor to Frederick D. Kohler & Co., a California state assayer. George C. Baldwin and Thomas S. Holman advertised Baldwin & Co. as assayers, refiners, and coiners who also did "all kinds of engraving." The boast, "our coin redeemable on presentation," was made. In 1850 there

1850 Baldwin & Co. "vaquero" $10.

was a shortage of gold coins of all kinds in circulation, and for a time the $5 pieces made in quantity by Baldwin that year circulated widely.

Early in 1851 private coinage was conducted at a furious pace in San Francisco. Baldwin coins fell from grace when assayer Augustus Humbert found that a sample of 13 $20 coins averaged an intrinsic value of $19.40, 10 $10 pieces averaged a value of $9.74, and 10 $5 pieces were valued at an average worth of $4.91. Later pieces issued by this firm changed hands at a 20% discount, a figure significantly less than their metallic value, thus enriching anyone who took them in trade and melted them.

Schultz & Co.
San Francisco, California (1851)

$5 gold coins were struck in 1851 by the firm of Schultz & Co. The design is a close copy of the contemporary United States $5 gold coin. The obverse bears on the coronet of Liberty the inscription SHULTS & CO., a misspelling eliminating the C and incorporating a final S instead of the Z.

1851 Schultz & Co. $5 gold piece.

Schultz & Co. was comprised of Judge G.W. Schultz and William Thompson Garratt. Garratt conducted a brass foundry on Clay Street, San Francisco, as early as 1850 and produced many of the dies used for private coinage by other firms in the city. His foundry was directly behind Baldwin's coining establishment. The partnership between Schultz and Garratt operated for just a short time and was dissolved in April 1851. Sample coins examined by Augustus Humbert averaged $4.87 in intrinsic value, which, while seemingly reasonable, was considered at the time to be lightweight. As a result the mintage stopped, and Schultz coins sold at a discount in commerce.

In later years William T. Garratt furnished a description of the early activities:

> We made a great many dies for private coining. Albert Küner, who is still in business here, would do the engraving and I the turning—that is, the machine work on the dies, for which at the time we would get $100 per day per man on that special job. After that, Schultz took a notion to go into coining for Burgoyne & Co. and Argenti & Co., who were bankers here at the time. They would buy the dust and we would do the coining. We ran for a while, and then Schultz and I separated, he taking the coining establishment and I the foundry, he keeping the room over the foundry for his business. He continued only a short time before the Legislature passed a law prohibiting private coining.

Dunbar & Co.
San Francisco, California (1851)

Edward E. Dunbar arrived in San Francisco in 1849, and not long thereafter he opened Dunbar's California Bank at 49 Washington Street, which became highly profitable. Through his bank, and also a separate office, Dunbar traded in bullion and specie, offering in December 1850 and early January 1851 to redeem at par the numerous gold issues of Baldwin & Co., this being before the Baldwin issues were discredited.

1851 Dunbar & Co. $5 gold coin.

In 1851 Dunbar commissioned Albert Küner to cut dies for a $5 piece. Probably at least a few thousand were struck, perhaps far more. However, today only a handful are known, despite the fact that in the same year the Philadelphia Mint assayed a single batch containing 111 pieces![5] Although no Dunbar coins were assayed by Humbert, they and all other coins except those of Moffatt & Co. and Humbert eventually fell under a cloud and were accepted only at a discount by merchants. In 1867 Dunbar's book, *The Romance of the Age; or, the Discovery of Gold in California*, was published in New York by D. Appleton and Company.

Wass, Molitor & Co.
San Francisco, California (1852–1855)

Among the private issuers of gold coins in San Francisco, Wass, Molitor & Co. was one of the most important, although their initial production did not begin until relatively late, in 1852. Count S.C. Wass and A.P. Molitor, Hungarians, were earlier engaged in refining and assaying in the same city. The *Daily Alta California* of January 8, 1852 noted:

Wass, Molitor & Co. 1852 $5 gold piece.

The day before yesterday we were shown a piece of the denomination $5 which Messrs. Wass, Molitor & Co. are preparing to issue from their assay office, Naglee's Building, in Merchant St. It has the head and stars like the American coin, with the letters WM & CO. in the place occupied by the word LIBERTY on our National currency. Below is the date, 1852. On the reverse is the eagle, with the words 'In California Gold—Five Dollars' around it. The coin has the pale yellow appearance which is peculiar to the private coinage of the State, and which is caused by the silver alloy natural to the gold, whereas the issues from the United States Mint are slightly alloyed with copper.

The Humbert assaying situation having been forgotten and the Wass, Molitor & Co. coins having enjoyed a good reputation, the firm prospered. Coins were made through 1855. In the last year round $50 coins were made.

Kellogg & Co.
San Francisco, California (1854–1855)

The last major entry in the field of private coinage in San Francisco was Kellogg & Co., which produced its first coins in 1854. John Glover Kellogg, of Auburn, New York, came to San Francisco on October 12, 1849. He secured a position with Moffat & Co. and remained with them during the operations of the United States

1855 Kellogg & Co. $20 gold coin.

Assay Office of Gold. When the latter institution discontinued business on December 14, 1853, and began the changeover of facilities that would lead to the opening of the San Francisco Mint, Kellogg formed a new partnership with G.F. Richter, who had earlier worked with the United States Assay Office as an assayer.

On January 14, 1854, a number of leading banking houses of San Francisco and Sacramento addressed a petition to Kellogg & Richter imploring them to produce coins, in the period after the United States Assay Office of Gold had ceased operations and before the United States Mint at San Francisco had begun production. The merchants indicated their willingness to receive any coins that would be produced. Kellogg & Richter jumped at the opportunity, and on February 9, 1854, the first Kellogg $20 coin was issued.

Following the opening of the San Francisco Mint, production at the government facility was quite limited. Kellogg & Co. therefore did a large business by continuing their private coinage. Toward the end of 1854 the firm of Kellogg & Richter was dissolved and a new firm, Kellogg & Humbert, took its place, with Augustus Humbert, formerly the United States assayer connected with Moffat, joining. This partnership continued until 1860, although the last coins were struck in 1855.

Oregon Exchange Co.
Oregon City, Oregon (1849)

Citizens of Oregon returning from the California gold fields in 1849 brought with them quantities of gold dust and nuggets. The Oregon Legislature on February 15, 1849, passed an act which provided for the establishment of a territorial mint. However, the governor declared this act to be in contravention to the laws of the United States, and plans for the mint were terminated.

1849 Oregon Exchange Co.
$5 gold coin.

To remedy the situation, a group of eight prominent merchants and citizens banded together to establish a private mint. The principals were W.K. Kilborne, Theophilus Magruder, James Taylor, George Abernathy, W H. Wilson, William H. Rector, J. G. Campbell, and Noyes Smith. The firm was designated as the Oregon Exchange Co. Hamilton Campbell, a Methodist missionary, was employed to cut dies for a $5 coin. Victor Wallace, machinist, engraved the dies for a coin of the $10 denomination. The coins produced were to be virgin gold without alloy.

The $5 gold dies bore on the obverse the initials K.M.T.A.W.R.G.S., representing the names of the company members. The G was an error and should have been C for Campbell. The obverse of the $5 piece pictured a beaver on a log, facing to the right, the same animal which, being a trademark of the Territory, was earlier used on the Northwest Co. tokens. Below was the designation T. O. for Territory of Oregon, and below that, the year 1849, with branches to the sides. On the reverse appeared the notation OREGON EXCHANGE COMPANY, 130 G. NATIVE GOLD 5 D. The pieces contained 130 grains of gold, or nearly 5-1/2 pennyweights.

Coinage amounted to approximately 6,000 of the $5 pieces and 2,850 $10 coins. These were accepted as legal tender throughout the Oregon Territory, which at that time included the present states of Oregon and Washington and all land east of these reaching to the Rocky Mountains. Oregon City had approximately 1,000 white citizens (as the population was counted), while the entire Territory contained about 9,000 inhabitants. Many of the Oregon issues were sent to California in payment for merchandise. Eventually nearly all were melted. By a decade after the original issue only a small number of Oregon coins existed.

Mormon Gold Coinage
Salt Lake City, Utah (1849–1860)

In the autumn of 1848, Mormons returning from California brought large quantities of gold dust into the Great Salt Lake area. In the settlement at Great Salt Lake, called the State of Deseret (*deseret* being another words for *honeybee*, a Mormon symbol of industriousness), Dr. Willard Richards, an official of the Mormon Church, weighed the gold dust and distributed it in paper packages which contained

1850 Mormon $5 gold coin.

from $1 to $20 in value. In November 1848 coinage designs were formulated. Each piece was to depict on one side the priesthood emblem, a three-pointed Phrygian crown over the all-seeing eye of Jehovah, with the phrase "Holiness to the Lord." The reverse was to bear the inscription "Pure Gold," clasped hands, and the denomination.

Coins of the $2.50, $10, and $20 denominations were minted with the date 1849, and $5 coins were made from 1850-dated dies, and then years later in 1860 from dies with a lion obverse and eagle reverse. Mormon coins had gold content significantly below their face value. In areas other than Salt Lake City they only circulated at a discount of 10% to 25% less than face value. Within Salt Lake City itself there were numerous questions raised, and in 1851 and 1852 many people were reluctant to accept the pieces, but the church applied pressures which made the coins circulate. Eventually Brigham Young, the Mormon leader, closed the mint, and the pieces disappeared from circulation.

Clark, Gruber & Co.
Denver, Colorado (1860–1861)

Note: This mint evolved into the federal Denver Mint (see chapter 10). For this reason, expanded coverage is given here.

GOLD IN THE ROCKY MOUNTAINS

In 1858 in Kansas Territory in the foothills of the Rocky Mountains at the western edge of the prairie, the town of Denver was established, taking its name from James W. Denver, the territorial governor. Not long afterward, gold nuggets and particles were discovered in nearby Cherry Creek. The time was propitious for individual fortune seekers, as California gold mining had tran-

1860 Clark, Gruber & Co.
Pikes Peak $10 gold coin.

sitioned into production by large companies. Word spread, and in 1859 the rush began. "Pikes Peak or Bust" was lettered on the side of Conestoga wagons, never mind that Cherry Creek was about 70 miles north of that mountain.

On March 23 the *Worcester* (Massachusetts) *Palladium* carried this:

Pike's Peak.

The *Lawrence* (Kansas) *Republican* publishes a letter from the gold mines at Pike's Peak, dated El Paso, Jan. 28, in which the writer states that he and others had just traced a vein . . . eight miles towards the mountains, and were well repaid for the time and expense. It was nothing uncommon to find lumps worth from $10 to $15. The people were flocking to the vein from all directions, and 600 men were then at work in the diggings. Speculation, of course, was rife, and town shares were changing hands rapidly. The tide of emigration was already setting in. The previous week a company of California miners, consisting of forty men and two women, arrived.

In 1859 an estimated 10,000 fortune hunters arrived to stake claims, including many from the California mines. Horace Greeley, the widely traveled editor of the *New York Tribune*, visited Denver and stated that the only hotel in town was a tent. Such was rough-and-ready frontier life. In May 1859 other gold discoveries were made in Gilpin County not far to the west, where Central City was established. In time many more strikes would be made.

In the *Annual Report of the Director of the Mint,* covering the fiscal year ended June 30, 1860, Director James Ross Snowden took notice of a private Mint in Denver in Jefferson Territory, the new name for the western part of Kansas that was split off and renamed Colorado in the next year:

> In the gold producing region of Kansas, namely at Denver City, a private minting establishment has been set in operation by Messrs. Clark, Gruber & Co., from which pieces of ten and five dollars are issued. They are of various grades of fineness; our assays [conducted at Philadelphia] show them to be from 815 to 838 thousandths, and the pieces are evidently made directly from native gold with its silver alloy, without any attempt to fix or maintain any exact standard. The weight is greater than in corresponding pieces of the national coinage in order to make up for the deficiency of fineness. The 10-dollar pieces vary from 273 to 283-1/2 grains. On the average, and adding the value of the silver alloy, and deducting the Mint charges, the pieces are found to be of professed value, or slightly over. The devices on the 10-dollar pieces are appropriate and distinctive; but on the five dollar piece they are made in close imitation of the legal coin, a reprehensible and illegal practice, countenanced by previous similar emissions in California.

From that beginning was to grow Colorado's largest mint. The partners in Clark, Gruber & Co. were Austin M. Clark, Milton Edward Clark, and Emanuel Henry Gruber.

The Bank and Mint of Clark, Gruber & Co., Denver, 1860.

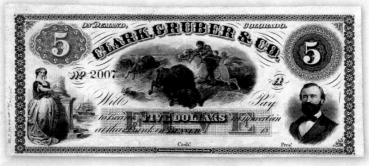

$5 note issued by Clark, Gruber & Co.

Framework for the minting business was begun in December 1859 when Milton E. Clark journeyed to Boston and Philadelphia to purchase assaying, refining, and minting equipment. Certain of the dies may have been made by Ernest G. Chormann, a Philadelphia engraver.

In the spring of 1860 Austin Clark and Emanuel H. Gruber arrived in Denver and purchased several lots on the northwest corner of McGaa and F Streets, later to become Market and 16th streets. An imposing two-story brick structure with a stone basement was set up. In April the machinery arrived by an ox-drawn wagon. George W. McClure came from Iowa to put the equipment in order. Later he was named as the assayer and coiner. By July 16 the building was complete inside and out, and coinage operations were ready to begin. On July 28 $10 coins were minted. Eventually $2.50 and $20 coins were added. The 1860 $10 and $20 coins had a volcanic-shaped peak on the obverse. Other issues of the two years had a Liberty Head in imitation of federal issues. A branch of the company was opened in Central City to receive gold and pay out coins.

In 1862 the business was sold to the United States Treasury Department and renamed the Denver Mint (see chapter 10).

The coining equipment from the earlier company was stored in the basement of the Denver Mint until 1898, when the Treasury Department transferred it to the rooms of the Colorado State Historical Society in the State Capitol Building.

John Parsons & Company
Tarryall, Colorado (1861)

John D. Parsons, or Parson, a medical doctor from Quincy, Illinois, went to Denver in 1858, where for a time he was in the real estate trade. By 1861 he was in the coining business at Tarryall Mines, striking $2.50 and $5 pieces bearing his imprint. The location of his mint was midway between Tarryall and Hamilton, near Como, on the way to Breckenridge.

John Parsons & Co.
$2.50 gold coin.

The obverse of each of these coins shows a large quartz stamping mill, J. PARSON [*sic*] & Co. (on the $2.50) or JNo PARSON [*sic*] & Co. (on the $5), and the word ORO, that being Spanish for *gold*. The reverse illustrates an eagle with inscriptions including PIKES PEAK GOLD. Today, only a few examples are known of each of the two denominations.

J.J. Conway & Company
Georgia Gulch, Colorado (1861)

In 1861, in a small settlement near Parkville (later to become a ghost town) in the Rocky Mountains at Georgia Gulch (not far from present-day Breckenridge), the jewelry and banking firm of J.J. Conway & Co. struck gold coins of the $2.50, $5, and $10 denominations. Bearing the firm's imprint, the dies may have been ordered through a jewelry supply house, as they closely resemble in style (but not denominations) certain trade tokens of that era, such as those made by John Stanton of Cincinnati.

J.J. Conway & Co.
$2.50 gold piece.

The pieces were intended to serve local miners, who at the time were receiving just $14 to $15 per ounce for gold dust of high purity, said to be worth $18 per ounce if coined. Conway turned out coins that proved to be debased when they were assayed in Denver. The coins had a fineness of .7725, considerably less than the federal standard of .900. Another assay, this conducted on behalf of Conway, showed a fineness of .882, which was better, but still below the standard.

As to whether these coins ever achieved utility in and around Parkville is not known, but the production must have been very small, as fewer than 20 coins are known totally of four varieties (one variety of $2.50 gold, two varieties of $5 gold, and one of $10 gold).

Manila Mint
Philippine Islands

This section is condensed and adapted from Dennis Tucker's history of Philippine coinage and the Manila Mint, first published in the Deluxe Edition *of the* Guide Book of United States Coins, *popularly referred to as MEGA RED (which includes more detailed series-by-series information on the coinage of the Philippines struck under U.S. sovereignty, a date-by-date study of typical prices, and grading instructions with additional photographs).*

In April 1899 control of the Philippine Islands was officially transferred from Spain to the United States, as a condition of the treaty ending the Spanish-American War. The U.S. military suppressed a Filipino insurgency through 1902, and in July 1901, the islands' military government was replaced with a civilian administration led by American judge William Howard Taft. One of the civilian government's first tasks was to sponsor a new territorial coinage that was compatible with the old Spanish issues, but also legally exchangeable for American money at the rate of two Philippine pesos to the U.S. dollar. The resulting coins bear the legend UNITED STATES OF AMERICA, which has sometimes caused them to be confused with standard U.S. coins. Otherwise they are quite different in design from regular federal coins. Today they can be found in many American coin collections, having been brought to the States as souvenirs by service members after World War II, or otherwise saved.

The coins, introduced in 1903, were designed by Filipino silversmith, sculptor, engraver, and art professor Melecio Figueroa, who had earlier worked for the Spanish *Casa de Moneda*, in Manila. They are sometimes called "Conant coins" or "Conants," after Charles Arthur Conant, an influential American journalist and banking expert who served on the commission that brought about the Philippine Coinage Act of March 2, 1903.

Following Spanish custom, the dollar-sized peso was decimally equivalent to 100 centavos. Silver fractions were minted in denominations of fifty, twenty, and ten centavos, and minor coins (in copper-nickel and bronze) included the five-centavo piece, the centavo, and the half centavo.

In addition to the name of the United States, the coins bear the Spanish name for the islands: FILIPINAS. The silver coins feature Miss Liberty standing in a flowing gown, holding in one hand a hammer that she strikes against an anvil, and in the other an olive branch, with the volcanic Mount Mayon (northeast of the capital city of Manila) visible in the background. The minor coinage shows a young Filipino man, bare-chested and seated at an anvil with a hammer, again with Mount Mayon seen in the distance. The first reverse design, shared across all denominations, shows a U.S. federal shield surmounted by an eagle with outstretched wings, clutching an olive branch in its right talon and a bundle of arrows in its left. This reverse design was changed in 1937 to a new shield emblem derived from the seal of the 1936 Commonwealth.

Dies for the coins were made at the Philadelphia Mint by the U.S. Mint's chief engraver, Charles E. Barber. Mintmarks were added to the dies, as needed, at the branch mints. From 1903 to 1908 the coins were struck at the Philadelphia Mint (with no mintmark) and the San Francisco Mint (with an S mintmark). From 1909 through 1919, they were struck only at San Francisco. In the first part of 1920, one-centavo coins were struck in San Francisco; later in the year a new mint facility, the Mint of the Philippine Islands, was opened in Manila, and from that point into the early 1940s Philippine coins of one, five, ten, twenty, and fifty centavos were struck there. The coins produced at the Manila Mint in 1920, 1921, and 1922 bore no mintmark. No Philippine coins were struck in 1923 or 1924. The Manila Mint reopened in 1925; from then through 1941 its coinage featured an M mintmark. The Denver and San Francisco mints would be used for Philippine coinage in the final years of World War II, when the islands were under Japanese occupation.

In addition to normal coins and paper currency, special token money in the form of coins and printed currency was made for use in the Culion Leper Colony. The token coinage saw six issues from 1913 to 1930, some produced at the Manila Mint. The leper colony was set up in May 1906 on the small island of Culion, one of the more than 7,000 islands comprising the Philippines, and the coinage was intended to circulate only there.

In 1935 the United States, responding to popular momentum for independence, approved the establishment of the Commonwealth of the Philippines, with the understanding that full self-governing independence would be recognized after a ten-year transition period. To celebrate this transfer of government, the Manila Mint in 1936 produced a set of three silver commemorative coins—one of the fifty-centavo denomination, and two of the one-peso. These were designed by Ambrosio Morales, professor of sculpture at the University of the Philippines School of Fine Arts.

Fifty Centavos.

The fifty-centavo coin and one of the set's pesos feature busts of Philippine president Manuel L. Quezon and the last U.S. governor-general, Frank Murphy, who served (largely ceremonially) as the first U.S. high commissioner to the Commonwealth of the Philippines. The other peso has busts of Quezon and U.S. president Franklin D. Roosevelt. The set's issue price was $3.13, or about 2.5 times the coins' face value expressed in U.S. dollars. Commemorative coins were popular in the United States at the time, but these sets still sold poorly, and thousands remained within the Philippine Treasury at the onset of World War II.

One Peso, Busts of Murphy and Quezon.

One Peso, Busts of Roosevelt and Quezon.

Following the bombing of Pearl Harbor on December 7, 1941, Japanese military forces advanced on the Philippines later that month and in early 1942, prompting the civil government to remove much of the Philippine treasury's bullion to the United States. Nearly 16 million pesos' worth of silver remained, mostly in the form of one-peso pieces of 1907 through 1912. These coins were hastily crated and dumped into Manila's Caballo Bay to prevent their capture by Japan. Later during World War II, in 1944 and 1945, the U.S. Mint struck coins for the Philippines at its Philadelphia, Denver, and San Francisco facilities. These coins were brought over and entered circulation as U.S. and Philippine military forces fought to retake the islands from the Japanese.

After the war the Commonwealth of the Philippines became an independent republic, on July 4, 1946, as had been scheduled by the Constitution of 1935. Today the Philippine coins of 1903 to 1945, including the set of commemoratives issued in 1936, remain significant mementoes of a colorful and important chapter in U.S. history and numismatics. They are a testament to the close ties and special relationship between the United States of America and the Republic of the Philippines.

THE OPENING OF THE MANILA MINT

During U.S. sovereignty, much of the civilian government of the Philippines was administered by the Bureau of Insular Affairs, part of the War Department. Most heads or secretaries of Philippine government departments were appointed by the U.S. governor general, with the advice and consent of the Philippine Senate. In 1919, the chief of the Bureau of the Insular Treasury (part of the Department of Finance) was Insular Treasurer Albert P. Fitzsimmons, formerly a mayor of Tecumseh, Nebraska, and member of the municipal board of Manila. Fitzsimmons, a surgeon who had served in the U.S. Army Medical Corps in Cuba and the Philippines, was active in civil affairs, and had been in charge of U.S. government bond issues in the Philippines during the Great War. On May 20, 1919, he was named director *ad interim* of the Mint of the Philippine Islands, which was then being organized.

The genesis of this new mint began on February 8, 1918, when the Philippine Legislature passed an appropriations bill for construction of its machinery. The war in Europe was interfering with shipments from the San Francisco Mint, where Philippine coinage was produced, and a local mint was seen as more expedient and economical. In addition, a mint in Manila would serve the United States' goal of preparing the Philippines for its own governance and infrastructure.

The mint was built in Manila in the Intendencia Building, which also housed the offices and hall of the Senate, and the offices and vaults of the Philippine Treasury. Its machinery was designed and built in Philadelphia under the supervision of U.S. Mint chief mechanical engineer Clifford Hewitt, who also oversaw its installation in Manila. The facility was opened, with formalities and machine demonstrations, on July 15, 1920. The fanfare included the production of an official commemorative medal, the first example of which was struck by Speaker of the House of Representatives Sergio Osmeña.

The medal has since come to be popularly known as the "Wilson Dollar" (despite not being a legal-tender coin), because of its size and its bold profile portrait of Woodrow Wilson on the obverse, surrounded by the legend PRESIDENT OF THE UNITED STATES. The reverse features the ancient Roman goddess Juno Moneta guiding a youth—representing the fledgling mint staff of the Philippines—in the art of coining. She holds a pair of metallurgical scales. The reverse legend is TO COMMEMORATE THE OPENING OF THE MINT / MANILA P.I., along with the date, 1920. The medal was designed by Hewitt, the mint's supervising engineer from Philadelphia. Its dies were made by U.S. Mint chief engraver George T. Morgan, whose initial, M, appears on the obverse on President Wilson's breast and on the reverse above the goddess's sandal.

The Manila Mint in 1920.

The issue was limited to 2,200 silver medals (2,000 of which were struck on the first day), sold to the public at $1 apiece; and 3,700 in bronze, sold for 50¢ each. In addition, at least five gold specimens were reportedly struck. These included one for presentation to President Wilson and one for U.S. Secretary of War Newton Baker. The other gold medals remained in the Philippines and were lost during World War II.

Interior of the new mint facility.

Of the medals unsold and still held by the Treasury in the early 1940s, some or all were dumped into Caballo Bay in April 1942 along with millions of silver pesos, to keep them from the approaching Japanese forces. The invaders learned of the coins, and in May they attempted to recover the sunken silver coins using the labor of Filipino divers. Although skilled divers, the Filipinos were not experienced in deep-sea diving, and the coins were at the bottom of the bay, 120 feet below the surface. After three deaths the Filipinos refused to participate in further recovery efforts. The Japanese then forced U.S. prisoners of war who were experienced deep-sea divers to recover the sunken treasure. The American divers conspired to salvage only small quantities of the sunken treasure. They repeatedly sabotaged the recovery process, and smuggled a significant number of recovered silver coins to the Philippine guerillas. Only about 2 to 3 percent of the dumped coinage was recovered before the Japanese ceased recovery operations.

Sergio Osmeña was Speaker of the House of Representatives of the Philippines in 1920, the year he struck the first medal commemorating the opening of the Manila Mint. He would later serve as the nation's fourth president.

Following the war the United States brought up much of the coinage that had been dumped into the sea. Many of the recovered silver and bronze Wilson dollars in grades VF through AU bear evidence of saltwater corrosion.

The Manila Mint Opening medal in gold.

The Manila Mint Opening medal in silver.

The Manila Mint Opening medal in bronze.

The Manila Mint Opening medal is popular with collectors of Philippine coins and of American medals. It is often cataloged as a *so-called dollar*, a classification of historic dollar-sized souvenir medals, some of which were struck by the U.S. Mint and some produced privately. The Manila Mint Opening medal is valued for its unique connections to the United States and to American numismatics.

A FEW NOTES ABOUT VARIOUS PHILIPPINE COINS

Half Centavos. In 1903 and 1904 the United States minted nearly 18 million half centavos for the Philippines. By March of the latter year, it was obvious that the half centavo was too small a denomination, unneeded in commerce despite the government's attempts to force it into circulation. The recommendation of Governor-General Luke Edward Wright—that the coin be discontinued permanently—was approved, and on April 18, 1904, a new contract was authorized to manufacture one-centavo blanks out of unused half-centavo blanks. In April 1908 Governor-General James Francis Smith received permission to

Half Centavo.

ship 37,827 pesos' worth of stored half centavos (7,565,400 coins) to the San Francisco Mint to be re-coined into one-centavo pieces. Cleared from the Philippine Treasury's vaults, the coins were shipped to California in June 1908, and most of them were melted and made into 1908 centavos.

Centavo. Unlike the half centavo, the bronze centavo was a popular workhorse of Philippine commerce from the start of its production. Nearly 38 million coins were minted in the denomination's first three years. In 1920 the centavo was struck for circulation by two different mints (the only year this was the case). San Francisco produced the coins during the first part of 1920; later in the year,

One Centavo.

the coins were struck at the Manila Mint, after that facility opened. From that point forward all centavos struck under U.S. sovereignty were products of the Manila Mint. Those dated 1920, 1921, and 1922 bear no mintmark identifying their origin. Manila produced no coins (of any denomination) in 1923 and 1924.

Coin collectors, notably educator, writer, and American Numismatic Association member Dr. Gilbert S. Perez, urged Manila Mint officials to include an M mark on their coinage—similar to the way that, for example, San Francisco coins were identified by an S—and this change was made starting with the coinage of 1925.

Five Centavos. Five-centavo coins were minted under U.S. sovereignty for the Philippines from 1903 to 1935, with several gaps in production over the years. Circulation strikes were made in Philadelphia in 1903 and 1904; then coinage resumed in 1916, this time at the San Francisco Mint. The newly inaugurated Manila Mint took over all five-centavo production starting in 1920, continuing through the end of direct U.S. administration, and under the Commonwealth government beginning in 1937.

The Manila Mint coins of 1920 and 1921 bore no mintmark indicating their producer, a situation noticed by coin collectors of the day. Gilbert S. Perez, superintendent of schools for Tayabas in the Philippines (and a member of the American Numismatic Association), wrote to Assistant Insular Treasurer Salvador Lagdameo in June 1922: "Several members of numismatic societies in Europe and America have made inquiries as to why the Manila mint has no distinctive mint mark. Some do not even know that there is a mint in the Philippine Islands and that the mint is operated by Filipinos." He recommended the letter M be used to identify Manila's coins. Lagdameo replied later that month, thanking Perez and informing him: "It is now

Five Centavos, Large Size (1903–1928, 20.5 mm).

Five Centavos, Reduced Size (1930–1935, 19 mm).

being planned that the new dies to be ordered shall contain such mark, and it is hoped that the coins of 1923 and subsequent years will bear the distinctive mint mark suggested by you." Coinage would not resume at the Manila facility until 1925, but from that year forward the mintmark would grace the coins struck in the Philippines.

Five-centavo coinage under U.S. sovereignty continued to 1935. From 1937 on, the Manila Mint's production of five-centavo coins would use the new Commonwealth shield design on the reverse.

Ten Centavos. The Philippine ten-centavo coin was minted from 1903 to 1935, in several facilities and with occasional interruptions in production.

Ten Centavos, Large Size (1903–1906, 17.5 mm).

In 1907 the silver ten-centavo coin's fineness was reduced from .900 to .750, and at the same time its diameter was decreased. This was in response to the rising price of silver, with the goal of discouraging exportation and melting of the silver coins. The net effect was nearly 40 percent less silver, by actual weight, in the new ten-centavo piece. The older coins continued to be removed from circulation, and by June 30, 1911, it was reported that only 35 percent of the ten-centavo pieces minted from 1903 to 1906 still remained in the Philippines.

Ten Centavos, Reduced Size (1907–1935, 16.5 mm).

The Manila Mint took over ten-centavo production from San Francisco in 1920. The ten-centavo coins of 1920 and 1921 bear no mintmark identifying them as products of Manila (this was the case for all Philippine coinage of those years, and of 1922). The efforts of Philippine numismatists, including American Numismatic Association member Gilbert S. Perez, encouraged mint officials to add the M mintmark when the facility reopened in 1925 after a two-year hiatus for all coinage.

Ten-centavo production after 1921 consisted of 1 million pieces struck in 1929 and just fewer than 1.3 million pieces in 1935. The next ten-centavo mintage would be under the Commonwealth, rather than U.S. sovereignty.

Twenty Centavos. In the early 1900s the rising value of silver was encouraging exportation and melting of the Philippines' silver twenty-centavo coins. As was the case with the ten-centavo piece, in 1907 the diameter of the twenty-centavo coin was reduced and its silver fineness decreased from .900 to .750. The net effect was about 40 percent less silver, by actual weight, in the new smaller coins. Attrition continued to draw the older coins out of circulation and into the melting pot, as their silver value exceeded their face value. A report of June 30, 1911, held that only about 25 percent of the twenty-centavo coins minted from 1903 to 1906 still remained in the Philippines.

Twenty Centavos, Large Size (1903–1906, 23 mm).

Twenty Centavos, Reduced Size (1907–1929, 20 mm).

Circulation strikes were made at the Philadelphia and San Francisco mints through 1919. The Manila Mint's output during the period of U.S. sovereignty included twenty-centavo pieces in 1920, 1921, 1928, and 1929. (Production of the coins later continued under the Commonwealth, with a slightly modified design.) The first two years of coinage did not feature a mintmark identifying Manila as the producer of the coins. As mentioned earlier, this was noticed by collectors of Philippine coins; they protested the oversight, and later coinage dies had an M mintmark added.

In 1928 a rush order for twenty-centavo coins was received at the Manila Mint—by that time the only producer of the denomination. Manila had not minted the coins since 1921, and the Philadelphia Mint had not shipped any new reverse dies (which would have featured their 1928 date). Under pressure to produce the coins, workers at the Manila Mint married a regular twenty-centavo obverse die with the 1928-dated reverse die of the five-centavo denomination, which was only .5 mm larger. As a result, the entire mintage of 100,000 1928 twenty centavos consists of these "mule" (mismatched-die) coins. The reverse of the 1928 coins, compared with others of 1907 to 1929, has a narrower shield and a larger date.

Fifty Centavos. After four years of fifty-centavo coinage, in 1907 the denomination's silver fineness was lowered from .900 to .750, and its diameter was reduced by ten percent. This action was in response to rising silver prices. The new smaller coins contained 38 percent less silver, by actual weight, than their 1903–1906 forebears, making them unprofitable to melt for their precious-metal content.

Fifty Centavos, Large Size (1903–1906, 30 mm).

The reduced-size coins of the U.S. sovereignty type were minted from 1907 to 1921. The Manila Mint took over their production from the Philadelphia and San Francisco mints in 1920, using coinage dies shipped from Philadelphia. The fifty centavos was the largest denomination produced at the Manila Mint. Neither the

Fifty Centavos, Reduced Size (1907–1921, 27 mm).

1920 nor the 1921 coinage featured a mintmark identifying Manila as its producer.

Production of the fifty-centavo denomination would again take place, in 1944 and 1945, in San Francisco, using the Commonwealth design introduced for circulating coins in 1937.

One Peso. The Philippine silver peso was struck under U.S. sovereignty from 1903 to 1912. The key date among those struck for circulation is the issue of 1906-S. Although the San Francisco Mint produced more than 200,000 of the coins that year, nearly all of them were held back from circulation. They were instead stored and then later sold as bullion.

Peso, Large Size (1903–1906, 38 mm).

By 1906 natural market forces were driving the Philippine silver pesos out of commerce and into the melting pot: the rising price of silver made the coins worth more as precious metal than as legal tender. In

Peso, Reduced Size (1907–1912, 35 mm).

1907 the U.S. Mint responded by lowering the denomination's silver fineness from .900 to .800 and reducing its diameter from 38 mm to 35. The resulting smaller coins had about one-third less silver, by actual weight, than those of 1903 to 1906, guaranteeing that they would stay in circulation. The older coins, meanwhile, were still profitable to pull aside and melt for their silver value. An official report of June 30, 1911, disclosed that less than ten percent of the heavier silver coins still remained in the Philippines.

The new smaller pesos were minted every year from 1907 to 1912, with the San Francisco Mint producing them for commerce and the Philadelphia Mint striking a small quantity of Proofs in 1907 and 1908. Millions of the coins were stored as backing for Silver Certificates (and, later, Treasury Certificates) in circulation in the Philippines. Although the Manila Mint started operations in 1920, the silver peso was never part of its production for circulation.

PHILIPPINE COMMONWEALTH ISSUES FOR CIRCULATION (1937–1945)

The Philippine Islands were largely self-governed, as a commonwealth of the United States, from 1935 until full independence was recognized in 1946. Coinage under the Commonwealth began with three commemorative coins in 1936. Circulating issues were minted from 1937 to 1941 (in Manila) and in 1944 and 1945 (in Philadelphia, Denver, and San Francisco).

The Commonwealth coinage retained the obverse motifs designed by Melecio Figueroa and used on the coinage of 1903 to 1936. Its new reverse design featured a shield derived from the official seal of the government of the Philippines.

World War II forced the Commonwealth government to operate in exile during the Japanese occupation of 1942 to 1945. A pro-Japan puppet government was set up in Manila in 1943; it issued no coins of its own, and in fact during the Japanese occupation many coins were gathered from circulation to be melted and remade into Japanese coins. Barter and low-denomination emergency paper money replaced coins in day-to-day commerce. (Much of the money used in the Philippines during World War II consisted of hastily printed "guerrilla" currency.) The United States military knew of the local need for circulating coins, and the U.S. and Philippine governments included new coinage in the plans to liberate the islands. The U.S. Treasury Department used its Philadelphia, San Francisco, and Denver mints to produce brass, copper-nickel–zinc, and silver coins in 1944 and 1945, to be shipped to the Philippines during and after the liberation.

One Centavo. The Manila Mint struck one-centavo coins for the Commonwealth of the Philippines every year from 1937 through 1941. This production was brought to an end by the Japanese invasion that started in December 1941, immediately after the bombing of Pearl Harbor. Part of the United States–Commonwealth plan to retake the islands included the San Francisco Mint's 1944 striking of 58 million one-centavo coins—a

One Centavo.
Bronze Alloy (1937–1941).
Brass Alloy (1944).

quantity greater than all of Manila's centavo output since 1937. Like the federal Lincoln cents of 1944 to 1946, these coins were made of *brass* rather than bronze—their alloy was derived in part from recycled cartridge cases, and their composition included copper and zinc, but no tin (a vital war material). The coins were transported to the islands to enter circulation as U.S. and Philippine military forces fought back the Japanese invaders. This would be the final mintage of centavos until the Republic of the Philippines, created on July 4, 1946, resumed the denomination's production in 1958.

The centavo was a popular coin that saw widespread circulation. As a result, many of the coins today are found with signs of wear or damage, exacerbated by corrosion and toning encouraged by the islands' tropical climate.

Five Centavos. The Manila Mint switched its coinage of five-centavo pieces to the Commonwealth reverse design in 1937. Production of the coins increased in 1938, then skipped two years. The 1941 output would be Manila's last for the type; the Japanese invasion at year's end stopped all of its coinage.

Five Centavos.
Copper-Nickel Alloy
(1937–1941).
Copper-Nickel-Zinc Alloy
(1944–1945).

Philippine commerce was starved for coins during the war. As part of the broader strategy for liberating the Philippines from Japanese occupation, the U.S. Treasury Department swung its mints into production of five-centavo coins in 1944 (Philadelphia and San Francisco) and 1945 (San Francisco alone). This effort dwarfed that of the Commonwealth's late-1930s coinage, producing in those two years more than ten times the combined output of 1937, 1938, and 1941. In order to help save copper and nickel for military use, the U.S. Mint reduced the proportions of those metals in the five-centavo coinage, making up for them with the addition of zinc. This substitution saved more than 4.2 million pounds of nickel and 3.2 million pounds of copper for the war effort. The Philadelphia and San Francisco coins were shipped to the islands during the combined American-Filipino military operations against Japan.

Ten Centavos. As with its production of other denominations, the Manila Mint under Commonwealth governance struck ten-centavo coins in 1937 and 1938, followed by a hiatus of two years, and a final coinage in 1941. Normal mint functions were interrupted in 1941 when Imperial Japan invaded the Philippines as part of its war with the United

Ten Centavos (1937–1945).

States. The Japanese puppet government of 1943–1945 would not produce any of its own coins, and the Manila Mint, damaged by bombing during the Japanese assault, was later used as part of the invaders' defensive fortifications on the Pasig River.

Japan's wartime exportation of Philippine coins resulted in a scarcity of coins in day-to-day commerce. The U.S. Treasury geared up the Denver Mint for a massive production of Philippine ten-centavo coins in 1944 and 1945, to be shipped overseas and enter circulation as American and Philippine troops liberated the islands. The 1945 coinage was particularly heavy: more than 130 million ten-centavo coins, compared to the Denver Mint's production of just over 40 million Mercury dimes that same year. This large mintage of silver coins continued to circulate in the Philippines into the 1960s.

Twenty Centavos. The twenty-centavo piece was the largest circulating coin struck by the Commonwealth of the Philippines at the Manila Mint. Production commenced in 1937 and 1938, followed by a hiatus of two years, and a final year of output in 1941 before Japan's December invasion put a halt to all

Twenty Centavos (1937–1945).

coinage. During their occupation, the Japanese pulled many twenty-centavo pieces out of circulation and melted them as raw material for new imperial coins.

Anticipating driving the Japanese military out of the islands, the United States and Commonwealth governments planned an impressive production of coinage for the

Philippines in 1944 and 1945. The Denver Mint was the source for twenty-centavo pieces, and its output was immense, in 1945 exceeding even the Philadelphia Mint's production of Washington quarters for domestic use. 111 million of the coins were shipped overseas to accompany the U.S. military as Americans and Filipinos fought to liberate the islands. The need was great, as legal-tender coins had largely disappeared from circulation. Most day-to-day commerce was transacted with small-denomination scrip notes and paper money issued by guerrilla military units, local governments, or anti-Japanese military and civilian currency boards.

Fifty Centavos. No fifty-centavo coins were struck at the Manila Mint for the Commonwealth of the Philippines. The denomination's first issue was a wartime production of the San Francisco Mint, in 1944, to the extent of some 19 million coins, or double that facility's production of Liberty Walking half dollars for the year. This was followed

Fifty Centavos (1944–1945).

by a similar mintage in 1945. These coins were intended to enter circulation after being shipped overseas with the U.S. military during the liberation of the Philippines from Imperial Japan's 1942–1945 occupation. They were readily accepted in the coin-starved wartime economy and continued to circulate in the islands into the 1960s.

This section on Philippine coins struck under U.S. sovereignty was condensed and excerpted from the **Guide Book of United States Coins, Deluxe Edition,** *popularly referred to as* **MEGA RED.** *For more information on these coins, including grading instructions, a date-by-date study of typical prices, and more photographs, please consult that reference.*

MINTS THAT NEVER WERE

While this book is about mints that produced coins, much could be said about non-mints: facilities that were proposed and sometimes even built, but never went into operation. Some such as in The Dalles are well documented, while others, such as the private mint that coined gold in Mount Ophir, California, are remembered only by legend and may or may not have existed. Over a period of years in the 19th century various bills were introduced in Congress to establish federal mints in New York City (in particular), Indianapolis (a large, presently vacant government building could have been used), Omaha, Louisville (proposed by Henry Clay in 1835), Cincinnati, St. Louis, and other cities, none of which ever received serious consideration.

Not included in our list of vaporous mints are private companies that have proposed coins for foreign countries, such as Duncan, Sherman & Company of New York, which in 1859 proposed to set up a mint in San Francisco to produce silver pesos similar in design, weight, and fineness to those produced under government authorities in Mexico.

Cincinnati Mining & Trading Company
(1849)

The Cincinnati Mining & Trading Company name indicates that this firm may have been an Ohio partnership which intended to produce coinage in California. Although few facts have been found concerning the operations of the firm, one of its $10 coins was overstruck on an 1849 J.S. Ormsby $10 made in Sacramento, lending credence to the thought that the company may have

Cincinnati Mining & Trading Co.

had limited minting operations in California, although no specific information has been located. Today, all Cincinnati Mining & Trading Co. coins are extreme rarities.

Massachusetts & California Co.
(1849)

A May 1849 account stated:

**Massachusetts &
California Company $5.**

> The Massachusetts & California Co., formed in Northampton, Mass., which originally contemplated a capital of only $6,000, has increased it to $50,000. Only a quarter of the amount, however, is to be paid in at the outset. Josiah Hayden of Haydenville is president of the company, and S.S. Wells of Haydenville, Miles G. Moies, and others, directors. Rev. Frederick P. Tracy goes out to California as its active agent.
>
> It is the intention of the Company to establish a private Mint at California, and, with the approbation of the government, to make coins of the same denominations as the coins of the United States, and of equal, if not a little higher, value. Mr. William H. Hayden goes out as assayer, having qualified himself for the purpose by a series of studies under Prof. Silliman[1] and by all the information that could be obtained at the United States Mint. Mr. Hayden is a graduate of Yale College and is son of the president of the company.
>
> The machinery will coin about $10,000 a day. It is the intention of the Company to purchase gold dust at the current price and transform it into coin for circulation. Should the government establish a Mint there, it will be worth its denominational value, or more, at the mint. The agent and those who accompany him will go by one of the land routes.[2]

It seems likely that the group did attempt to take the overland route west but there is no indication that they ever arrived in California. Like many other adventurers of the day they probably fell on hard times, or split up with everyone going their own way.

Prior to heading to the land of gold, the company made or commissioned some $5 pattern coins. No account has been found of the company's success, or lack thereof, nor has any information been found about a mint being established.

Current evidence indicates that all of the known trial pieces made in 1849 were likely given to prospective investors in the company and probably never circulated far from Northampton, where some have been discovered. Best estimates are that about two dozen pieces were coined in gold, silver, and copper. They almost certainly were made in the East as the gold alloy is unlike that of any other California coins. If Hayden's equipment ever reached California there is no evidence of it, and all dreams of the venture were dashed early in 1850 when private gold coins and mints were declared illegal.[3]

Mount Ophir Mint
Mount Ophir, California (1851)

Legends, probably building on other legends and folklore, have it that a mint was established in Mount Ophir in Mariposa County in the gold district of California, where $50 and other gold coins were struck. There are several examples of erroneous "history" of the mint in print:

On the road between Mount Bullion and Bear Valley may be seen the ruins of the first mint in California. Fifty-dollar slugs were coined there.[4]

Mount Ophir has one of the state's most interesting relics of the gold days—the Mount Ophir Mint, built in 1850, privately owned by sanctioned by the government, which had to provide some way for the issuance of money until the official Mint was built in San Francisco.[5]

Now completely abandoned, Mt. Ophir is sometimes said to be the site of California's first mint. Although it is probably true that here were coined the famous octagonal fifty dollar gold slugs, the Pacific Company in San Francisco seems to have minted the first gold coins. . . . The foundations of the Mt. Ophir mint, about 30 feet square, may still be seen. Like the other ruins in Mt. Ophir, they are made of quarried slabs of schist set in mud mortar.[6]

The Mount Ophir Mint turned out hexagonal [*sic*] gold slugs of fifty-dollar denomination. Today these slugs have a value of many times that amount among coin collectors. It has been said that a sack of two hundred such coins lies buried somewhere in the vicinity of these ruins. A gold digger's dream![7]

As of May 2016, Wikipedia proposed that Mount Ophir was the site of the first authorized mint in California, and that it was operated by Moffat and Co. under the authority of Augustus Humbert. All of this is incorrect.

These theories contravene numismatic knowledge that no $50 coins were minted other than in San Francisco. One explanation is that, somehow, a notation about the Mount Ophir *Mine* was mistranslated to Mount Ophir Mint.

We learn from geological records that the Mount Ophir vein, an outcrop of gold-bearing white quartz, was worked extensively in the 1850s and yielded an average of 6 dwt of gold per long ton of 2,240 pounds.[8] No record has been found of any coinage operation.

The Dalles Mint
Dalles City, Oregon (1864)

The *Annual Report of the Director of the Mint* for the fiscal year ended June 30, 1863, included this:

New Mining Regions

Idaho at this time especially claims our attention. It is emphatically the land of promise and of gold. This region was set off as a separate territory at the last session of Congress . . . The localities where workings have commenced are numerous, but many of them have not been reported or described. They must be various and widely separated, judging from the characteristic varieties in the quality of their production. Among the deposits [of gold] we have had grades of fineness from 795 to 949 thousandths; the latter in considerable quantity from Salmon River, a tributary to the Columbia. The quality of the gold produced from the mines of Idaho is equal to that in the older gold regions of our country, and the quantity appears to be inexhaustible.

Not less promising are the mines opening in Oregon and Washington Territory. The workings are numerous and constantly increasing in number. The characteristic energy of our people will, no doubt, soon develop the mineral wealth of these far distant [from Philadelphia where the Mint director's office was located] regions, and thus invite and stimulate immigration to our North Pacific territories. In Oregon the fineness of the gold seems to be tolerably regular and steady, and nearly equal to the average of California. In the gold from the Washington Territory the variation is great, ranging from 650 to 938 thousandths.

The returns from Oregon, Idaho, and Washington Territories are as yet imperfect, but enough is known to warrant the statement that in quantity and quality the gold of those regions will rival, if not surpass, the productions of the California mines.

These comments turned into reality. By an act of Congress on July 4, 1864, a federal branch mint was to be established in the town of Dalles City (today known as The Dalles), Oregon, on the Columbia River, for the coinage of gold and silver. The sum of $100,000 was appropriated for its construction. On March 20, 1869, a plot of land was purchased. Construction commenced in 1869. In 1871 the partially completed building was gutted by fire. Work resumed, but the outlook was uncertain. Native gold and silver was being found in reduced quantities. In 1875 the project was abandoned. Had the mint gone into operation the coins probably would have had a DC mintmark. In 1889 the premises were sold to a private buyer.[9]

The Henning "Mint"
Erial, New Jersey (1954)

This is a mint that was, in a way, but never should have been.[10]

No story about our nation's mints would be complete without some mention of the dark side of what has often unscrupulously been passed off as "coin of the realm". Counterfeiting is one of the oldest scams and has plagued both governments and the public

A 1944 Henning "Mint" nickel.

since the beginning of coinage. That easy pathway to wealth has taken many forms, though fortunately many of the would-be makers were amateurs whose work was quickly and easily detected.

Prior to the middle of the last century nearly all of the counterfeit coins in circulation were of denominations ranging from dimes through minor gold coins. Cents were not unheard of but were hardly worth the effort to replicate. Larger coins were costly to make and subject to greater scrutiny. In more recent times where coins are merely tokens and contain little precious metal, circulating counterfeits are rarely a threat to society.

Statistically, probably not one in a thousand readers of this narrative has ever received a bad coin in change. The Secret Service does a good job of seeing to that, and it has become unprofitable to even attempt the crime of counterfeiting coins. Further, it is far easier to duplicate high-denomination paper money. Regardless of all this there was one extraordinary exception to the norm that changed history, and it was only because of a slight numismatic blunder made by the counterfeiter that the deception was exposed.

The story of the astonishing fake five-cent coins known as "Henning nickels" began in 1954. The thought of producing fake nickels may seem like a waste of time but Francis LeRoy Henning, a 62-year-old mechanical engineer from Erial, New Jersey, knew it was a brilliant scheme. Because there was no precious metal in the genuine coins the cost of making them was less than face value and through mass production he could nearly double his manufacturing cost and likely avoid detection with a near perfect product. In fact, his nickels were so well made that they easily passed alongside genuine pieces.

It is estimated that more than 100,000 Henning nickels were placed in circulation before being detected. Many of them were deposited directly to New Jersey local banks by the maker under the guise of being a vending machine operator. His downfall came when members of the Camden County Coin Club discovered some unusual nickels in circulation that were made of nickel instead of the normal 1944 war-time silver alloy, and they lacked the important large "P" mintmark on the reverse. When authorities were alerted the Secret Service and Philadelphia Mint initially thought the coins were real, but later determined that they were being made somewhere in the vicinity of Camden, New Jersey. At that point Henning left town after dumping over 400,000 coins into Cooper Creek and the Schuylkill River. He was eventually arrested in Cleveland, Ohio, in October 1955, went to court, and was convicted of counterfeiting. He was sentenced to six years in jail and fined $5,000, which put an end to his further plan of manufacturing his own $5 bills.

Henning admitted to making six obverse and six reverse dies for his coins that were dated 1939, 1944, 1946, 1947, and 1953. All are deceptive when studied today, and it is only the 1944 coin, made in the wrong alloy and without the P mintmark, that is easily identified. All of his nickels passed easily along with genuine coins of that time. In recent years his coins have been regarded by some as a coveted numismatic item and have sold for more than any other lightly-worn Jefferson nickel variety. They are acknowledged in *A Guide Book of United States Coins* as a warning to collectors.

MINT DIRECTORS AND SUPERINTENDENTS

THE COINAGE ACT OF 1873

The Coinage Act of February 12, 1873, also known as the Mint Act of 1873, provided for changes in coin specifications as well as for moving the office of the director of the Mint from Philadelphia, where it had been located since 1792, to the Treasury Building in Washington, D.C.:

SEC. 1. Be it enacted by the Senate and House of Representatives of the United States of America in Congress assembled, That the Mint of the United States is hereby established as a Bureau of the Treasury Department, embracing in its organization and under its control all Mints for the manufacture of coin, and all assay-offices for the stamping of bars, which are now, or which may be hereafter, authorized by law. The chief officer of the said Bureau shall be denominated the Director of the Mint, and shall be under the general direction of the Secretary of the Treasury. He shall be appointed by the President, by and with the advice and consent of the Senate, and shall hold his office for the term of five years, unless sooner removed by the President, upon reasons to be communicated by him to the Senate.

SEC. 2. That the Director of the Mint shall have the general supervision of all Mints and assay-offices, and shall make an annual report to the Secretary of the Treasury of their operations, at the close of each fiscal year, and from time to time such additional reports, setting forth the operations and condition of such institutions, as the Secretary of the Treasury shall require, and shall lay before him the annual estimates for their support. And the Secretary of the Treasury shall appoint the number of clerks, classified according to law, necessary to discharge the duties of said Bureau.

SEC. 3. That the officers of each Mint shall be a superintendent, an assayer, a melter and refiner, and a coiner, and for the Mint at Philadelphia, an engraver, all to be appointed by the President of the United States, by and with the advice and consent of the Senate.

SEC. 12. That there shall be allowed to the Director of the Mint an annual salary of four thousand five hundred dollars, and actual necessary travelling expenses in visiting the different Mints and assay-offices, for which vouchers shall be rendered, to the superintendents of the Mints at Philadelphia and San Francisco, each four thousand five hundred dollars. . . .

The Treasury Building in 1890.

SEC. 66. That the different Mints and assay-offices authorized by this act shall be known as "the Mint of the United States at Philadelphia," "the Mint of the United States at San Francisco," "the Mint of the United States at New Orleans," "the Mint of the United States at Carson," "the Mint of the United States at Denver," "the United States assay-office at New York," and "the United States assay-office at Boise City, Idaho," "the United States assay-office at Charlotte, North Carolina"....

MINT DIRECTORS

It was standard practice for the president to make an appointment for director, furnishing one date of entry, and for the Senate to confirm the appointment at a later date. Often, when a new appointment was made or a new confirmation, the incumbent remained in office, creating a transition period. In many instances there has been a gap between the services of two directors. In modern times, an interim, deputy, or acting director has served.

After 1856 the Mint's fiscal year ran from July 1 of the first year into June 30 of the second. Most accounts and reports were on a fiscal-year basic, supplemented by some numbers calculated into calendar years, the last necessary for numismatists interested in mintage figures (the number of dies used in a calendar year) and related aspects. *The Annual Report of the Director of the Mint* was compiled by staff, starting after the close of the fiscal year and published months later.

David Rittenhouse: Appointed by President George Washington. Served from April 1792 to June 1795 (Senate confirmation April 14, 1792).

Henry William de Saussure: Appointed by President George Washington. Served from July 1795 to October 1795 (Senate in recess).

Elias Boudinot: Appointed by President George Washington. Served from October 1795 to July 1805 (Senate confirmation December 1, 1795).

Robert Patterson: Appointed by President Thomas Jefferson. Served from January 1806 to July 1824 (Senate confirmation December 25, 1805).

Samuel Moore: Appointed by President James Monroe. Served from July 1824 to July 1835 (Senate confirmation January 3, 1825).

Robert Maskell Patterson: Appointed by President Andrew Jackson. Served from May 1835 to July 1851 (Senate confirmation January 5, 1835).

George N. Eckert: Appointed by President Millard Fillmore. Served from July 1851 to April 1853 (Senate confirmation August 30, 1852).

Thomas Pettit: Appointed by President Franklin Pierce. Served from April 1853 to May 1853. Died in office.

James Ross Snowden: Appointed by President Franklin Pierce. Served from June 1853 to May 1861 (Senate confirmation February 4, 1853).

James Pollock: Appointed by President Abraham Lincoln. Served from May 1861 to September 1866 (Senate confirmation July 15, 1861).

William Millward: Appointed by President Andrew Johnson. Served from October 1866 to April 1867 (not confirmed by the Senate).

Henry Richard Linderman: Appointed by President Andrew Johnson. Served from April 1867 to May 1869 (Senate confirmation April 1, 1867).

James Pollock: Appointed by President Ulysses S. Grant. Served from May 1869 to March 1873 (Senate confirmation April 20, 1869).

Director's office moved from the Philadelphia Mint to the Treasury Building in Washington, D.C., in 1873.

Henry Richard Linderman (second term): Appointed by President Ulysses S. Grant. Served from April 1873 to December 1878 (Senate confirmation December 8, 1873).

Horatio C. Burchard: Appointed by President Rutherford B. Hayes. Served from February 1879 to June 1885 (Senate confirmation February 19, 1879).

James P. Kimball: Appointed by President Grover Cleveland. Served from July 1885 to October 1889 (Senate confirmation May 6, 1886).

Edward O. Leech: Appointed by President Benjamin Harrison. Served from October 1889 to May 1893 (Senate confirmation December 19, 1889).

Robert E. Preston: Appointed by President Grover Cleveland. Served from November 1893 to February 1898 (Senate confirmation January 14, 1894).

George E. Roberts: Appointed by President William McKinley. Served from February 1898 to July 1907 (Senate confirmation January 26, 1898).

Frank A. Leach: Appointed by President Theodore Roosevelt. Served from September 1907 to August 1909 (Senate confirmation February 12, 1898).

Abram Piatt Andrew: Appointed by President William Howard Taft. Served from November 1909 to June 1910 (Senate confirmation August 5, 1909).

George E. Roberts (second term): Appointed by President William Howard Taft. Served from July 1910 to November 1914 (Senate confirmation December 14, 1910).

Robert W. Wooley: Appointed by President Woodrow Wilson. Served from March 1915 to July 1916 (Senate confirmation March 3, 1915; legislative day of February 19[1]).

Friedrich J.H. von Engelken: Appointed by President Woodrow Wilson. Served from September 1916 to March 1917 (Senate confirmation August 17, 1916; legislative day of August 16).

Raymond T. Baker: Appointed by President Woodrow Wilson. Served from March 1917 to March 1922 (Senate confirmation March 15, 1917).

F.E. Scobey: Appointed by President Warren G. Harding. Served from March 1922 to September 1923 (Senate confirmation March 7, 1922).

Robert J. Grant: Appointed by President Calvin Coolidge. Served from November 1923 (at which time he was superintendent of the Denver Mint) to May 1933 (Senate confirmation December 18, 1923).

Nellie Tayloe Ross: Appointed by President Franklin D. Roosevelt. Served from May 1933 to April 1953 (longest term of any director) (Senate confirmation April 28, 1933; legislative date of April 17).

William H. Brett: Appointed by President Dwight D. Eisenhower. Served from July 1954 to January 1961 (Senate confirmation July 1, 1954).

Eva Adams: Appointed by President John F. Kennedy. Served from October 1961 to August 1969 (Senate confirmation September 23, 1961).

Mary Brooks: Appointed by President Richard M. Nixon. Served from September 1969 to February 1977 (Senate confirmation August 8, 1969).

Stella Hackel Sims: Appointed by President Jimmy Carter. Served from November 1977 to April 1981[2] (Senate confirmation November 5, 1977).

Donna Pope: Appointed by President Ronald Reagan. Served from July 1981 to August 1991 (Senate confirmation July 13, 1981).

David J. Ryder: Appointed by President George H.W. Bush. Served from September 1992 to November 1993 (Never confirmed by the Senate).

Philip N. Diehl: Appointed by President Bill Clinton. Served from June 1994 to March 2000 (Senate confirmation June 24, 1994).

Jay Johnson: Appointed by President Bill Clinton. Served from May 2000 to August 2001 (Confirmed by the Senate on May 24, 2000).

Henrietta Holsman Fore: Appointed by President George W. Bush. Served from August 2001 to August 2005 (Confirmed by the Senate on August 3, 2011).

Edmund Moy: Appointed by President George W. Bush. Served from September 2006 to January 2011 (Confirmed by the Senate July 26, 2008).

Richard A. Peterson: Appointed by President Barack Obama as acting deputy director. Served from January 2011 to January 2015 (Never confirmed by the Senate).

Rhett Jeppson: Appointed by President Barack Obama as the first deputy director. Served from January 2015 to date (Never confirmed by the Senate).

MINT SUPERINTENDENTS (AND PLANT MANAGERS) AND TERMS

Each of the mints active in 1873 and later was under the direction of an appointed superintendent, an office usually granted as a reward for good deeds to the party in control of Congress or to the president. Superintendents reported to the director of the Mint in Washington, D.C. The superintendents had charge of operations, including the processing of metal and planchets, coining to meet the requirements requested by the Treasury Department, and all other activities and departments.

The most important superintendency was at the Philadelphia Mint, where the Engraving Department was located and where dies for all the mints were made until Denver began making some. The Philadelphia superintendent managed these important departments and corresponded much more frequently with the director than did those of the other mints. Each year through 1978 he helped organize the U.S. Assay Commission to review the previous year's coinage of precious metals.

Since 1999 the superintendent positions are no longer political appointments, and the role has been renamed as "plant manager."

Note: Government records are often inconsistent. Some entries represent dates of service in a position and others are dates of official appointments, although an individual may have served before that time.

CARSON CITY MINT SUPERINTENDENTS, 1870–1878

Abram Van Santwood Curry: April 1869 to September 1870.[3]

Henry F. Rice: September 1870 to June 1873.

Frank D. Hetrich: July 1873 to August 1874.

James Crawford: September 1874 to March 1885.

William Garrard: March 1885 to June 1889.

Samuel Coleman Wright: July 1889 to August 1892.

Theodore R. Hofer:[4] August 1892 to May 1894 (coinage stopped in 1893).

Jewett W. Adams: May 1894 to September 1898.

Roswell K. Colcord: September 1898 to July 1899.[5]

CHARLOTTE MINT SUPERINTENDENTS, 1835–1861

General R.M. Saunders: March 1835 to 1837.

John Hill Wheeler: January 1837 to August 1841 (overlapping the following).

Burgess Sidney Gaither: July 1841 to June 1843.

Green Washington Caldwell: June 1843 (appointed earlier; took over the office on June 15) to 1847 (resigned to serve in the Mexican-American War).

The Mint was completely destroyed by fire in July 1844.

William Julius Alexander: March 1847 to August 1849 (overlapping the following).

James Walker Osborne: June 1849 to April 1853.

Green Washington Caldwell: April 1853 to April 1861 (second term).[6]

From 1885 to 1915 the facility was used as an assay office and repository with an officer in charge. No coinage took place.

DAHLONEGA MINT SUPERINTENDENTS, 1837–1861

Joseph J. Singleton: January 1837 to March 1841.

Paul Rossignol: March 1841 to May 1843.

James Fairlie Cooper: May 1843 to July 1849 (served for a while after his successor had been appointed).

Anderson W. Redding: July 1849 to June 1853 (began serving about October 1, 1849).

Jacob R. Davis: April 1853 to May 1853 (appointed by President Franklin Pierce and confirmed by the Senate to replace Redding, but his appointment was canceled after protests; he never served, nor was Redding ever dismissed).

Julius M. Patton: June 1853 to October 1860.

George Kellogg: October 1860 to May 1861.

DENVER MINT SUPERINTENDENTS AND PLANT MANAGERS, 1905 TO DATE

Frank M. Downer: July 1907 to August 1913.[7]

Thomas Annear: August 1913 (took office on August 29) to July 1921.

Robert J. Grant: May 1921 (took office on July 1) to November 1923 (on November 12 he was appointed director of the Mint with an office in Washington, D.C.).

Frank E. Shepard: December 1923 to June 1933 (officially resigned on April 26, 1933, to become advisor for the central Chinese government mint in Shanghai).

Mark A. Skinner: June 1933 to December 1942.

Moses E. Smith: March 1943 to April 1952.

Gladys P. Morelock: August 1953 to February 1953.

Alma K. Schneider: February 1953 to May 1961.

Fern V. Miller: May 1961 to July 1967.

Marian N. Rossmiller: August 1967 to March 1969.

Betty Higby: March 1969 to March 1978.

Evelyn Davidson: March 1978 to August 1981.

Nora Hussey: August 1981 to July 1987.

Cynthia Grassby Baker: January 1988 to May 1989.

Barbara McTurk: October 1989 to July 1993.

Raymond J. DeBroeckert (acting): July 1993 to November 1999.[8]

Jay Neal: November 1999 to 2001.

Tim Riley: January 2001 to 2008 (plant manager).

David Croft: 2008 to 2014 (plant manager).

Randall Johnson: August 2014 to date (plant manager).

NEW ORLEANS MINT SUPERINTENDENTS, 1835–1909

Martin Gordon: March 1835 to 1837.

David Bradford: March 1837 to October 1839 (removed from office).

Joseph M. Kennedy: October 1839 to September 1850.

Robert M. McAlpine: September 1850 to May 1853.

Charles Bienvenu: May 1853 to December 1857.

Logan McKnight: January 1858 to May 1858.

John H. Alpuente (acting): May 1858 to May 1858.

Howard Millspaugh (acting): May 1858 to July 1858.

William A. Elmore: July 1859 to January 1861.[9]

The Mint closed from 1861 to 1874.

M.F. Bonzano (assayer in charge; superintendent[10]): 1874 to 1878.

Michael Hahn: 1878 to December 1878.

Henry S. Foote: December 1878 to May 1880 (died on this May 19).

Martin V. Davis: June 1880 to August 1882.

Andrew W. Smythe: August 1882 to July 1885.

Gabriel Montegut: July 1885 to March 1889.

Andrew W. Smythe (second term): March 1889 to July 1893 (fired for lack of oversight of a major theft).

Overton Cade: July 1893 to 1898.

Charles W. Boothby: July 1898 to 1902.

Hugh S. Sithon: 1902 to 1911 (last coinage was in 1909).

From 1909 to June 30, 1942, the facility was managed by various appointees, but no coinage took place.

PHILADELPHIA MINT SUPERINTENDENTS AND PLANT MANAGERS, 1873 TO DATE

James Pollock: 1873 to 1879.

Colonel A. Loudon Snowden: 1879 to 1885.

Daniel M. Fox: 1885 to 1889.

Colonel Oliver C. Bosbyshell: May 1889 to March 1894.

Dr. Eugene Townsend: March 1894 to May 1895.

Major Herman Kretz: May 1895 to January 1898.

Henry Boyer: January 1898 to April 1902.

John H. Landis: April 1902 to July 1914.

Adam M. Joyce: July 1914 to March 1921.

Freas M. Styer: April 1921 to January 1934.

A. Raymond Raff: January 1934 to April 1935.

Edwin H. Dressel: April 1935 to July 1953.[11]

Rae V. Biester: 1953 to 1961.

Michael H. Sura: 1961 to 1969.

Nicholas G. Theodore: 1969 to 1977.

Shallie M. Bey Jr.: 1978 to 1981.

Anthony H. Murray Jr.: 1981 to July 1988.

John T. Martino: October 1989 to July 1993.

Augustine A. Albino: August 1993 to October 1996.[12]

Stephen Kunderewicz: November 1998 to 2002 (plant manager title beginning in 1999).[13]

Richard Robidoux: 2002 to 2008 (plant manager).

Marc Landry: March 2008 to date (plant manager).

SAN FRANCISCO MINT SUPERINTENDENTS AND PLANT MANAGERS, 1853 TO DATE

Lewis Aiken Birdsall: June 1853 to July 1855.[14]

Peter Lott: June 1855 to July 1857.

Charles H. Hempstead: June 1857 to June 1861.

Robert Julius Stevens: June 1861 to June 1863.

Robert Bunker Swain: June 1863 to July 1869.

Oscar Hugh LaGrange: August 1869 to December 1877.

Henry Lee Dodge: January 1878 to June 1882.

Edward Freeman Burton: July 1882 to July 1885.

Israel Lawton: July 1885 to July 1889.

William Henry Dimond: August 1888 to July 1893.

John Daggett: July 1893 to July 1897.

Frank A. Leach: August 1897 to November 1907.[15]

Edward Sweeney: November 1907 to August 1912.[16]

Frank A. Leach (second term): August 1912 to August 1913.

Thaddeus W.H. Shanahan: August 1913 to June 1921.

M.J. Kelly: July 1921 to June 1933.

Peter J. Haggerty: July 1933 to December 1944.

N.H. Callaghan: May 1945 to November 1947.

G.B. Gillin: July 1948 to November 1951.

J.P. McEnery: June 1952 to November 1952.

R.P. Buell: July 1953 to June 1958.

A.C. Carmichael: July 1955 to July 1958.

No mintmarked coinage occurred from 1956 to 1967.

J.R. Carr: August 1958 to November 1968 (officer in charge, U.S. Assay Office, San Francisco).

John F. Brekle: October 1968 to June 1972 (officer in charge, U.S. Assay Office, San Francisco).

Bland T. Brockenborough: August 1972 to November 1980 (officer in charge, U.S. Assay Office, San Francisco).

Thomas H. Miller: November 1980 to March 1988 (officer in charge, U.S. Assay Office, San Francisco).

Carol Mayer Marshall: March 1990 to January 1993 (officer in charge, U.S. Assay Office, San Francisco).

Donald T. Butler: 1993 to 1996 (acting superintendent).

Dale B. DeVries: September 1996 to December 1999 (plant manager).[17]

Larry Eckerman: December 1999 to May 2015 (plant manager).

David Jacobs: June 2015 to date (plant manager).

West Point Mint Superintendents and Plant Manager, 1984 to Date

The West Point Mint was known as the U.S. Bullion Depository from 1938 to 1988.

Clifford M. Barber: March 1988 to November 1990.

Bert W. Corneby: November 1990 to July 1993.

Bradford E. Cooper: August 1996 to March 1999.

Ellen McCullom: December 1999 to date (plant manager).

APPENDIX

MINTING PROCEDURES IN 1854

New York Industrial Exposition: Special Report of Professor Wilson, London, 1854, includes a lengthy article, "Mints of the United States," which is reprinted below, with subtitles added and light edits. This gives an exceptionally detailed view of the operations and procedures in effect in early 1854 when the author, Professor John Wilson, visited the Philadelphia Mint:

Mints of the United States

The transmissions of gold from the new State of California have caused a corresponding increase in the gold currency of the States, and have invested the Mint operations with more general interest than under the previous ordinary circumstances they possessed. The same condition of things exists in this country; and as it is intended to establish a mint in the gold producing colony of Australia, I thought it desirable to obtain as much information as I could in reference to the organization and working details of those in the United States.

The head establishment is at Philadelphia, and is called "the Mint;" there are also three "branch mints;" at New Orleans, in Louisiana; at Charlotte, in North Carolina; and at Dahlonega, in Georgia, respectively. The Branch Mint in California, and the Assay Office in New York, are not yet completely organized.

At the Mint in Philadelphia, gold, silver, and copper, are coined; at New Orleans, gold and silver are coined; while the branches at Charlotte and Dahlonega coin gold only. At "the Mint," the executive staff consists of a director, treasurer, chief coiner, melter and refiner, engraver, assayer, and assistant assayer. At the New Orleans Branch Mint, the staff consists of a superintendent, treasurer, melter and refiner, and coiner; and at each of the other two branch mints there are but three officers,—superintendent and treasurer (combined), assayer and coiner.

At the United States Mint at Philadelphia, the salaries are fixed as follows: — Director, $3,500; treasurer, $2,000; chief coiner, $2,000; melter and refiner, $2,000; assayer, $2,000. At the New Orleans Branch Mint the salaries are, to the superintendent, $2,500, and $2,000 each to the other officers; and at the other branch mints the superintendents receive $2,000, and the other officers $1,500 respectively. In each of the establishments, the appointment of assistants, subordinate officers and servants, is left entirely in the hands of the chief of the different departments.

Receiving Deposits

In visiting the Mint at Philadelphia I had the advantage of being taken through the several departments by the chief coiner, Mr. Franklin Peale, and the melter and refiner, Professor J.O. Booth, who kindly furnished me with the following details of their operations. As the gold is brought to the Mint in various quantities and in a crude state, it passes necessarily through the department of the refiner before it reaches that of the chief coiner; I therefore give the actual details of the refining operations upon sundry deposits of gold, amounting in the aggregate to $2,000,000.

The deposits are immediately weighed and a certificate of their gross weight issued. The fires having been lighted in the five furnaces of the deposit melting room at four or five o'clock, A.M., all the deposits, amounting, perhaps, to seventy or eighty, are melted before noon; assay slips are then taken off, and the assays finished the next morning, after which their values are calculated by the weight after melting, care being taken to include all the grains that can be procured from the flux, pots, &c, by grinding them up under a pair of small chasers, sifting, and washing.

There is a clerk and his assistant and one hand wholly engaged in performing all the weighings for the treasurer, such as weighing deposits before and after melting, ingots for coinage, fine bars, and the clippings after cutting out the planchets. There are five men in the deposit melting room, two of whom attend to two furnaces each at the same time, one to one furnace and washing grains, and the remaining two are laboring assistants. The whole deposit of $2,000,000 is melted in three or four days in the deposit-room, and assayed by from the third to the seventh day.

Assaying and Refining

As soon as the first deposits are assayed, say on the third day (if expedition is necessary), or always on the fourth, they are granulated in the proportion of one part of gold to two parts of silver. The pots contain 50 lbs. of gold and 100 lbs. of silver, equal to 1,800 oz., and each melt requires about an hour. With four furnaces (attended by four melters and two aides), there are ordinarily made 32 melts per day, but when hurried 48 melts can be made, making from one-third of a million to one-half of a million of dollars per day. Two days' work, or about $650,000 worth of gold, equal in weight to one ton (avoirdupois weight), are granulated for a single setting with acid.

The granulated metal is charged into large pots, together with pure nitric acid of 39° Beaume, between the hours of seven and nine A.M. on the sixth day, and steamed for five hours. The pots, made in Germany, are two feet in diameter by two feet in depth, set in plain wooden vats, lined with three-sixteenth-inch sheet-lead; a single coil of copper pipe passing around the bottom of the vat blows the steam directly into the water, in which the pots are set to about half their depth.

The vats are arranged in a small house in the middle of the room with a large flue connecting with the chimney-stack, so that when in action the odor of nitrous fumes is scarcely perceptible in the building. The $2,000,000 require about 60 such pots; they are stirred about once each hour, say altogether five times, with simple wooden paddles; the next day (seventh) the acid solution of nitrate of silver is drawn off by a gold-syphon into wooden buckets, and transferred to the large vat, in which it is precipitated by salt (chloride of sodium), and fresh acid added to the metals, now containing very little silver. Steaming for five hours on the seventh day completes the refining of $650,000. Early on the eighth, one pot is drawn off, washed with a little warm water, and the gold-powder transferred to a filter. Fresh granulations are put into this empty pot, and the acid of the adjoining pot baled over upon them, and thus through the series, the whole being re-charged in from two to two and a half hours. After steaming for five hours, the acid which contained but little silver from the preceding day becomes a nearly saturated solution of nitrate of silver. The mode of assaying is according to the "wet proves" of Gay Lussac. By this arrangement 4-1/2 lbs. of nitric acid are consumed altogether for each pound of gold refined, and the latter is brought up to 990 to 998 in fineness,—rarely below 990. Thus every two days, 13,000 lbs. of nitric acid are used. In the course of the last year 1,000,000 lbs. of pure nitric acid, at seven cents per pound, equal to $70,000, were consumed.

The gold is washed with hot water on the filter during the eighth day, and until it is sweet, (say by 7 P.M.). The filter consists of two layers of tolerably stout coarse muslin, with thick paper between, in a tub with a false bottom, 2-1/2 feet in diameter and 2-1/2 feet deep, and mounted on wheels. One of the men remains, after washing hours, until 7 P.M., when the watchman of the parting-room continues washing the gold and silver until sweet, i.e., until the wash-water ceases to color blue litmus paper.

Early on the ninth day the wet gold is pressed with a powerful hydraulic press, and the cakes then thoroughly dried on an iron pan, at a low red heat. This process saves wastage in the melting-pot, since there is no water remaining in the pressed metal to carry off gold in its steam. The same day (ninth) the gold is usually melted with a less proportion of copper than is requisite to make standard metal, and cast into bars, which are assayed by noon on the tenth. They are then melted with the proper quantity of copper, partly on the same day, partly early on the eleventh, and assayed and delivered to the coiner the same day. On the fourteenth day they are ready for delivery to the treasurer as coins.

The silver solution drawn off from the pots is precipitated in a large wooden vat of 10 feet diameter by 5 feet deep, and the chloride of silver immediately run out into large filters (6x3x14) where it is washed sweet. The filter is covered with coarse muslin, and the first turbid water thrown back; the filter, which is on wheels, is then run over to the reducing vats, and the chloride shoveled into them. There are four such vats (7x4x12) made of wood and lined with lead, one inch thick in the bottom. A large excess of granulated zinc is thrown on the moist chloride in the vats, without the addition of acid; the reduction is very violent, and when it slackens, oil of vitriol is added to remove the excess of zinc. The whole reduction occupies a few hours, and after a night's repose the solution of mixed sulphate and chloride of zinc is run off into the sewer.

About two tons of zinc per $1,000,000 of gold are employed; the silver however, in this amount, say 10 per cent, by weight, should only take, by equivalents, about 2,400 lbs., so that nearly two equivalents of zinc for one equivalent of silver are used. This is found to be advantageous, as both time and space are greatly economized by this excess.

The day after the reduction the reduced silver is washed, and the second day it is pressed and dried by heat, the same hydraulic press as for gold being used, but with different drying-pans. The same silver is used again for making fresh granulations, but as it accumulates from the Californian gold, 10,000 or 20,000 ounces are now and then made into coin, great care being taken in this case to avoid getting gold in it when drawing off the silver solution, and in the press.

Such are the actual working details in refining a specified amount ($2,000,000) of gold, the first third of which is delivered as coin in fourteen days after its arrival, and the third in eighteen days.

But as there is a bullion-fund of $5,500,000 allowed by government, depositors are paid from the third to the fifth day after an arrival, *i.e.*, as soon as the gold is melted, assayed, and its value calculated. When two heavy arrivals occur in close succession, the time of refining and coining can be shortened from 14 to 10 days.

The number of men engaged in the refining department is 14: 1 foreman, 8 for the parting process, 3 for reducing, and 2 for pressing and drying. In the gold melting-room there are 3 melters and 2 assistants. The total number of hands in the melting and refining departments is 34, including a melting and parting foreman, and 3 in the place for grinding, sifting, washing, and sweeping. The last place or sweep, embraces all pots, ashes of fires, trimmings of furnaces, ashes of all woodwork, &c.

The late law for reducing the weight of silver coin necessitated an increase of force, and 15 more were in consequence employed for this purpose. While $50,000,000 in a year have been parted with the above force, they could with the same force and apparatus refine $80,000,000 if it were required.

After many experiments upon anthracite, Professor Booth stated that he had at length fully succeeded in employing it for melting both gold and silver in the same furnaces, slightly modified, in which he had been accustomed to melt with charcoal. This change had been accompanied by great economy in the cost of material and labor, and by greater comfort to the workmen, from being less exposed to heat. The cost of charcoal (of the best quality—hard pine-knot coal) is 16 cents per bushel, delivered at the Mint; and while the cost of this fuel for all their operations in 1852, when gold was chiefly refined and melted, was about $7,000, the cost of anthracite will be from $600 to $1,000. In using the anthracite, he found that a simple draft of air, without a blast, was quite sufficient to sustain combustion.

Californian gold frequently contains the alloy "iridosmine," which is not always detected by the assay. In order to remove it as far as possible without actually dissolving gold, it is allowed to subside, first in the granulating crucibles, and then in the crucibles for toughening (melting fine gold and copper). If the assayers report its presence in the toughened bars, they are again melted, and the iridosmine allowed to subside. By these three, and often four successive meltings, the gold is separated from its troublesome companion as far as practicable. The gold thus refined, and reduced to the proper standard:

Section 8: "And be it further enacted, that the standard for both gold and silver coins of the United States shall hereafter be such, that of 1,000 parts by weight, 900 shall be of pure metal and 100 of alloy; and the alloy of silver coins shall be of copper, and the alloy of gold coin shall be of copper and silver; provided that the silver does not exceed one half of the whole alloy," is delivered over to the chief coiner in the form of bars or ingots of a certain weight, to be divided and shaped into pieces required for the currency of the country.

ROLLING MILLS AND PLANCHET PREPARATION

The *Coining* department of the establishment is of a power and efficiency sufficient to perform all the mechanical processes incidental to the issue of nearly 70,000,000 of pieces during the past year; and I was assured by Mr. Franklin Peale, the chief coiner, that it could have executed much more if it had been steadily employed, or fully supplied with material during the whole of that period. It is not necessary to go through the whole course of operations in this department, but to notice only such as possess novelty or present special characteristics.

The necessary power for working the machinery is obtained from a large steam-engine of the form usually known as the steeple-engine; it is a double vertical high-pressure engine, with cranks at right angles, the power being carried off by a caoutchouc belt, 2 feet wide, from a drum of 8 feet in diameter; the estimated power is equal to 90 horses. At times, this is all required, at others much less is sufficient, and in uncertain proportions; to meet this irregularity, and to ensure that steadiness of motion so necessary in such delicate operations, a governor and throttle-valve of a peculiar construction have been devised, which have now been in use for some time, and have produced most satisfactory results, fully effecting the purpose for which they were designed.

The rolling mills, four in number, are driven by belts, at the rate of six revolutions per minute; the distances between the rollers being adjusted by double wedges, moved by a train of wheels which are connected with a dial plate and bands, divided and numbered into hours and minutes, so as to indicate the proper thickness of the strips of metal without the use of gauges. It is not necessary to enter into details of their construction, as a full and minute description is given in the *Journal of the Franklin Institute* for July, 1847. Gold strips are heated in an iron heater by steam, and waxed with a cloth dipped in melted wax, and the silver strips are coated with tallow by means of a brush. The draw bench is used for both metals, and trial pieces are cut from every strip and their weight tested, preparatory to the cutting of the whole.

The cutting processes are very simple and efficient, consisting of a shaft moved by pulley and a 2-1/2 inch belt, with a fly-wheel of small diameter but sufficient in momentum to drive the punch through the slip of metal by means of an eccentric of 3/8 inch, at the rate of 250 pieces per minute, which skilled hands can readily accomplish and continue until the slip is exhausted. The annealing during the rolling of the ingots into slips is performed in copper cases, in muffles of fire-clay and brick, heated by anthracite coal, three muffles or hearths being kept at a bright red heat by one fire-grate or furnace, and the distribution and intensity regulated by dampers. These annealing furnaces are recent in their construction and very satisfactory in operation; they are heated by anthracite at the cost of about one fourth the expense of the wood previously employed.

The whitening of planchets is performed as usual by inclosing the gold in luted boxes, and by exposing the silver in an open pan, to the heat of a simple furnace with wood fuel; the drying and sifting after the action of dilute sulphuric acid, is rapidly and effectually accomplished by a rolling screen—one portion of which consisting of a pair of closed concentric cylinders, between which high-pressure steam is admitted. The blanks, with a sufficient quantity of light wood saw-dust (linden or bass wood is the best), being introduced into the interior cylinder, a revolving motion is given to it by the engine for a certain time; the door is then opened and the blanks and saw-dust gradually find their way into the wire screen, by which they are separated, the movement being continued until the separation is complete, when the blanks are discharged at the end of the machine. An arrangement exists by which a slight inclination is given to the machine so as to direct the motion of the blanks towards the discharging end.

The milling machines are, I was informed, peculiar to this mint, and are in a great measure original, the operation being performed by a continuous rotary motion, with great rapidity and perfect efficiency, varying in rate according to the denomination of the coin, between 200 and 800 pieces per minute, and at the same time separating any pieces that are notably imperfect.

It must be understood that the operation here termed "milling," is merely for the purpose of thickening and preparing the edge, so as to give a better and more protective border to the coin, the ornament or reed, commonly known I believe in this country as "milling," being given to the piece by the reeded collar of the die in which the piece is struck.

COINING PRESSES

The coining presses, 10 in number, and milling machines are worked by a high-pressure horizontal steam-engine, made from the design and under the direction of the present chief coiner, in the workshops of the establishment, in 1838.

The presses are three sizes, the largest applicable to the striking of silver dollars and double eagles:—the second to pieces of medium value:—and the smallest to the dime, half dime, and three cent pieces. The first is usually run at the rate of 80 per minute, the last at 104 per minute,—the average rate of the whole is 82 per minute. This rate can be increased if required.

If all the presses were employed in coinage at the usual rate, they would strike in one day (9 working hours) 439,560 pieces; and if employed upon gold, silver, and copper, in the usual manner, and on the usual denomination of coin, they would amount in value to $966,193.

During the past year, on one occasion, 8 of the presses were run 22 out of 24 consecutive hours, and coined in that time 814,000 pieces of different denominations of coin.

These presses have been made principally in the workshops of the Mint. They possess, in common with the presses of Uhlhorn, in Germany, and Thonnelier, in Paris, the advantage of "the progression lever," "le genou" or "toggle joint," a mechanical power admirably adapted to this operation; but in almost every other particular they are original in arrangement, being the result of experience, beginning as far back as 1836.

In order to supply these presses, various means have been devised: among them, and not the least important, is the "shaking box," in which advantage is taken of a disposition observable in similar bodies, or bodies of similar form, to arrange themselves in similar positions. This is a box, whose bottom is constructed with parallel grooves adapted to the size of the blanks of planchets to be arranged. A quantity of them is thrown indiscriminately into the box, which is then quickly shaken in the direction of the grooves, the pieces immediately lay themselves side by side, in parallel rows, from which they can easily be lifted in rouleaux as required to be passed to the feeding tubes of the mills or presses.

COUNTING THE COINS

It is very evident to all visiting the establishment, that such a large number of pieces could not be coined and manipulated by such a limited number of hands without the aid of some labor-facilitating arrangements, one of the most worthy of remark of which is, the method of counting the pieces coined—if counting it can be called, for in principle it is a measuring machine. The arrangement of this counting frame, or tray, may be understood from the following sketch of its construction:

A board or tray of such dimensions as may be required, is divided by a given number of parallel metallic plates dissected into its plane and slightly elevated above it, the edges of which rise no higher than the thickness of the coin for which it is intended. The board is of such a length as will admit of a few more than the required number of pieces to be laid longitudinally in the rows, and is divided across and at right angles with the rows, and hinged at a point opposite to a given number. One of those employed by this department, counted 1,000 pieces, that is to say, it had 25 parallel grooves or rows sufficiently long to receive 45 pieces. Now, having thrown on this board a large excess of pieces, it is agitated by shaking until all the grooves are filled, and then inclined forwards until all the surplus pieces have slid off, one layer only being retained by the metallic ledge; the hinged division is then suffered to fall, which at once throws off all but the 45 pieces in the length of each row. This operation, somewhat difficult and tedious to describe, is performed in a few seconds, and results in retaining on the board 1,000 pieces, each piece exposed to inspection, and the whole accurately counted without the wearisome attention—so likely to result in error—required under usual circumstances.

WOMEN AT THE MINT

The very large number of pieces coined during the last year has been counted exclusively by two female manipulators, assisted by a man who had the duty of weighing them in addition as a testing check. The same amount of labor by ordinary means could not have been performed with fewer than thirty or forty hands, to say nothing of inferior accuracy. This machine was originally arranged and patented by the late R. Tyler, coiner of the New Orleans Branch Mint, but materially improved in its application and construction by Mr. Franklin Peale.

The balances of the Mint of the United States have received the attention necessary to an instrument of such importance in mint operations. They have been arranged and made generally in the workshops of the establishment, and operate entirely to the satisfaction of the department. I, perhaps, ought to mention that since that appeared, some slight improvements have been made by inclosing all but the stirrups and pans in glass, by these means excluding dust and protecting them from the influence of air currents.

In concluding this brief sketch of the practical working of the two most important departments of the United States Mint, I cannot omit a reference to the very excellent remarks of the chief coiner on the employment of females in some of the operations in his department. This, he informed me, had generally excited the surprise of, and had been commented upon, by foreigners, who had visited the Mint. His experience, however, had led him to believe, that in places of trust, where no great physical exertion was called for, but where accuracy and strict integrity were of first importance, the moral perceptions of the female, generally stronger and of a higher standard than in the man, would qualify her as his substitute, and thus, while opening a new field of labor for the occupation of females, would strengthen their claims to it by the superior accuracy and economy of their work.

MEDALS AND OTHER COLLECTIBLES OF THE UNITED STATES MINT

In chapter 9 of *American Gold and Silver: U.S. Mint Collector and Investor Coins and Medals, Bicentennial to Date*, numismatist Dennis Tucker explores modern national medals such as the Valley Forge medal minted at Philadelphia for the United States Capitol Historical Society (1978); the Benjamin Franklin Firefighters medal (1993); the 9/11 silver medal; and others. He writes:

> The mint of the United States has a long history of issuing medals for commemorative and historical purposes, dating back to the early 1800s. Even before that—before there existed a national mint, as far back as during the American Revolution—official medals were authorized to be designed and struck (typically in France) for the fledgling nation.
>
> In 1776 the Continental Congress voted to award the thanks of the Thirteen United Colonies to General George Washington, his officers, and his soldiers, after their victorious bombardment of Boston and the resulting British evacuation. Along with the congressional gratitude a gold medal was presented to General Washington. Medals (separate from military awards and decorations) have been granted to other commanders from the Revolutionary War and the War of 1812 to the present day. Beyond the military, the nation has struck medals to celebrate everything from acts of heroic lifesaving to famous entertainers to wildlife conservation.

Self-referentially, the Mint has struck medals honoring people, places, and events in its own history. Hobbyists can assemble significant, attractive, and historically important collections of these medals.

Private producers have also created souvenirs of the U.S. Mint, including medals, postcards, and other collectibles. This appendix illustrates many types of Mint collectibles, both official and private. This is just the tip of the iceberg, and an active collector will be able to find many more examples by searching online and inquiring among coin dealers, ephemera and book shops, and similar sources.

FIRST STEAM COINAGE MEDALET (1836)

Today at the American Numismatic Association museum in Colorado Springs, you can see on display a remarkable piece of machinery—the first steam-powered coinage press installed at the Philadelphia Mint. This press was used in 1836 to strike souvenir medals, 28 mm in diameter, with the legends UNITED STATES MINT and FIRST STEAM COINAGE MAR. 23. 1836, although the piece was

First Steam Coinage medalet.

meant to be struck on George Washington's birthday, February 22 (with the die matching that date). Today, nearly 200 years after their production, these are popular little pieces of early Mint memorabilia. A Proof-65 example (Julian MT-21) sold in June 2016 for $2,115.

The first of these medals that rolled off the press was engraved with the words THE VERY FIRST on its edge. It was presented to Mint Director Robert Maskell Patterson, who wrote to Treasury Secretary Levi Woodbury: "I also send, by this mail, some copper pieces struck at the mint today on our new press by steam. They are the first ever struck by this power in America. We must consider this day, therefore, as marking an epoch in our coinage." This unique "very first" medal sold for $20,563 in August 2014.

PHILADELPHIA MINT LORD'S PRAYER MEDALET (LATE 1800s)

Philadelphia Mint employee George B. Soley acquired the Mint's first steam-powered coinage press as surplus scrap in 1875, using it to strike medals sold as souvenirs at expositions (including the 1876 Centennial Exposition) and at the Mint in the late 1800s. Russell Rulau lists this piece as PA-Ph394 in his *Standard Catalog of*

Lord's Prayer medalet.

United States Tokens, 1700–1900. One side of the medal shows the Philadelphia Mint, with the date 1832, and the other side reproduces the Lord's Prayer in tiny lettering.

The medal measures 13 mm and is typically found holed and with a ring and/or ribbon for suspension. It is relatively common and can be found for $20 or so in worn conditions. Higher-grade examples can be collected for $100 to $200.

U.S. ASSAY COMMISSION MEDALS (1860–1977)

Many of the medals minted for the annual meeting of the U.S. Assay Commission feature portraits of Mint directors, façades of Mint buildings, and similar design elements. This makes perfect sense, since the Commission was directly involved in American coinage. From *American Gold and Silver*:

> A special annual assembly of government officials (and, later, invited citizens) started in the late 1790s, not long after the first U.S. Mint was established in Philadelphia. These dignitaries comprised the United States Assay Commission. Their charge was to review silver and gold coins selected at random from the previous year's mintage. Their goal: to confirm that the nation's coins were of full weight and fineness, conforming with established laws. . . . Today there exist tangible and collectible mementoes of this obscure and largely forgotten ceremony. Starting in 1860, the U.S. Mint struck special presentation medals for those attending the annual assay—federal officials, dignitaries, and the public, as well as members of the commission. . . .[1]

All Assay Commission medals are scarce. Copper issues dating from the 1800s are the ones most often seen in the marketplace, and therefore the most affordable. Silver and aluminum pieces are rarer and more expensive. Specialists seek out unusual die varieties and combinations in various metals. The final medal was minted in 1977, in pewter, and the Commission was formally disbanded in 1980.

"No Assay Commission medals were issued from 1862 through 1866; the 1936 medal was actually a Roosevelt presidential medal obverse with a Washington medal reverse; and none were issued in 1954. Collectors eagerly await their appearance at auction or in private sales, as they represent a colorful and historic part of U.S. coinage history."[2] The scarcest pieces from the 1940s and 1950s can fetch prices up to $6,000 or more. A review of auction results from 2015 and 2016 show many Assay Commission medals selling for $400 to $800, with some rarer pieces bringing $2,000 or more. Ten years earlier, typical prices were in the $200 to $400 range.

All of the Assay Commission medals feature either allegorical or direct references to the U.S. Mint. Here are some examples. The first medal, of 1860, with James Longacre's French Liberty Head, bears the legend MINT OF THE UNITED STATES and PHILADELPHIA. The 1872 medal shows the goddess Juno Moneta in front of minting and assaying machinery, with a cornucopia of coins guarded by a dog; the name of Philadelphia Mint director James Pollock is in the exergue. The 1879 medal shows a portrait of Mint Director Henry R. Linderman. Several medals including that of 1892 show Juno Moneta instructing a youth in the science of coinage. Some, including the medal of 1893, show the façade of the Treasury Building in Washington, D.C., while others, such as 1902, 1948, 1967, and 1977, show various incarnations of the Philadelphia Mint (including Edwin Lamasure's fantastical "Ye Olde Mint" rendition of the original Philadelphia buildings). Some draw on historical scenes from the early Mint, such as the medal of 1955 (showing men working with the first coining press) and that of 1975 (with a rendition of John Ward Dunsmore's painting "Inspection of the First Coins").

1860.　　　　　　1872.

1879.

1892.

1893.

1902.

1940.

1948.

1955.

1967.

1977.

MINT DIRECTOR MEDALS (1700s TO DATE)

"The U.S. Mint produces national medals to commemorate significant historical events of the Nation, or to honor those persons whose superior deeds and achievements have enriched our history, or the world," the Mint's product catalog notes. "'List Medals' are also produced, some as part of a continuing series, such as the Presidents of the United States, Secretaries of the Treasury, Directors of the Mint, and Historic Buildings of the United States and others, when approved by the secretary of the Treasury."

The Mint describes one of its continuing series wherein medals are issued for Mint directors, noting that "it is not a complete record of all those appointed." Their obverses bear a portrait of the director, and their reverses include Treasury and Mint buildings and career symbols. Some of those honored include David Rittenhouse (the first Mint director); Nellie Tayloe Ross; Eva Adams; Mary Brooks; Stella Hackel Sims; Donna Pope; and Henrietta Holsman Fore.

Auction records of the past decade show older medals selling for hundreds of dollars. A Henry Linderman medal (76 mm, bronze, MS-63) sold for $920 in 2008. A

Henry R. Linderman.

James Kimball.
(shown at 80%)

James Ross Snowden.
(shown at 80%)

James Kimball medal (76 mm, bronze, MS-63) sold for $403 the same year. Modern restrikes of the 1859 James Ross Snowden medal can be purchased for less than $100, a fraction of the going rate for originals. Three-inch bronze medals of more recent Mint directors such as Donna Pope sell for $20 or so in online auctions.

PHILADELPHIA MINT SUPERINTENDENT MEDALS (1879–1917)

In the June 2016 issue of *The Numismatist* (monthly magazine of the American Numismatic Association, online at www.money.org/the-numismatist), Michael B. Costanzo describes the superintendent medals of the U.S. Mint. "Only six were created from 1879 to 1917 in special recognition of Philadelphia Mint officials," he notes. "The office of superintendent came into being with the Mint Act of 1873. Though superintendents were appointed to each mint, the individual assigned to the Philadelphia facility wielded greater influence. He closely conferred with the mint director and dealt with a large variety of coinage issues, from new designs and denominations to die production. He also oversaw the Medals Department, which operated out of the Philadelphia Mint."

Archibald Loudon Snowden.

The first Philadelphia superintendent was James Pollock, but no medal was issued for him, nor for Henry K. Boyer. The six pieces minted (and their mintages, based on the research of Robert W. Julian published in *Medals of the United States Mint: The First Century, 1792–1892)*, were for A. Loudon Snowden (1879; mintage of 53), Daniel M. Fox (1885; 21), Oliver C. Bosbyshell (1889; 24), Eugene Townsend (1894; 67), Herman Kretz (1895; 151), and Adam M. Joyce (1917; unknown). Varieties exist, and some medals were struck later than their year dates indicate.

The superintendent medals are all scarce, as their mintages attest, but also so obscure and specialized that their market prices today are not unreasonable. Their appeal is obvious, however, given their large size (76 to 80 mm), their historical importance, and their designs by famous Mint engravers such as Charles Barber and George Morgan. Professional numismatist John Kraljevich offered a Barber-designed Snowden medal (Julian MT-14) in 2016 for $1,150.

MANILA MINT OPENING MEDAL (1920)

The medals struck at the Manila Mint to celebrate its opening in 1920 are actively sought by collectors of So-Called Dollars—a class of U.S. medals about the size of a silver dollar, struck from the 1800s to the mid-1900s to commemorate historical subjects. Numismatist Jeff Shevlin describes So-Called Dollars in appendix

Manila Mint Opening medal.

F of the second edition of *Mega Red* (the *Guide Book of United States Coins, Deluxe Edition*): "There are more than 750 different design types . . . struck in virtually every metal composition conceivable. . . . Many of the most famous engravers of U.S. coins also engraved So-Called Dollars."

Bronze examples of the Manila Mint medal are the most common of the type, with a mintage of 3,700, and are valued at $35 to $100 in lower circulated grades; up to $200 or more in About Uncirculated; and $4,500 in MS-65. Silver examples were minted to the extent of 2,200 pieces, and are worth around $900 in MS-60, or double that in MS-63. Only five examples were minted in gold, and when they come to market they sell for $40,000 or more.

MINT EMPLOYEE MEDALS FOR SERVICE, RECOGNITION, ETC. (1800s TO DATE)

Collectors can find many kinds of medals struck by and/or for the Philadelphia Mint in the 1800s and beyond.

In 1879 Mint employees made a medal for President Ulysses Grant, in commemoration of a parade held in Philadelphia in his honor on December 16. Brass examples of the medal were handed out for free to the public during the parade; today they can be collected for $100 or less. A rare gold example (Julian CM-18, in deep prooflike MS-63, pictured) sold for $2,585 in December 2015.

A unique 1899–1941 retirement medal in gold was offered by Stack's Bowers Galleries in February 2016. (It went unsold.) The auction-lot description: "This rare medal is 50.6 mm, 1,407.43 grains, and features a view of the front entrance to the third Philadelphia Mint building (1901–1969). Inscription MINT OF THE UNITED STATES / PHILADELPHIA around, signed A. PIETZ at the bottom. This is the same die used on the 1934 Assay Commission medal." The reverse was struck from a specially prepared die, unsigned, with a laurel wreath surrounding the legend, "To H. Dudley Coleman from his associates in the United States Mint in sincere appreciation of his long and distinguished service," with the dates June 17, 1899, and October 21, 1941. The medal was housed in a burgundy-leather and cream-velour presentation case.

Another retirement medal features the same view of the Philadelphia Mint, with the reverse inscription, "To William R. Siner from your associates at the U.S. Mint as a tribute of good will and in appreciation of long and faithful service," with the dates 1899 and 1935. This was struck in silver, about 50 percent larger than a normal U.S. Assay Commission medal of the similar 1935 type. It sold for $1,840 in 2008.

Grant parade medal.

Coleman retirement medal.

Siner retirement
medal.

Since 2008 the Mint has awarded exceptional employees with the Rittenhouse Medal for Excellence, a silver medal struck in Proof on one-ounce American Silver Eagle planchets. It features a portrait of the first director of the Mint, David Rittenhouse. This award is pictured and described in detail in *American Gold and Silver: U.S. Mint Collector and Investor Coins and Medals, Bicentennial to Date*. "The bureau's goal is to be recognized as the finest mint organization in the world through excellence in its people, products, customer service, and workplace," writes author Dennis Tucker. "The Rittenhouse Medal for Excellence recognizes employees across the organization who have significantly furthered and ultimately improved Mint programs, operations, and services. Through the Rittenhouse Medal awards program the Mint honors the legacy of its famous first director."

Because of the personal nature of most such awards, and the fact that they are either unique or struck in very small quantities, they rarely are seen in the marketplace. However, opportunities do occasionally arise for collectors to seek them out and add them to their holdings of Mint-related memorabilia.

MINT BUILDINGS (1900S TO DATE)

In recent decades the Mint has produced commemorative medals for current and historic facilities, including for the new mint at Philadelphia; the Denver Mint; the U.S. Bullion Depository at Fort Knox; the West Point Mint; the San Francisco Mint; and the New Orleans Mint. Many of these medals, struck in bronze, are common and can be collected for a few dollars.

Older medals connected to Mint buildings and facilities are more valuable. For example, in 1905 the recently constructed Denver Mint struck a bronze medal to test its new machinery. Fewer than 75 are believed to have been made. An MS-63 example sold for $2,070 in January 2009; an MS-62 sold for $1,058 in February 2014.

Fort Knox Bullion
Depository medal.

San Francisco
Mint medal.

New Orleans
Mint medal.

Denver Mint
Opening medal.

Philadelphia Mint Coinage Bicentennial Medal (1993)

Dennis Tucker catalogs this silver medal in *American Gold and Silver*:

"A truly unique collector's item. Presenting THE PHILA-DELPHIA SET. Commemorating U.S. Mint history." Thus did the Mint advertise a new numismatic package in 1993, part of which was a unique silver medal. "Specially created for 1993 is The Philadelphia Set. It includes the one-half, one-quarter, and one-tenth ounce American Eagle Proof gold coins; the one-ounce silver coin; and a special silver medal commemorating the 200th anniversary of the first official U.S. coins at the Philadelphia Mint—all bearing the 'P' mint mark. We plan to offer this unique collection only in 1993, the first year the silver one-ounce coin is being minted in Philadelphia. Next year, the minting of the gold fractional coins will move from the Philadelphia Mint to the West Point Mint. You won't want to miss this special numismatic opportunity."

Philadelphia Mint Coinage Bicentennial medal.

Nearly 13,000 collectors took the Mint up on its offer. Packaged in a green velvet box with a certificate of authenticity, the set was issue-priced at $499. Today the silver medal alone is worth a good portion of that original cost, and the entire set sells for nearly $2,000.

The Philadelphia Mint Bicentennial silver medal was struck on a .900 fine planchet the size of a silver dollar. The obverse shows a rendering of the famous John Ward Dunsmore painting of Martha Washington inspecting the first coins from the Mint. (For the full story of this painting, see the award-winning *Secret History of the First U.S. Mint: How Frank H. Stewart Destroyed—and Then Saved—a National Treasure*, by Joel Orosz and Leonard Augsburger.) The reverse features a spiraling montage of U.S. coins dating from the 1790s to the1990s.

Many of the silver medals (and the accompanying American Silver Eagle) have toned over the years from exposure to the heavy green cardboard of the set's packaging. The resulting coloration can be dramatic and visually appealing—of course with beauty being in the eye of the coin holder.

The medal is worth $125 in unimpaired Proof.

San Francisco Old Mint Commemorative Coins (2006)

In 2006 two commemorative coins were struck by the San Francisco Mint to mark the 100th anniversary of the second mint (nicknamed "The Granite Lady") having survived the Bay Area earthquake and fires of 1906. The silver dollar shows the Old Mint as viewed off the left front corner. Its reverse features the reverse of the Morgan dollar of the era. Uncirculated examples are worth $45, and Proofs $35. The gold $5 coin shows the front portico of the Old Mint, with a portion of the building visible to each side, modeled after an 1869 construction drawing by Supervising Architect A.B. Mullet. The reverse is a rendition of the reverse of the Liberty Head half eagle of 1866 to 1907. The gold commemoratives are worth $325 in either Uncirculated or Proof.

San Francisco Old Mint silver dollar.   San Francisco Old Mint gold $5.

A portion of the sales of these coins went to the San Francisco Museum and Historical Society, for the purpose of rehabilitating the Old Mint as a city museum and an American Coin and Gold Rush Museum. (See chapter 8 on the San Francisco Mint for more on that effort.)

MINT LEDGERS, PAPERWORK, AND SIMILAR EPHEMERA (1792 TO DATE)

From time to time U.S. Mint letters, ledgers, and similar paperwork have entered the private-arena marketplace, despite their public-domain origin as property of the United States. While not common enough to be considered popularly collectible, this type of material is sought by researchers and numismatic specialists.

For example, in October 2000 an 1880s payroll ledger from the Philadelphia Mint's Engraving Department was auctioned for $920. The leather-covered booklet contained salary information for Mint employees including George T. Morgan, William Key, and George B. Soley (who privately produced the Lord's Prayer medalets mentioned on page 397, among others).

In the same auction, a tattered and unbound Philadelphia Mint ledger dating from 1857 to 1868 sold for $2,415. Its contents included information on die steel, date and letter punches, building materials, and various personnel matters.

In September 2001 an 1883 San Francisco Mint receipt for the shipment of 5,000 silver dollars by Wells Fargo & Co. sold for $44.

Autographs of famous Mint directors and engravers are collectible. A letter signed by David Rittenhouse in 1790 (two years before he became the first director of the Mint) sold for $1,700 in April 2012.

MINT-RELATED POSTCARDS

Postcard collecting was a popular craze in the late 1800s and early 1900s. Among the thousands of different postcards made during this era are numismatic creations by various private printers. Their scenes include coin and medal production, displays of the Mint Collection, workers going about their business, Mint buildings, and the like. The more common cards in nice condition sell for $10 to $20 for real-photo postcards, and $5 to $10 for later linen and glossy examples.

MINT-RELATED BOOKS

The bibliography herein lists some of the many Mint-related books published over the years. Some of these are as collectible as the Mint's coins themselves. An entire organization has been formed for collectors of numismatic literature—the Numismatic Bibliomania Society, online at www.coinbooks.org—and there are rare-book dealers who specialize in the field, such as Kolbe & Fanning Numismatic Booksellers (www.numislit.com), Charles Davis Numismatic Literature (www.numisbook.com), and Lake Books (www.lakebooks.com).

Vintage postcards showing the Philadelphia Mint.

A postcard view of the Denver Mint.

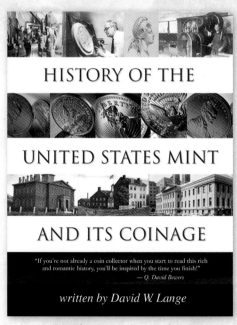

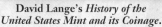

David Lange's *History of the United States Mint and its Coinage*.

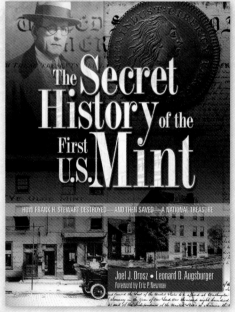

Joel Orosz and Leonard Augsburger's *The Secret History of the First U.S. Mint*.

HIGHLIGHTS OF THE U.S. MINT'S 2015 ANNUAL REPORT

The following appendix is composed of selected sections pulled from the 2015 Annual Report of the Director of the Mint, *lightly edited. As of publication, back issues as far back as 2001 can be viewed on the U.S. Mint's Web site at www.usmint.gov.*

PRINCIPAL DEPUTY DIRECTOR'S LETTER

Year after year, the men and women of the United States Mint have consistently delivered results for the American people, fulfilling our mission of facilitating commerce and protecting critical national assets.

In 2015, our work is as important as ever. In response to rising demand for circulating coins, we produced 16.2 billion circulating coins in FY 2015—an increase of 23.9 percent from the previous year. We have not seen this rate of circulating coin production since the pre-recessionary levels of 2006. To meet this increased demand, we added a third shift to the Philadelphia and Denver Mints and hired additional personnel.

Thanks to the ingenuity of our workforce, we have continued to reduce our fixed production costs and our selling, general, and administrative expenses while sustaining a very demanding production schedule. In fact, the cost of producing each of our circulating coin denominations went down for the fourth year in a row. Compared to FY 2014, we reduced the per unit cost of producing the penny by 13.9 percent; the nickel by 8.0 percent; the dime by 9.5 percent; and the quarter by 5.7 percent. We also returned $550 million in seigniorage to the Treasury General Fund in FY 2015, more than double the amount we returned last fiscal year.

The demand for American Eagle Silver Bullion Coins increased throughout the year and set new records again. This year we sold 47,870,000 silver bullion ounces. In 2013—a record-breaking year—we sold 43,559,000 ounces. We use an allocation process when demand exceeds available inventory, and we used this process for 36 weeks in FY 2015.

For our gold bullion products, we were able to keep up with demand and did not have to go on allocation—unlike several other mints around the world.

As gold and silver spot prices fell to the lowest levels since FY 2009, global demand for bullion rose rapidly in FY 2015, putting a strain on the world's suppliers. Employees of the West Point Mint, which produces the Mint's bullion coins, worked tirelessly at all hours of day and night on a three shift schedule to satisfy the increased demand. We sold 49.7 million ounces of gold and silver bullion this year—an increase of 25.4 percent from last year—for revenue totaling $2.1 billion. This represents a nearly five-fold increase in bullion sales since 2007, from 8.2 million ounces in 2007 to 49.7 million ounces this year.

Matthew Rhett Jeppson
United States Mint
Principal Deputy Director

While earnings from numismatic products increased 31.5 percent to $66.8 million, revenue and unit sales shrunk by 10.2 and 6.0 percent to $453.2 million and 5.4 million units, respectively. We believe the decline in revenue and unit sales is primarily driven by the absence of some of last year's most popular products, such as the Baseball Hall of Fame Commemorative Coins, and the shrinking customer base of numismatic hobbyists.

One of the things I have stressed from the outset of my tenure is providing all our customers only the highest quality products and services. As we continue our efforts to retain current customers and reach new audiences, we have to listen to the people who buy our products. That means making it easier for our customers to shop however, wherever, and whenever they want. To do that, we launched a new e-commerce site last year, which has made the shopping experience faster, simpler, and more reliable. Building on that progress, we released the Mint's first-ever mobile app this year, enabling smartphone users to learn about the history of the Mint and purchase our products using their devices. As one of the nation's largest online retailers, we need to have this type of capability if we are to be relevant in the marketplace.

Even as technology is driving significant changes in the way we operate, our employees remain the lifeblood of our organization. A significant portion of our workforce will soon retire, making it essential that we take the steps necessary today to prepare for our future human capital needs. To continue to meet the requirements of our mission in a resource-constrained environment, we must invest in training opportunities such as Lean Six Sigma and the Manufacturer Certification and Apprenticeship Program (MCAP) to enable our employees to grow into roles with increased responsibility.

For more than 200 years, we have told the story of our country's history, culture, traditions, and values by connecting America through coins. In January 2015, I was privileged to join this dedicated and talented group of individuals. While our core mission remains the same, we have embraced a culture of continuous improvement, never tiring in our endeavor to find safer, cheaper, and faster ways to produce the highest quality products in service of the American people. As we look toward our 225th anniversary in 2017, I am confident that we will continue to make great strides toward those goals.

Sincerely,

Matthew Rhett Jeppson, Principal Deputy Director

Principal Deputy Director Rhett Jeppson and Treasurer Rosie Rios cut
the unveiling ribbon for Treasury's display of U.S. Mint coin artwork.

THE UNITED STATES MINT AT A GLANCE

UNITED STATES MINT (MINT)

The men and women of the Mint manufacture and distribute circulating coins, precious metal and collectible coins, and national medals to meet the needs of the United States. Our vision is to become the finest mint in the world, through excellence in our people, products, customer service, and workplace.

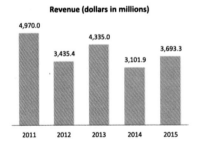

Revenue (dollars in millions)

Year	Revenue
2011	4,970.0
2012	3,435.4
2013	4,335.0
2014	3,101.9
2015	3,693.3

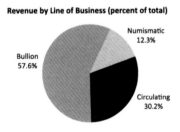

Revenue by Line of Business (percent of total)

- Bullion 57.6%
- Numismatic 12.3%
- Circulating 30.2%

CIRCULATING COINAGE

The Mint is the sole manufacturer of legal tender coinage in the United States. The Mint's highest priority is to efficiently and effectively mint and issue circulating coinage.

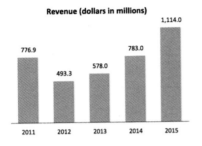

Revenue (dollars in millions)

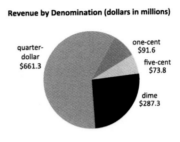

Revenue by Denomination (dollars in millions)

BULLION COINS

The Mint is the world's largest producer of gold and silver bullion coins. The bullion coin program provides consumers a simple and tangible means to acquire precious metal coins. Investors purchase bullion coins for the intrinsic metal value and the United States Government's guarantee of each coin's metal weight, content, and purity.

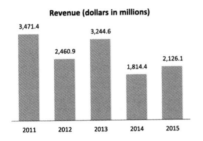

Revenue (dollars in millions)

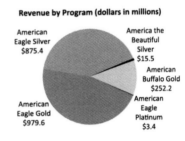

Revenue by Program (dollars in millions)

NUMISMATIC PRODUCTS

The Mint prepares and distributes numismatic products for collectors and those who desire high-quality versions of coinage. Most of our recurring products are required by federal statute. Others are required by individual public laws.

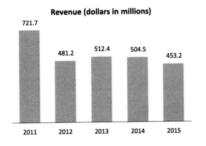

Revenue (dollars in millions)

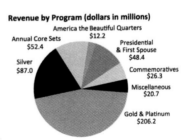

Revenue by Program (dollars in millions)

SEIGNIORAGE AND NET INCOME

Seigniorage is the difference between the face value and the cost of producing circulating coinage. The Mint transfers seigniorage to the Treasury General Fund to help finance national debt. Net income from bullion and numismatic operations can also fund government programs.

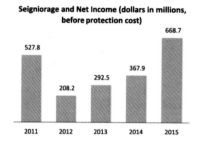

Seigniorage and Net Income (dollars in millions, before protection cost)

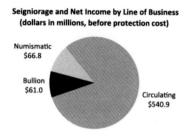

Seigniorage and Net Income by Line of Business (dollars in millions, before protection cost)

2015 PERFORMANCE

FY 2015 revenue was $3,693.3 million, an increase of 19.1 percent compared to last year. Cost of goods sold (COGS) increased 11.5 percent to $2,880.8 million. Selling, general and administrative (SG&A) expenses decreased 5.0 percent from last year. Total seigniorage and net income before Protection expenses increased 81.8 percent to $668.7 million compared to last year, reflecting the impact of increased circulating coinage shipments.

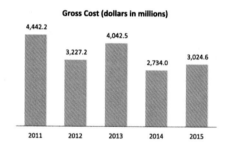

Gross Cost (dollars in millions)

CIRCULATING COINAGE

Circulating coin shipments increased 23.9 percent to 16,151 million coins in FY 2015. Quarter-dollar coin shipments experienced the strongest annual percentage growth at 57.2 percent compared to last year. Circulating revenue increased 42.3 percent to $1,114.0 million. Seigniorage increased 87.1 percent to $540.9 million. Seigniorage per dollar issued increased to $0.49 from $0.37 last year.

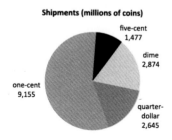

Shipments (millions of coins)

BULLION COINS

Demand for bullion coins increased in FY 2015 compared to last year. The Mint sold 49.7 million ounces of gold, silver, and platinum bullion coins, an increase of 10.1 million ounces from last year. Total bullion coin revenue increased 17.2 percent to $2,126.1 million in FY 2015, primarily due to a 32.8 percent increase in gold bullion coin demand. Bullion coin net income increased 117.9 percent to $61.0 million and bullion coin net margin increased to 2.9 percent compared to 1.5 percent last year.

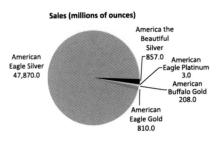

Sales (millions of ounces)

NUMISMATIC PRODUCTS

Numismatic sales decreased 5.3 percent to 5.4 million units in FY 2015. Numismatic revenues decreased 10.2 percent to $453.2 million. Numismatic net income increased 31.5 percent to $66.8 million (before Protection expenses). Numismatic net margin increased to 14.7 percent compared to 10.1 percent last year.

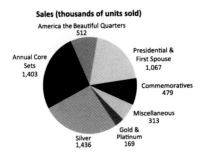

Sales (thousands of units sold)

TRANSFER TO THE GENERAL FUND

In FY 2015, the Mint transferred $561 million to the Treasury General Fund from the United States Mint Public Enterprise Fund. The Mint transferred $550 million of seigniorage as a non-budget transfer. In the first quarter, the Mint made a budget transfer of $11 million from FY 2014 numismatic program results.

Transfer to the Treasury General Fund
(dollars in millions)

MOBILE APP LAUNCHES

The United States Mint released MyUSMint, the bureau's new mobile application (app) for Android™, iPhone® and iPod touch® on July 15. Since then, 12,326 people have downloaded the app.

"We're constantly striving to find new ways to serve our customers and the American public. The new app comes on the heels of introducing a new order management system and reflects our commitment to be a world-class organization," said Rhett Jeppson, Principal Deputy Director of the U.S. Mint.

The free mobile app gives users easy access to United States Mint products (including the ability to place and track orders) as well as educational information. So far, the most popular features are the customized settings and historical information. The Mint updates content every month.

MyUSMint

STYLE NOTES

Certain quoted material has been lightly edited, but in no instances have meanings been changed. Certain now-archaic spellings such as planchettes (now planchets), deposites (deposits), etc., have been changed to modern usage.

NOTES

PREFACE BY DENNIS TUCKER

1. The U.S. Mint was born in 1792 as part of the Department of State, was made an independent agency reporting directly to the president in 1799, and in 1873 became a bureau of the Department of the Treasury.

CHAPTER 1

1. John Eadie, "An Historical Survey of the Origin and Changes of Money, Coin Values, Etc.," *Banker's Magazine*, April 1855.

2. Much information concerning commodities and goods as money is adapted from S.S. Crosby, *Early Coins of America*, 1875.

3. Joseph B. Felt, *An Historical Account of Massachusetts Currency*, 1839, pp. 82, 83.

4. Certain information is from Q. David Bowers, *Whitman Encyclopedia of Colonial and Early American Coins*, 2009.

5. By 1797 when the Soho Mint in Birmingham produced "cartwheel" pennies the use of British coppers had declined greatly. British pennies never circulated to any extent in the United States.

6. Spanish halves, or four-real pieces, circulated but were few in number. Judging from coins surviving today, the most popular fractional denomination was the two-real or quarter piece.

7. Colonial and Continental notes are delineated in Eric P. Newman's magisterial *Early Paper Money of America*.

CHAPTER 2

1. In 1667 the Dutch officially relinquished their claim to the colony in a diplomatic exchange which gave the Dutch control of the Spice Islands. No minting activity took place in New Amsterdam.

2. Contribution to *Whitman Encyclopedia of Colonial and Early American Coins*, 2008.

3. Discovered by Howland Wood and published in *The Numismatist*, July 1913. Unique.

4. Among the leading students of Vermont coppers over a long period of years have been Sylvester S. Crosby (1875), Hillyer C. Ryder (1918), John M. Richardson (1947), and in modern times (alphabetically) Q. David Bowers, Walter Breen, Kenneth Bressett, Tony Carlotto, and Eric P. Newman.

5. Addressed to Charles I. Bushnell; cited by Crosby, *Early Coins of America*, 1875.

6. "A Recently Discovered Coin Solves a Vermont Numismatic Enigma" in *The Centennial Publication of the American Numismatic Society*, New York, 1958.

7. Cited by Everett T. Sipsey in *The Colonial Newsletter*.

8. In *The Colonial Newsletter*, January 1976, Edward R. Barnsley, in "The Problem of James F. Atlee," took issue with the popular idea that James F. Atlee cut dies for Machin's Mills, Connecticut, Vermont, New Jersey, and certain other coins. A mis-reading of Crosby may have led to this. Michael J. Hodder's masterful "The 1787 'New York' Immunis Columbia; A Mystery Re-Ravelled," in *The Colonial Newsletter*, January 1991, also debunks this myth. Related commentary

can be found in John Lorenzo's "Atlee Broken 'A' Letter Punch," in the proceedings from the Coinage of the Americas Conference, 1995. Also see indexed Atlee items in *The Whitman Encyclopedia of Colonial and Early American Coins*, 2008, which contains findings by many modern scholars.

9. Updates can be found in issues of *The Colonial Newsletter* and the *C4 Newsletter* (published by the Colonial Coin Collectors Club).

10. https://www.whitehouse.gov/1600/constitution.

11. These names are from research by William Swoger and John Kraljevich.

12. Brasher was not the only metalsmith with EB initials, so hallmarks need to be studied for proper attribution.

13. Maris knew of only one, but modern day scholar Michael J. Hodder has seen 24—an interesting reflection of how rarity estimates can change.

14. Not all pieces show this feature. WM, for Walter Mould, can be seen under the left and right leaves below the horse's head on early die state examples in high grades.

CHAPTER 3

1. Morris did not address the 1776 Continental dollar and either was not familiar with its origin (whatever that may have been) or did not consider it to be of federal issue.

CHAPTER 4

1. Letter from Short to Jefferson, October 14, 1791: "Drost called on me yesterday and after some hesitation told me that several circumstances had taken place in his private affairs which rendered it necessary that he should decline going to America."

2. The popular story that President Washington supplied the silver has no foundation in fact.

3. On October 4, 1794, property at 629 Filbert Street was purchased for $1,200, and later three other properties were rented.

15. Counterfeit obverse die muled with an apparently discarded genuine reverse die.

16. The reason for the dramatic change of direction of the horse's head for this and the next two obverse dies is not known today, but it must have been significant in its time. This orientation matches the state arms, as does the intricate design of the plow, while the head facing right does not.

17. Dies completely hand-engraved without the use of punches.

18. Same obverse die as 18 of 1788.

19. Same die used for other 1787 cents and for obverse 17 of 1788.

20. Same die used for obverse 12 of 1788.

21. The Campbell study is by Kenneth Bressett. Sources include: Frances A. Westervelt, "The Final Century of the Wampum Industry in Bergen County New Jersey." New Jersey: 1916; *The British Numismatic Journal. Proceedings of the British Numismatic Society*. London: 1910; Don Taxay, *Money of the American Indians*. New York, New York: Nummus Press, 1970.

4. Andrew Mason Smith. *Coins and Coinage, The United States Mint, Philadelphia*, 1881, p. 5.

5. Leonard Augsburger, communication, January 19, 2016.

6. Frank H. Stewart, *History of the First United States Mint*, pp. 147, 157, here edited. Washington's presence at the foundation laying seems to be from oral tradition in the Kates family, who purchased the old Mint property in 1836 (per letter from J. Lewis Kates to Frank H. Stewart, April 19, 1915, quoted in Stewart's *History*).

7. Frank H. Stewart, *History of the First United States Mint*, p. 125: On January 7, 1793, Henry Voigt paid $3 for a watchdog which was a "savage brute named Nero."

8. Stephen Decatur Jr., *Private Affairs of George Washington From the Records and Accounts of Tobias Lear, Esquire, his Secretary* Boston, 1933.

9. As cited by R.W. Julian in "The Mint Investigation of 1795, *Numismatic Scrapbook Magazine*, July 1961.

10. Citation from Craig Sholley, manuscript for "Adam Eckfeldt, Mint Engraver?" refuting multiple published statements, including by Walter Breen and Don Taxay, that Eckfeldt engraved dies for coinage in 1792. RG104, *Records of the U.S. Mint at Philadelphia*, Entry 3 – Letters Sent.

11. Gardner did not work for a full year.

12. This would not happen again until 1840 (comment from William R. Eckberg, March 7, 2016).

13. On March 3, 1803, the authorization was extended for five years; on April 1, 1808, for five more years; on December 2, 1812, until March 4, 1818; on January 4, 1818, to March 4, 1823; on March 3, 1823, to March 4, 1828; on March 10, 1828, until otherwise provided by law.

14. Neil Carothers, *Fractional Money*, 1930, p. 75.

15. Sources are many, including Secretary of the Treasury William H. Crawford's report to the House of Representatives, February 24, 1820.

16. Erastus Root letter to the Mint, January 24, 1815, on behalf of the Select Committee on Sundry Resolutions in Relation to the Copper Coinage of the United States.

17. A very rare mention of gold; the amount of gold was not stated, but was probably small. 99 percent of the financial and newspaper accounts of the period refer to the importation of silver specie.

18. Leonard Augsburger, communication, January 19, 2016.

19. Adapted from *The American Machinist*, 1893, by Raymond H. Williamson, and reprinted by him in *The Numismatist*, January 1951; here excerpted. Also, Eugene S. Ferguson (editor), *Early Engineering Reminiscences (1815-40) of George Escol Sellers*. Washington, D.C: Smithsonian Institution, 1965. Sellers was a frequent visitor for a long period of years. In another reminiscence he told of watching cents struck in 1812.

20. Leonard Augsburger communication to Joel J. Orosz, January 19, 2016: This is the new middle building of the Mint (which Frank Stewart mistakenly thought to be the coinage building), which replaced the horse mill that had burned to the ground in 1816).

21. *Niles' Weekly Register*, June 30, 1821.

22. Adapted from *The American Machinist*, 1893, by Raymond H. Williamson, and reprinted by him in *The Numismatist*, January 1951; here excerpted. Also, Eugene S. Ferguson (editor), *Early Engineering Reminiscences (1815-40) of George Escol Sellers*. Washington, D.C: Smithsonian Institution, 1965.

23. Per a comment from Edward D. Cogan in the *American Journal of Numismatics*, March 1868. Cogan had no first-hand knowledge of the activities of the Mint in the 1820s and 1830s, for he did not become involved in numismatics until the late 1850s.

24. From research presented by John Kleeberg at the ANS at the Coinage of the Americas Conference, December 4, 1999.

25. *Travels Through North America During the Years 1825 and 1826*, 1828.

26. *Annual Report of the Director of the Mint*, 1828.

27. Document No. 51, 20th Congress, 2nd Session, House of Representatives. To accompany a bill HR 356 dated January 25, 1829, Mint of the United States.

28. *Mrs. Royall's Pennsylvania, Or Travels Continued in the United States*, 1829.

29. Sale advertisement, *Philadelphia Inquirer*, July 4, 1832. The same notice ran multiple times and in other papers.

30. Joel J. Orosz and Leonard Augsburger, separate communications, January 19, 2016; also see their study, *The Secret History of the First United States Mint*, pp. 49–69.

31. R.W. Julian from Mint records (communication to the author, March 8, 2009).

32. Adapted from *The American Machinist*, 1893, by Raymond H. Williamson, and reprinted by him in *The Numismatist*, January 1951; here excerpted. Also, Eugene S. Ferguson (editor), *Early Engineering Reminiscences (1815-40) of George Escol Sellers*. Washington, D.C: Smithsonian Institution, 1965. The accuracy of certain of Sellers' recollections has been questioned. Certainly, at age four in 1812 he could not have been a very qualified observer. Also see earlier Sellers commentary in the present narrative.

33. Statement issued by the Mint, "Small Change," printed in *Niles' Weekly Register*, June 17, 1820.

34. R.W. Julian, "1815 Quarter First of the Capped Bust Coinage," *Coins Magazine*, December 2013.

35. This figure, equal to 2,580 per hour, seems to be too high. In 1800 Director Boudinot said a hand press could turn out about 14,000 coins in a 10-hour day. Of course, there would have been some down time during the day.

36. In *The Granite Monthly*, November 1881, Hon. George Stark contributed an account from the 1828 journal one of his forebears.

37. Don Taxay, *U.S. Mint and Coinage*, p. 96, citing "A View of Philadelphia in 1829." *Pennsylvania Magazine*, July 1954.

38. In the 1830s Strickland also designed the Charlotte (North Carolina) Mint building.

39. *Niles' Register* on August 8, 1829, told more: "We have seen a specimen of the half dime lately issued by the mint, and must pronounce it one of the most beautiful coins our country has produced." Years later, in May 1903, *The Numismatist* reported that the cornerstone of the second Mint had been found and opened. The stone block, weighing about 300 pounds, concealed what was described as an old-fashioned candy jar, inside of which were found three coins and two newspapers. Description of two of the three coins was left to the imagination, but one was a dime of 1829. The newspapers were Philadelphia periodicals printed on July 3 and 4, 1829.

40. R.W. Julian, communication, February 25, 2009.

41. Frank H. Stewart, *History of the First United States Mint*, p. 123.

42. "Message from the president of the United States transmitting a report of the operations of the Mint during the year 1832," January 19, 1833.

43. Excerpted from a lengthy article on Mint history printed as an exchange item in the *Alexandria Gazette*, Alexandria, Virginia, July 27, 1833.

44. As copied by the *New Bedford Mercury*, Massachusetts, August 16, 1833.

45. The merchants at Canton (the only Chinese port open to American traders) preferred payment in silver coins, with Spanish-American dollars being the coins of choice. For a short time, drafts on the Bank of the United States were honored there, this being especially true in the early 1830s. It is interesting that editor Hezekiah Niles, for years an outspoken critic of the Bank of the United States, now seems to miss it!

46. Probably the engine built 1829–1830 by Rush & Muhlenberg, 30 h.p. (see George Escol Sellers, *Early Engineering Reminiscences*, pp. 72, 75). Adam Eckfeldt disliked the horizontal type of steam

engine, which was thought by many technicians to be superior. The Charlotte, Dahlonega, and New Orleans mints used horizontal engines.

47. Certain information, including citation, from the report of Franklin Peale to the Franklin Institute as reported in the *Journal of the Franklin Institute*, September 1846, pp. 307–310.

48. In April 1831 at the request of Mint Director Moore, President Andrew Jackson removed the old restriction on coining dollars, but none were made at that time.

49. Citation provided by Craig Sholley.

50. Nicholas Biddle, president of the bank, formed the deceptively named Bank of the United States of Pennsylvania, but dropped "of Pennsylvania" from most publicity. Periodicals and other publications stated that it was the old Bank of the United States now operating under a Pennsylvania charter. In fact, it had no connection with the former bank. The venture ultimately failed, bringing disgrace to Biddle, who had been a pillar of Philadelphia society.

51. For a unique detailed study of the assay offices see Roger W. Burdette, *From Mine to Mint*, 2013.

52. Text lightly edited.

53. *Portsmouth Journal*, New Hampshire, July 21, 1855, exchange item from the *Philadelphia Enquirer*.

54. As quoted in *Banker's Magazine*, June 1854.

55. Details were later published in *Report of the Secretary of the Treasury, on the State of the Finances, for the year ending June 30, 1854*. Washington, D.C. Report to the House of Representatives, Thirty-third Congress, 2nd Session. Ex. Doc. No. 3.

56. Rene Bache, "Picking Uncle Sam's Pocket," *San Francisco Chronicle*, May 2, 1920.

57. Don Taxay, *U.S. Mint and Coinage*, pp. 181, 182, 190.

58. *Annual Report of the Director of the Mint* for 1855.

59. His predecessor, Superintendent A. Loudon Snowden, amassed a vast private reserve of thousands of patterns, including great rarities, that were transferred to the private ownership of William H. Woodin in 1910.

60. The object is to avoid repeating information given in early descriptions of Mint processes.

61. In general, Mint directors of the era favored the Philadelphia Mint and, to an extent, San Francisco, and were critical of other Mints, either actual (that at New Orleans and the newly opened Carson City facility) or proposed new Mints.

62. Most small transactions were accomplished by using paper Fractional Currency notes, often dirty and tattered, disdainfully called "stamps" by the public. This name was derived from the original Postage Currency issue of 1862, which depicted postage stamps and occasionally came with perforated edges like those of actual stamps. These in turn had replaced postage stamps of 1861 as small change.

63. Report No. 851, U.S. Senate, February 9, 1881 (the same information was also printed elsewhere).

64. R.W. Julian from Mint documents in the National Archives.

65. Craig Sholley, letter to the author, January 18, 2001. Also, early in 1857 large copper cents were paid out to the extent of 100,000 coins, again until February 28.

66. R.W. Julian, "U.S. Half Cent Thrives Despite Second Rate Treatment," *Numismatic Scrapbook Magazine*, October 1972.

67. Craig Sholley, communication, January 18, 2001.

68. A few additional coins were copper-nickel (not relevant to the copper version). *United States Patterns* lists the copper strikings as Judd-611, although they are not really patterns.

69. Retrospective commentary in the *Annual Report of the Director of the Mint*, 1879; the June 1874 figure was given as an estimate.

70. At the World's Columbian Exposition in Chicago in 1893, the "widow's mite" was often mentioned as a highlight of Mint Cabinet coins on display there.

71. *The Numismatist*, August 1903.

72. For details see William T. Gibbs's article "Who bought patterns?" in *Coin World*, May 24, 2004.

73. *The Numismatist*, March 1919.

74. Later de Francisci stated that his wife, the former Teresa Cafarelli, was the model, which could have been at least partially the case. Many portraits of Miss Liberty over the years have been composites.

75. For details see David W. Lange, *The Complete Guide to Lincoln Cents*, 1995.

76. Images of the modern issues can be found in *A Guide Book of United States Coins*. Coinage ongoing at time of press.

77. Certain information is from Paul Gilkes, Mint places second call for artists; Seeking 16 new members for Artistic Infusion Program," *Coin World*, November 22, 2004.

CHAPTER 5

1. Detailed in *Niles' Weekly Register*, November 14, 1835.

2. Clay, from Kentucky, was against nearly everything supported by President Martin Van Buren and his predecessor, Andrew Jackson.

3. Per a note from coiner Rufus Taylor published years later in 1894; the 1838-O was not listed in the *Annual Report*.

4. Director Robert M. Patterson, *Mint Report*, January 30, 1840.

5. The three depositions are given in *The 1838-O Half Dollar: An Alignment of the Stars*, Kevin Flynn and John Dannreuther, 2015.

6. Certain information concerning 1845 is from J.L. Riddell, "The Mint at New Orleans with an Account of the Process of Coinage," 1845, a pamphlet discovered by Eric P. Newman.

7. Certain information about Riddell's problems is from *John Leonard Riddell*, a 1977 study by Karlem Reiss, especially pp. 82–104. Riddell's life dates: February 20, 1807-October 7, 1865.

8. This text was copied in *Hunt's Merchants' Magazine and Commercial Review*, November 1848.

9. *Hunt's Merchants' Magazine and Commercial Review*, Vol. 22, January to June, 1850, pp 320–322:

10. This wording is identical to that in Norman's guide quoted earlier.

11. The extensive rest of the article was essentially an uncredited copy of the text in the *Encyclopædia Americana*, 1831 and other editions, and states that the equipment in the New Orleans Mint was made by Boulton & Watt! Among other wrong information was that New Orleans was producing copper coins.

12. This information is incorrect, as after seizure by Southern forces, coinage of 1861-O half dollars was continued for a short time.

13. This information is incorrect, as after seizure by Southern forces, coinage of 1861-O half dollars was and double eagles continued for a short time.

14. *History of the Mint*, 1895.

15. Coinage in Carson City was deemed to be costly and inefficient, and the production of dollars there was only 750,000 pieces.

CHAPTER 6

1. *Annual Reports of the Director of the Mint*, congressional records, National Archives, and contemporary newspaper accounts were the main sources of contemporary information for this chapter; Clair M. Birdsall, *The United States Branch Mint at Charlotte: Its History and Coinage*, was especially valuable among modern sources.

2. Thomas Ligon and Reuben Perry.

3. The letter quotation from Peale is from a presentation Dr. Richard Doty a program given in the Coinage of the Americas Conference, American Numismatic Society, 1989, later printed in the *Proceedings;* also Mint correspondence.

4. The prime source for the following is "An Onerous and Delicate Task: Franklin Peale's Mission South, 1837," proceedings of the 1989 Coinage of the Americas Conference at the American Numismatic Society.

5. "Reports on Public Expenditures," March 31, 1842, excoriated the branch Mint in New Orleans for excessive expenses and also point out that minting coins in Charlotte and Dahlonega was far more costly per coin that minting them in Philadelphia. This news was widely circulated.

6. "Reminiscences of a Coin Collector, No. 5," *Mason's Coin & Stamp Collector's Magazine*, November 1867.

7. Dispatch of that date from Charlotte published in the *Macon Telegraph*, April 24, 1861.

CHAPTER 7

1. Major sources include Mint reports; biographical information on Senator John C. Calhoun; contemporary news accounts; Thomas L. Elder, "A Trip to Dahlonega, Georgia's Mint, Town, *The Numismatist*, March 1939; Penelope Prewitt Cunningham, "The Dahlonega Mint Story," *The Numismatist*, January 1962; Clair M. Birdsall, *The United States Branch Mint at Dahlonega: Its History and Coinage*, 1984 (important source general history); Silvia Gaily Head and Elizabeth W. Etheridge, *The Neighborhood Mint: Dahlonega and the Age of Jackson*, 1986 (important for extensive details on Mint personnel, problems, and the like; the title is somewhat of a misnomer as Jackson left office on March 4, 1837, before coinage began, and was succeeded by Martin Van Buren.).

2. In the late 1850s some miners from this area went to Jefferson Territory (later Denver Territory) and established a town they called Auraria that later became Denver.

3. In nearby Auraria (formerly called Nuckollsville) the Pigeon Roost Bank was operated by the Pigeon Roost Mining Company (chartered in December 1834). It issued paper money in quantity, which soon became worthless.

4. On December 21, 1835, the Cherokee Mining Company headquartered in Dahlonega was chartered.

5. He later returned to Athens when the matter was largely forgotten and Singleton had assumed the important office of Mint superintendent.

6. In *The Annual Report of the Director of the Mint* for 1838, which did not mention the Peale visit.

7. The prime source for the following is "An Onerous and Delicate Task: Franklin Peale's Mission South, 1837," proceedings of the 1989 Coinage of the Americas Conference at the American Numismatic Society.

8. Featherstonehaugh, *A Canoe Voyage Up the Minnay Sotor*, 2 volumes, London, 1847.

9. This and certain information about the Calhoun mine are from various sources including W.S. Yeates, *A Preliminary Report on a Part of the Gold Deposits of Georgia*, 1896.

10. *Hunt's Merchants' Magazine*, February 1844.

11. This shipment involved three pairs each of quarter eagle and half eagle dies.

12. At the court house in Atlanta in December 1862, 32,076 shares of the company were auctioned, amounting to control of the business, with no mention of Nachoochee in the offering. Over

$100,000 had been spent on operations it was stated. Some shares remained privately held and were disposed of in various ways into the late 1860s.

13. *Charleston Mercury*, May 14, 1861, exchange item from the *Dahlonega Signal*.

14. *Report of the Commissioner of Agriculture for the Year 1873*. Washington: Government Printing Office, 1873.

CHAPTER 8

1. R.W. Julian, "The Original 'S' Mint."

2. Nancy Y. Oliver and Richard G. Kelly. *The Inconspicuous Gold Rush Mint: San Francisco 1854–1874*, 2014, p. 56.

3. Edited excerpts.

4. Hubert Howe Bancroft, *History of Oregon*, Vol. II. 1848–1888, p. 641, citing the *Portland Oregonian*, August 30, 1864.

5. *Across the Continent*, 1865, pp. 345–347.

6. John S. Hittell, *The Resources of California*, 2nd edition, 1866, pp. 337–338.

7. Rene Bache, "Picking Uncle Sam's Pocket," *San Francisco Chronicle*, May 2, 1920.

8. V.L. Arrington, "A Trip Through the San Francisco Mint," *The Numismatist*, November 1949.

CHAPTER 9

1. The railroad was financed by California businessmen: Collis P. Huntington (stepfather of Archer M. Huntington, who in the early 1900s funded the new headquarters of the American Numismatic Society in New York City), Leland Stanford, Charles Crocker, and Mark Hopkins.

2. Thompson and West's *History of Nevada*, 1881, p. 557.

3. This is a rare laudatory remark for a superintendent of the Mint, whose office was in Philadelphia, to make about Carson City. Most later *Annual Reports of the Director of the Mint* had remarks critical of the Nevada operation.

4. Rene Bache, "Picking Uncle Sam's Pocket," *San Francisco Chronicle*, May 2, 1920.

5. Upon his death in 1906 the collection passed to his son, John H. Clapp. In 1942 the Clapp Collection was sold by agents Joseph and Morton Stack to Baltimore collector Louis E. Eliasberg for $100,000, tying for value the 1923 sale of the Ellsworth Collection as the largest American numismatic transaction. The Eliasberg Collection coins were auctioned in 1982, 1996, and 1997 when Bowers supervised the sales.

6. Thanks to Rusty Goe for many of the mintage estimates. The narrative information by Bowers in the narrative sections of the Battle Born Collection catalog (Stack's Bowers Galleries, August 2013) was a source for certain other information. *The Gobrecht Journal* published by the Liberty Seated Collectors Club was an important source as well.

CHAPTER 10

1. *Annual Report of the Director of the Mint*, 1906, p. 6; other sources.

2. Jacob G. Willson, "Denver Mints," *Numismatic Scrapbook*, June 1940.

3. By 1961 it had become dangerous and slippery, and in that year it was replaced by the same design in mosaic tiles.

4. *Denver Post*, January 1, 1921.

5. *Annual Report of the Director of the Mint*, 1917.

6. May 11, 1919.

7. Adapted from Frances Wayne's story in the *Denver Post*, February 5, 1920, and other coverage in the same issue.

8. *Richmond Times Dispatch*, Virginia, December 2, 1934.

9. Founded as a shop in Bridgeton, New Jersey, in 1863, the Ferracute Machine Company became prominent in the making of presses for tin cans, automobile parts, ammunition cases, and other

products in addition to coining presses. It closed in 1968.

10. By the early 1980s all of these presses were sold and replaced by newer models. Source: Michael P. Lantz, "Ammunition Presses Worked for Coins Too," *Numismatic News*, February 23, 1999.

11. Uncirculated coins from Philadelphia are shipped to Denver, where they're combined with Denver Mint coins in special packaging for each year's Uncirculated Mint sets.

CHAPTER 11

1. "West Point Mint marks its 75th anniversary," *Coin World*, July 1, 2013.

CHAPTER 12

1. George William Featherstonhaugh, *A Canoe Voyage Up the Minnay Sotor*, 2 volumes, London, 1847.

2. *New-York Herald*, May 11, 1849; C.W. Haskins, *The Argonauts of California*, p. 399; Edgar H. Adams, *Private Gold Coinage of California 1849–1855*, p. 88; 1850 federal census.

3. Information from Longacre's diary in the National Archives, as quoted by Rick Snow in the magazine *Longacre's Ledger*, June 2001. Perhaps the dies given to Cross were hubs. Cross was born in England circa 1820 and at a later time practiced die cutting in that country. By 1845 he was in America and did work for

the Philadelphia Mint as an assistant to Longacre. In 1862 he was listed in *McElroy's Directory:* "Engraver, 2011 Girard Avenue, Philadelphia."

4. Edgar H. Adams, Private Gold Coinage of California, 1849–1855, p. 71: "That he did strike gold coins in 1849 is extremely probable."

5. Jacob R. Eckfeldt and William E. Dubois, *New Varieties of Gold and Silver Coins, Counterfeit Coins and Bullion, With Mint Values*. 1851, p. 9: "Five-dollar piece of Dunbar & Co., in imitation of United States coin. A lot of 111 pieces averages 131 grains weight, 883 fineness, value $4.98."

CHAPTER 13

1. Benjamin Silliman (1779–1864) was associated with Yale College and was well known for his studies in the natural sciences.

2. Newspaper name or day date not given, probably published in Massachusetts, reprinted in the *American Journal of Numismatics*, January 1898.

3. Certain information is from Kenneth Bressett, communication, February 22, 2016.

4. *Report XVII of the State Mineralogist*, 1921, p. 98.

5. *San Francisco Chronicle*, July 27, 1937.

6. "Survey of Building Structures of the Sierran Gold Belt—1848–1870," by Robert F. Heizer and Franklin Fenenga, included in *Geologic Guidebook Along Highway 49*, 1948, p. 99.

7. Otheto Weston, *Mother Lode Album*, 1949, pp. 18–19.

8. J. Arthur Phillips, *The Mining and Metallurgy of Gold and Silver*. London: 1867, p. 50.

9. Sources include James J. Padden, "The Mint We Almost Had," *The Numismatist*,

April 1959; and Ralph R. Keeney, "The Mint That Never Was," *The Numismatist,* June 1983. The building attracted casual numismatic interest including a visit by a group of American Numismatic Associa-

tion members who were at a Portland convention in 2015.

10. This entry is by Kenneth Bressett. "The Counterfeit 1944 Jefferson Nickel" by Dwight Stuckey, 1982, was the source of certain information.

APPENDIX A

1. A "legislative day" starts when the Senate meets after an adjournment and ends when the Senate next adjourns. Hence, a legislative day may extend over several calendar days or even weeks and months. Definition taken from http://www.senate.gov/reference/glossary.htm.

2. Often referred to as Stella Hackel (no Sims) in print.

3. For biographies of all Carson City Mint superintendents, see the "Superintendent Shrine" at www.carsoncitycoinclub.com.

4. For a short time in 1885 he was temporary superintendent.

5. On July 1, 1899, the status of the facility was changed from mint to assay office. No coinage had taken place in recent years (since 1893, during Theodore R. Hofer's superintendency).

6. On April 20, 1861, the Charlotte Mint was seized by Confederate forces, who continued minting operations with Caldwell in charge until May 20, 1861.

7. Downer helped supervise the Denver Mint's construction and outfitting.

8. DeBroeckert was named acting superintendent on July 26, 1993; on October 17, 1996, he became the first superintendent to serve under Civil Service rules rather than having the possibility of termination by political action.

9. The New Orleans Mint served as assaying and custodian facilities from 1875 to 1878. In 1879 the Mint com-

menced coinage again. Various superintendents were in office plant manager until the next, when Liberty Seated coinage took place for a short time.

10. Bonzano served as assayer in charge (or caretaker in charge or special agent; records vary) from 1874 to 1876; in 1876 and 1877 he served as assayer in charge and superintendent; in 1877 and 1878 he was superintendent.

11. Dressel served the longest term of any mint superintendent.

12. Albino was appointed to the post on August 2, 1993, but did not officially have his position become a career post until October 17, 1996.

13. Kunderewicz was named acting superintendent on November 8, 1998. The position was renamed in late 1999 under the Mint's restructuring.

14. Certain information was provided by Nancy Oliver and Richard Kelly on May 26, 2015.

15. Leach resigned his superintendency to become director of the Mint in the same month.

16. Sweeney died in office.

17. DeVries was named deputy superintendent on September 12, 1996, replacing acting superintendent Donald T. Butler. DeVries's formal appointment was confirmed on October 17, 1996, along with the appointments of Raymond J. DeBroeckert at Denver and Augustine A. Albino at Philadelphia.

APPENDIX C

1. Tucker, Dennis. *American Gold and Silver: U.S. Mint Collector and Investor Coins and Medals, Bicentennial to Date,* pp. 333–334.

2. Ibid, p. 335.

BIBLIOGRAPHY

American Journal of Numismatics. New York, NY, and Boston, MA: Various issues 1866 to 1912.

Asylum, The. Various issues 1980s onward. Published by the Numismatic Bibliomania Society.

Augsburger, Leonard and Joel J. Orosz. *The Secret History of the First U.S. Mint.* Atlanta, GA: Whitman Publishing, LLC, 2011.

Banker's Magazine and Statistical Register, The. Boston. Various issues 1846–1860s.

Birdsall, Clair M. *The United States Branch Mint at Dahlonega: Its History and Coinage.* Easley, SC: Southern Historical Press, Inc., 1984.

——, *The United States Branch Mint at Charlotte: Its History and Coinage.* Easley, SC: Southern Historical Press, Inc., 1988.

Bowers, Q. David. *The History of United States Coinage as Illustrated by the Garrett Collection.* First printing, Los Angeles, CA: Published for The Johns Hopkins University, Baltimore. 1979.

——, *Whitman Encyclopedia of Colonial and Early American Coins.* Atlanta, GA: Whitman Publishing LLC, 2009.

——, *United States Gold Coins: An Illustrated History.* Los Angeles, CA: Bowers and Ruddy Galleries, Inc., 1982.

——, *The History of American Numismatics Before the Civil War, 1760-1860.* Wolfeboro, NH: Bowers and Merena Galleries, 1998.

——, *The Treasure Ship S.S. New York.* Atlanta, GA: Whitman Publishing LLC, 2009. Wolfeboro, NH and New York, NY: Stack's LLC, 2009.

Breen, Walter H. *Walter Breen's Complete Encyclopedia of U.S. and Colonial Coins.* New York, NY: Doubleday & Co., 1988.

——, *Walter Breen's Encyclopedia of U.S. and Colonial Proof Coins, 1792–1977.* Albertson, NY: FCI Press, 1977; Updated, Wolfeboro, NH: Bowers and Merena Galleries, 1989.

Bressett, Kenneth E. (editor). *A Guide Book of United States Coins.* Racine, WI: Various modern editions. Earlier editions edited by Richard S. Yeoman.

——, "The Vermont Copper Coinage," part of *Studies on Money in Early America.* New York, NY: American Numismatic Society, 1976.

——, *Walter Breen's Encyclopedia of Early United States Cents 1793–1814.* Wolfeboro, NH: Bowers and Merena Galleries, 2000.

Burdette, Roger W. *From Mine to Mint.* Great Falls, VA: Seneca Mill Press LLC, 2013.

Cain, Andrew W. *History of Lumpkin County for the First Hundred Years 1832–1932.* Atlanta, GA: Stein Printing Company, 1932.

Carlotto, Tony. *The Copper Coins of Vermont and Those Bearing the Vermont Name.* Colonial Coin Collectors Club (C4), 1998.

Carothers, Neil. *Fractional Money.* New York, NY: John Wiley & Sons, Inc., 1930.

Cohen, Roger S. Jr. *American Half Cents, the Little Half Sisters.* Baltimore, MD: Published by the author, 1971; second edition, 1982.

Coin Collector's Journal, The. New York, NY: J.W. Scott & Co., 1870s and 1880s.

Coin Week. Internet site. 2013 to date.

Coin World. Sidney, OH: Amos Press, *et al.,* 1960 to date.

Coin World Almanac. Seventh edition. Sidney, OH: Amos Press, 2000.

Coinage Laws of the United States 1792–1894. Modern foreword to reprint by David L. Ganz. Wolfeboro, NH: Bowers and Merena Galleries, Inc., 1991.

COINage magazine. 1964 to date.

Coinage of Gold and Silver. Collection, amounting to 491 printed pages, of documents, testimonies, etc., before the House of Representatives, Committee on Coinage, Weights, and Measures, 1891. Washington D.C.: Government Printing Office, 1891.

Colonial Coin Collectors Club Newsletter, a.k.a. *C4 Newsletter.* Various issues 1993 to date.

Colonial Newsletter, The. Established in 1960 by Al Hoch, continued for a long time by James C. Spilman and The *Colonial Newsletter* Foundation, and currently published by the American Numismatic Society, New York.

Crosby, Sylvester S. *The Early Coins of America.* Boston, MA: Published by the author, 1875.

Dickeson, Montroville W. *American Numismatical Manual.* Philadelphia, PA: J.B. Lippincott & Co., 1859.

Doty, Richard, "An Onerous and Delicate Task: Franklin Peale's Mission South, 1837," proceedings of the 1989 Coinage of the Americas Conference at the American Numismatic Society.

——, *America's Money, America's Story.* Iola, WI: Krause Publications, 1998.

——, *The Soho Mint & The Industrialization of Money.* London: Spink and The British Numismatic Society, 1998.

DuBois, William E. *Pledges of History: A Brief Account of the Collection of Coins Belonging to the Mint of the United States, More Particularly of the Antique Specimens.* Philadelphia. PA: C. Sherman, Printer, (1st Edition) 1846; New York, NY: George P. Putnam, (2nd Edition) 1851.

Dunlap, William. *History of the Rise and Progress of the Arts of Design in the United States.* New York, NY: George P. Scott & Co., Printers, 1834.

Eadie, John. "An Historical Survey of the Origin and Changes of Money, Coin Values, Etc.," *Banker's Magazine,* April 1855.

eSylum Internet site of the Numismatic Bibliomania Society, edited by Wayne Homren.

Evans, George G. *Illustrated History of the United States Mint.* Philadelphia, PA: published by the author, editions of 1883, 1885, 1889, 1893.

Featherstonhaugh, G.W. *A Canoe Voyage Up the Minnay Sotor.* 2 volumes. London, England: Richard Bentley, 1847.

Felt, Joseph B. *An Historical Account of Massachusetts Currency.* Boston, MA: Printed by Perkins & Marvin, 1839.

Ferguson, Eugene S. (editor). *Early Engineering Reminiscences (1815–1840) of George Escol Sellers.* Washington, D.C.: Smithsonian Institution, 1965.

Flynn, Kevin and John Dannreuther. *The 1838-O Half Dollar: An Alignment of the Stars.* Lamberton, NJ: Kevin Flynn, 2015.

Gilkes, Paul, "West Point Mint marks its 75th anniversary," *Coin World,* July 1, 2013.

Hall, Frank. *History of the State of Colorado.* Rocky Mountain Historical Co., 1889.

Head, Silvia Gaily and Elizabeth W. Etheridge. *The Neighborhood Mint: Dahlonega and the Age of Jackson.* Macon, GA: Mercer University Press, 1986.

Heaton, Augustus G. *A Treatise on the Coinage of the United States Branch Mints..* Washington, D.C.: published by the author, 1893.

Historical Magazine, The. Morrisania, NY. Issues in Series 1, 2, and 3, 1850s and 1860s.

Hodder, Michael J. "More on Benjamin Dudley, Public Copper, Constellatio Novas and Fugio Cents." *The Colonial Newsletter*, No. 97, June 1994.

Hull, John. "The Diaries of John Hull, Mint-Master and Treasurer of the Colony of Massachusetts Bay." *Archæologica Americana: Transactions and Collections of the American Antiquarian Society*, vol. III. Cambridge, MA: Bolles and Houghton, 1850.

Johnston, Elizabeth Bryant. *A Visit to the Cabinet of the United States Mint at Philadelphia.* Philadelphia, PA: J.B. Lippincott & Co., 1876.

Jordan, Louis E. Robert H. Gore Jr., Numismatic Endowment, University of Notre Dame, Department of Special Collections, Website compiled maintained by Louis E. Jordan. Anthology of published information on various series.

Julian, R.W., "The Mint Investigation of 1795," *Numismatic Scrapbook Magazine*, July 25, 1961.

——, "The Mint in 1792," *Numismatic Scrapbook Magazine*, April 25, 1962.

——, "The Copper Coinage of 1794," *Numismatic Scrapbook Magazine*, March 25, 1963.

——, "The Beginning of Coinage—1793," *Numismatic Scrapbook Magazine*, May 25, 1963.

——, "The Harper Cents," *Numismatic Scrapbook Magazine*, September 25, 1964.

——, "The Copper Coinage of 1795," *Numismatic Scrapbook Magazine*, December 25, 1964.

——, "The 1796 Copper Coinage," *Numismatic Scrapbook Magazine*, December 25, 1965.

——, "The Early Half Cents 1793–1799," *Numismatic Scrapbook Magazine*, January 25, 1973.

——, "The Cent Coinage of 1793," *Numismatic Scrapbook Magazine*, December 25, 1974.

——, "Cent Coinage of 1794–1795," *Numismatic Scrapbook Magazine*, January 25, 1975.

——, "Cent Coinage of 1796," *Numismatic Scrapbook Magazine*, March 25, 1975.

——, "Cent Coinage of 1797," *Numismatic Scrapbook Magazine*, April 25, 1975.

——, "The Philadelphia Mint and Coinage of 1814–1816," *American Numismatic Association Centennial Anthology*, 1991.

——, "Aspects of the Copper Coinage, 1793–1796," chapter in *America's Large Cent*, American Numismatic Society, Coinage of the Americas Conference, November 9, 1996, published in 1998.

Lange, David. *History of the United States Mint and Its Coinage.* Atlanta, GA: Whitman Publishing, LLC, 2005.

Martin, Sydney F. *The Hibernia Coinage of William Wood (1722–1724).* Colonial Coin Collectors Club, 2007.

Mason's Monthly Illustrated Coin Collector's Magazine. Philadelphia, PA, and Boston, MA: Ebenezer Locke Mason. Various issues from the 1860s onward. Titles vary.

McCusker, John J. *Money and Exchange in Europe and America, 1600–1775: A Handbook.* Chapel Hill, NC: University of North Carolina Press, 1978.

Miller, Henry C., and Hillyer C. Ryder, *The State Coinages of New England.* New York, NY: American Numismatic Society, 1920.

Mossman, Philip L. *Money of the American Colonies and Confederation: A Numismatic, Economic & Historical Correlation.* New York, NY: American Numismatic Society, 1993.

——, "The American Confederation: The Times and Its Money." *Coinage of the American Confederation Period.* New York, NY: American Numismatic Society, 1996.

Newman, Eric P. "The 1776 Continental Currency Coinage." *The Coin Collector's Journal*, July–August 1952.

——, "Varieties of the Fugio Cent." *The Coin Collector's Journal*, July–August 1952.

Niles' Weekly Register. Baltimore, MD: H. Niles, 1811–1847.

Noe, Sydney P. *The Silver Coinage of Massachusetts.* Compilation of earlier studies with addenda. Lawrence, MA: Quarterman Publications, Inc., 1973.

Norton's Literary Letter. New York, NY: 1857–1860.

Numisma. House organ published by Édouard Frossard, 1877–1891.

Numismatic News. Iola, WI: Krause Publications, 1952 to date.

Numismatist, The. George F. Heath. Monroe, MI: American Numismatic Association; Colorado Springs, CO (and other addresses), various issues 1888 to date.

Oliver, Nancy Y. and Richard G. Kelly. *The Inconspicuous Gold Rush Mint: San Francisco 1854–1874.* Hayward, CA: O.K. Associates, 2014.

Riddell, J.L. "The Mint at New Orleans with an Account of the Process of Coinage," 1845.

Goe, Rusty. *The Mint on Carson Street.* Reno, NV: Southgate Coins and Collectibles, 2003.

——, *James Crawford: Master of the Mint at Carson City.* Reno, NV: Southgate Coins and Collectibles.

Runkel, William M. *The United States Mint.* Philadelphia, PA: Published for the author by Turner & Co., 1870.

Ryder, Hillyer C., and Henry C. Miller. "The Colonial Coins of Vermont." Part of *State Coinages of New England.* New York, NY: American Numismatic Society, 1920.

Schilke, Oscar G., and Raphael E. Solomon. *America's Foreign Coins: An Illustrated Standard Catalogue with Valuations of Foreign Coins with Legal Tender Status in the United States 1793–1857.* New York, NY: Coin and Currency Institute, Inc., 1964.

Sheldon, William H., Dorothy I. Paschal, and Walter Breen. *Penny Whimsy.* New York, NY: Harper & Brothers, 1958.

Siboni, Roger S., John L. Howes, and A. Buell Ish. *New Jersey State Coppers.* New York, NY: American Numismatic Society and the Colonial Coin Collectors Club, 2013.

Sipsey, Everett T. "New Facts and Ideas on the State Coinages: A Blend of Numismatics, History, and Genealogy." *The Colonial Newsletter*, No. 13, October 1964.

Snowden, James Ross. *A Description of Ancient and Modern Coins in the Cabinet of the Mint of the United States.* Philadelphia, PA: J.B. Lippincott, 1860.

Spilman, James C. "An Overview of Early American Coinage Technology." *The Colonial Newsletter*, Nos. 62–65, April and July 1982, March and July 1983.

Stewart, Frank H. *History of the First United States Mint, Its People and Its Operations.* Philadelphia, PA: Frank H. Stewart Electric Co., 1924.

Stone, William Fiske. *History of Colorado.* Vol. 1. Chicago, IL: The S.J. Clarke Publishing Company, 1918.

Sumner, William Graham. *A History of American Currency*. New York, NY: Henry Holt & Co., 1874.

——, *The Financier and the Finances of the American Revolution*. Vols. I and II. New York, NY: Dodd, Mead and Co., 1891.

Taxay, Don. *Counterfeit, Mis-Struck, and Unofficial U.S. Coins*. New York, NY: Arco Publishing, 1963.

——, *U.S. Mint and Coinage*. New York, NY: Arco Publishing, 1966.

Token and Medal Society Journal. Various addresses: 1960s to date.

Tompkins, Daniel Augustus. *History of Mecklenburg County and the City of Charlotte from 1740 to 1903*. Charlotte, NC: Observer Printing House, 1903.

Tucker, Dennis. *American Gold and Silver: U.S. Mint Collector and Investor Coins and Medals, Bicentennial to Date*. Atlanta, GA: Whitman Publishing, LLC, 2016. Wikipedia.

Witham, Stewart. *Johann Matthaus Reich, Also Known as John Reich*. Canton, OH: Published by the author, November 1993.

Young, James Rankin. *The United States Mint at Philadelphia*. Philadelphia, PA: "For sale by Capt. A.J. Andrews," 1903.

CREDITS AND ACKNOWLEDGMENTS

The **American Antiquarian Society** furnished images. **Leonard Augsburger** reviewed the manuscript and made suggestions. **Richard August** permitted photography of his collection. **Anne Bentley**, curator at the Massachusetts Historical Society, provided images and information. **Kenneth Bressett**, editor of *A Guide Book of United States Coins*, made suggestions. Mint Director **Mary T. Brooks** arranged Mint visits and provided documents. **Patrick Brown** of the Denver Mint furnished selected images. **Roger Burdette** supplied several early images of the various mints. **Patrick Case** furnished the image of the West Point Mint on page TK. **John Dannreuther** assisted during mint visits and helped in other ways. **Tom DiNardi**, deputy manager of the West Point Mint, assisted during a visit. **Rachel Dobkin** of the Heritage Assets Project helped in several ways. **William R. Eckberg** reviewed parts of the manuscript and provided information on the first Philadelphia Mint. **Larry Eckerman**, plant manager of the San Francisco Mint, facilitated research, photography, and other aspects of a 2005 visit. **Don Everhart**, sculptor-engraver at the U.S. Mint, helped in many ways. **Tom Fesing** of the Denver Mint provided historical information. **Francis B. "Barry" Frere** of the U.S. Mint facilitated access to Mint correspondence and documents in the 1980s. **Kathryn Fuller** provided a file of archival information about the New Orleans Mint.

Frank Gasparro, chief engraver of the U.S. Mint, helped in many ways from the 1960s to the 1990s. **Robert Goler**, Mint historian, cataloged many galvanos and plasters that were later available for study and has been a very important factor with the extensive Mint archives. **Tim Grant** of the Philadelphia Mint helped with several visits to that institution. **Stella B. Hackel**, director of the Mint 1977–1981, made arrangements for visits to the Philadelphia, Denver, and San Francisco mints and to take photographs and do research there. **Phebe Hemphill**, medallic sculptor at the U.S. Mint, helped in many ways. Mint Director **Jay Johnson** provided much information about "golden dollars" and facilitated research circa 1999–2001. **Elizabeth Jones**, chief engraver of the U.S. Mint 1981–1991, assisted the author with various research in Mint records and made available various historical galvanos and other items from the Engraving Department. Certain information supplied by **R.W. Julian** for *Silver Dollars and Trade Dollars of the United States: A Complete Encyclopedia*, 1993, has been adapted here. **Thomas Jurkowsky** of the U.S. Mint provided extensive help including arranging visits to the four mints, directing inquiries, and providing archival information: he was key to the creation of this book. **John Kraljevich** provided images. **Michael Levin**, historian at the San Francisco Mint, helped in several ways.

Joseph Menna, medallic sculptor at the U.S. Mint, provided images and information. **John Mercanti**, former lead engraver at the U.S. Mint, discussed various aspects of coinage. Mint Director **Edmund C. Moy** helped in several ways. Certain pictures taken by **Tom Mulvaney** at the **National Numismatic Collection** in the **Smithsonian Institution** have been used. The **National Archives and Records Administration (NARA)** provided much Mint data over a long period of years. **Eric P. Newman** provided information and images. **Norphotos.com** supplied the image of Paul Balan. **Joel J. Orosz** provided information on the first Philadelphia Mint and proofread certain of the manuscript. **Jaclyn Penny** of the American Antiquarian Society helped with an inquiry. Deputy Mint Director **Richard Peterson** helped in several ways, including arranging mint visits and facilitating research. **The Phillips Library** of Salem, Massachusetts, provided research opportunities. Certain images are by **Douglas Plasencia**.

Donna Pope, director of the Mint 1981–1991, arranged for several visits to do research at various U.S. mints: additionally, through her office in Washington she and her staff helped in many ways.

Jane Samuels of the Heritage Assets Project helped in several ways. **Tracy Scelzo** of the U.S. Mint supplied valuable information about mint directors and superintendents. **Lateefah Simms** of the U.S. Mint helped with information. **Roger Siboni** furnished illustrations of certain colonial and early American coins, some used in *The Whitman Encyclopedia of Colonial and Early American Coins*. **Adam Stump** of the U.S. Mint provided information. **David M. Sundman**, who accompanied the authors on several trips to mints, furnished photographs. Over a long period of years officers and employees of the **United States Mint** and the **United States Treasury Department** have facilitated visits to mints and access to archives and have helped in many ways. **Stephanie Westover** of Littleton Coin Company supplied many images of modern coins. **Michael J. White** of the U.S. Mint gathered many documents, assisted with Mint visits and photography, and helped in other ways.

About the Author

Q. David Bowers has been in the rare-coin business since he was a teenager, starting in 1953. He is a founder of Stack's Bowers Galleries and is numismatic director of Whitman Publishing. He is a recipient of the Pennsylvania State University College of Business Administration's Alumni Achievement Award (1976); he has served as president of the American Numismatic Association (1983–1985) and president of the Professional Numismatists Guild (1977–1979); he is a recipient of the highest honor bestowed by the ANA (the Farran Zerbe Award); he was the first ANA member to be named Numismatist of the Year (1995); and he has been inducted  into the ANA Numismatic Hall of Fame maintained at ANA headquarters. He has also won the highest honors given by the Professional Numismatists Guild. In July 1999, in a poll published in *COINage*, "Numismatists of the Century," Dave was recognized as one of six living people in this list of just 18 names. He is the author of more than 50 books, hundreds of auction and other catalogs, and several thousand articles, including columns in *Coin World* (now the longest-running by any author in numismatic history), *The Numismatist*, and other publications. His books have earned more "Book of the Year Award" honors bestowed by the Numismatic Literary Guild than have those of any other author. He and his firms have presented the majority of the most valuable coin collections ever sold at auction. Dave is a trustee of the New Hampshire Historical Society and a fellow of the American Antiquarian Society, the American Numismatic Society, and the Massachusetts Historical Society. He has been a consultant for the Smithsonian Institution, the Treasury Department, and the U.S. Mint, and is research editor of *A Guide Book of United States Coins*. For many years he was a guest lecturer at Harvard University. This is a short list of his honors and accomplishments. In Wolfeboro, New Hampshire, he is on the Board of Selectmen and is the town historian.

INDEX